Origins and Varieties of Logicism

This book offers a plurality of perspectives on the historical origins of logicism and on contemporary developments of logicist insights in philosophy of mathematics. It uniquely provides up-to-date research and novel interpretations on a variety of intertwined themes and historical figures related to different versions of logicism.

The essays, written by prominent scholars, are divided into three thematic sections. Part I focuses on major authors like Frege, Dedekind, and Russell, providing a historical and theoretical exploration of such figures in the philosophical and mathematical milieu in which logicist views were first expounded. Part II sheds new light on the interconnections between these founding figures and a number of influential other traditions, represented by authors like Hilbert, Husserl, and Peano, as well as on the reconsideration of logicism by Carnap and the logical empiricists. Finally, Part III assesses the legacy of such authors and of logicist themes for contemporary philosophy of mathematics, offering new perspectives on highly debated topics—neo-logicism and its extension to accounts of ordinal numbers and set-theory, the comparison between neo-Fregean and neo-Dedekindian varieties of logicism, and the relation between logicist foundational issues and empirical research on numerical cognition—which define the prospects of logicism in the years to come.

This book offers a comprehensive account of the development of logicism and its contemporary relevance for the logico-philosophical foundations of mathematics. It will be of interest to graduate students and researchers working in philosophy of mathematics, philosophy of logic, and the history of analytic philosophy.

Francesca Boccuni is Associate Professor at Vita-Salute San Raffaele University, Milan. She has published articles in scientific journals and collected volumes and edited collections in the philosophy of mathematics and logic.

Andrea Sereni is Associate Professor at the School for Advanced Studies IUSS Pavia. He has published essays and books as well as edited special issues and collections in the epistemology and philosophy of mathematics and logic.

Routledge Studies in the Philosophy of
Mathematics and Physics
Edited by Elaine Landry
University of California Davis, USA

and

Dean Rickles
University of Sydney, Australia

The Logical Foundation of Scientific Theories
Languages, Structures, and Models
Décio Krause and Jonas R. B. Arenhart

Einstein, Tagore and the Nature of Reality
Edited by Partha Ghose

A Minimalist Ontology of the Natural World
Michael Esfeld and Dirk-André Deckert

Naturalizing Logico-Mathematical Knowledge
Approaches from Philosophy, Psychology and Cognitive Science
Edited by Sorin Bangu

The Emergence of Spacetime in String Theory
Tiziana Vistarini

Origins and Varieties of Logicism
On the Logico-Philosophical Foundations of Mathematics
Edited by Francesca Boccuni and Andrea Sereni

For more information about this series, please visit: www.routledge.
com/Routledge-Studies-in-the-Philosophy-of-Mathematics-and-Physics/
book-series/PMP

Origins and Varieties of Logicism

On the Logico-Philosophical Foundations of Mathematics

Edited by Francesca Boccuni
and Andrea Sereni

Routledge
Taylor & Francis Group

NEW YORK AND LONDON

First published 2022
by Routledge
605 Third Avenue, New York, NY 10158

and by Routledge
2 Park Square, Milton Park, Abingdon, Oxon, OX14 4RN

Routledge is an imprint of the Taylor & Francis Group, an informa business

© 2022 Taylor & Francis

The right of Francesca Boccuni and Andrea Sereni to be identified as the authors of the editorial material, and of the authors for their individual chapters, has been asserted in accordance with sections 77 and 78 of the Copyright, Designs and Patents Act 1988.

All rights reserved. No part of this book may be reprinted or reproduced or utilised in any form or by any electronic, mechanical, or other means, now known or hereafter invented, including photocopying and recording, or in any information storage or retrieval system, without permission in writing from the publishers.

Trademark notice: Product or corporate names may be trademarks or registered trademarks, and are used only for identification and explanation without intent to infringe.

Library of Congress Cataloging-in-Publication Data
A catalog record for this book has been requested

ISBN: 978-0-367-23005-0 (hbk)
ISBN: 978-1-032-15910-2 (pbk)
ISBN: 978-0-429-27789-4 (ebk)

DOI: 10.4324/9780429277894

Typeset in Sabon
by Apex CoVantage, LLC

To Bob Hale and Eva Picardi

Contents

Preface and Acknowledgments

The chapters collected in this volume, as well as the ideas behind it, originated in two homonymous workshops we organized in 2015 and 2017: *Origins and Varieties of Logicism* (School for Advanced Studies IUSS Pavia, March 16, 2015) and *Origins and Varieties of Logicism II—Further Explorations* (Vita-Salute San Raffaele University, Milan, February 20–21, 2017). The workshops offered (splendid, to our eyes) occasions to gather established and younger scholars in the philosophy of mathematics working on several aspects of the logicist tradition and to witness how lively the debate surrounding that tradition still is, even beyond some regularly frequented quarters. We hope reading this volume will convey part of the stimulating atmosphere of those two meetings.

Edited volumes are known to take time and effort, and having to cope with the exhausting consequences the pandemic has caused globally in the last year and a half could not but make things even harder on everyone involved. We are most grateful to all the contributing authors, as well as to Routledge, for their continued support and patience during the whole editing process.

Neither the workshops nor the volume could have been possible without participation and cooperation from institutions, authors, and reviewers.

First of all, we want to thank the School of Advanced Studies IUSS Pavia, Vita-Salute San Raffaele University, joint IUSS/San Raffaele PhD Program in Cognitive Neuroscience and Philosophy of Mind, San Raffaele PhD Program in Philosophy, NEtS (Neurocognition, Epistemology and theoretical Syntax) Center at IUSS, and CRESA (Research Center in Experimental and Applied Epistemology) at San Raffaele University for sponsoring the workshops and making them possible, as well as the then-Rector of IUSS, Michele Di Francesco, and the coordinator of the San Raffaele PhD Program in Philosophy, Massimo Reichlin, for supporting the two workshops, respectively. Finally, thanks to the Italian Network for the Philosophy of Mathematics (FilMat) for granting its auspices.

All participants in the workshops helped in providing fruitful discussions and enhancing the ideas later to be published in this volume. Beyond the contributing authors to be found in these pages, we want to thank other speakers, including Salvatore Florio, Eva Picardi, and Michael Potter, as well as all participating discussants: Massimiliano Carrara, Ciro De Florio, Miriam Franchella, Alessandro Giordani, Daniele Molinini, Francesco Orilia, and Claudio Ternullo. Marco Panza and Erich Reck have been supportive of our project from the start, and we owe them for many helpful suggestions.

The volume also includes chapters from authors who did not take part in the workshops and were later reached for their expertise in areas that were not originally covered. Some authors could not contribute in the end, but we want nonetheless to thank them for their efforts: Salvatore Florio, Stefan Rosky, and Matthias Schirn.

A number of colleagues have greatly helped in providing comments and reviews of the collected chapters, and we are grateful for their assistance: Bahram Assadian, Joan Bertran-San Millán, Philip Ebert, José Ferreirós, Sebastien Gandon, Ansten Klev, Gabriele Lolli, Markus Pantsar, Matteo Plebani, Gabriel Uzquiano, and Sean Walsh.

Our gratitude naturally goes to the editors of the Routledge series *Studies in the Philosophy of Mathematics and Physics*, Elaine Landry and Dean Rickles, for considering our proposal; to two anonymous reviewers for their helpful comments; to Andrew Weckenmann at Routledge for helping us with the initial submission; and especially to Alexandra Simmons for her continuous and valuable support (and, again, patience) in all phases of the production process.

This volume is dedicated to two scholars who took active part in the workshops but sadly are not with us anymore: Bob Hale and Eva Picardi.

Bob Hale greatly contributed (among many other things) to the study of Frege's philosophy of mathematics and language and to the development of Scottish neo-logicism. His influence in these areas cannot be overestimated. Bob sent us his chapter for this volume shortly before passing away. Together with Jessica Leech, who was editing a collection of his posthumous writings, we thought that his contribution should rightfully appear in both volumes. It was originally published in Bob Hale and Jessica Leech, *Essence and Existence*, Oxford University Press, Oxford, pp. 240–255, and is reprinted here with Margaret Lovell Hale's permission. We want to thank Jessica Leech and Margaret (Maggie) Hale for their support in the reprint process. Bob was an inspiring thinker and an exquisite person to talk with, not just about philosophy. His loss was a shock to us, and to the entire scientific community.

Eva Picardi was a renowned Frege (and Dummett) scholar who was an illuminating and passionate teacher to (some of us and) many Italian students. Eva profoundly influenced many generations of philosophers by contributing to the dissemination of Frege's philosophy and more generally analytic philosophy in Italy. Above and beyond that, she was an internationally esteemed thinker, as the vast scope of her publications makes clear (some of her papers are now collected in *The Selected Writings of Eva Picardi: From Wittgenstein to American Neo-Pragmatism*, ed. Annalisa Coliva, Bloomsbury, 2020; a second volume, *Frege on Language, Logic and Psychology*, ed. Annalisa Coliva, is forthcoming for Oxford University Press). We sorely miss her for her philosophical insights, her professional passion, and her wit, both within and outside academia.

We hope the present volume will be a worthy homage to the legacy of Bob and Eva.

Milan, Pavia
May 2021

Francesca Boccuni
Andrea Sereni

1 Origins and Varieties of Logicism

An Overview

Francesca Boccuni and Andrea Sereni

WHAT IS LOGICISM?

The basic idea behind logicism in the philosophy of mathematics is disarmingly easy to state: mathematics is reducible to logic. Equally disarming, however, is the amount of subtleties, complexities, and variations that exploring that basic idea entails.

Logicism, as it was originally introduced or as it is still perceived today, is a foundational theory, and appeal to "reduction" is a pointer to its foundational role. How foundation and reduction are to be conceived of, however, is a vastly debated matter, combining issues in both the semantics, ontology, epistemology, and formal reconstruction of mathematics. Ideally, one may expect a foundation to roughly entail the following: unclear terms in a disputed area of discourse are explained, via definitions, on the basis of clearer or more familiar primitives, thus yielding clarity to the disputed concepts; such semantic reduction brings about an ontological reduction, for objects apparently referred to in, or anyway featuring in the subject-matter of, the target discourse are identified with objects referred to in the basic discourse, which are also taken to be more familiar or more easily characterized; finally, semantic and ontological reductions bring about epistemic reduction: disputed objects are now shown to be more easily accessible, and statements in the disputed discourse can be derived from statements in the basic language expanded with suitable definitions, thus inheriting part of the certainty the basic statements enjoy. Following Quine (1969), we can call the process of attaining clarity by reduction of obscure concepts to basic ones a *conceptual* study and the process of attaining certainty by deduction of disputed statements from basic ones a *doctrinal* study; to these, an *ontological* and an *epistemological* study clearly follow suit. A logicist foundation would then see in the language of mathematics (or some of its branches) the target disputed discourse, in some sort of logical objects those with which mathematical objects are to be identified, and logical vocabulary as the basic vocabulary with which, via definitions, we can recover the originally obscure concepts and objects, explain the content and truth-conditions of mathematical statements, and deduce such statements from purely logical ones.

DOI: 10.4324/9780429277894-1

This idealized picture, however, is hostage to a number of different variables. A standard presentation of logicism, mainly reflecting the views expounded by the German mathematician and philosopher Gottlob Frege, would roughly deliver the following outline when limited to arithmetic.

Despite its undeniable usefulness in applications, which gives us an indirect, inductive, and *a posteriori* warrant in its truth, arithmetic stands in need of a much more robust justification. Such justification will guarantee a number of alleged properties of arithmetical statements: that they are necessary, that they enjoy universal applicability and are not tied to particular applications, that they do not assert anything of particular material objects or their properties, that the reasons for holding them true make no appeal to empirical evidence, and that they have a form of objectivity that makes their truth independent of subjective human activities. An adequate semantic analysis of arithmetical statements characterizes numerical expressions as singular terms, whose semantic role is to denote individual, self-subsistent objects. These objects can be identified with particular logical objects (the ontological study). Frege considered these extensions of concepts, that is, loosely speaking, the sets of objects falling under (sortal) concepts. The existence of such logical objects is established by basic logical laws, delineating the essential vocabulary of our basic logical language. Via such vocabulary, through appropriate definitions, we can define the target arithmetical notions (the conceptual study): for instance, a certain law will characterize extensions; via extensions we can define the notion of cardinal number; other logical definitions will introduce the notions of following-in-a-series and immediate successor and allow definitions of individual numbers, beginning with the number 0; together these definitions will lead to defining in turn natural numbers as those finite cardinals that follow in the successor-series starting with 0. This will show how it is possible to derive all arithmetical truths from basic logical truths and definitions (the doctrinal study) once a proper formal language is set and legitimate inference rules are specified. Such derivation will also show that whatever epistemic access we have to logical objects and laws will transmit to mathematical objects and statements (the epistemological study). In an ideal scenario like the one just depicted, this access will be purely *a priori* and will show arithmetical statements to enjoy the same necessity and generality of logical statements: indeed, arithmetical statements will just turn out to be a sub-class (or a notational variant) of logical statements. Hence, they will be *ipso facto* provable by logic and definitions and thus analytic in Frege's sense (cf. [GLA], §3).

Even without looking at the complexities such a picture requires when placed in Frege's overall mathematical, semantical, and philosophical setting, it is easy to see from how many angles it can be attacked or simply modified.

To begin with, we need a clarification of what logic is, what its basic concepts and laws are, and how we acquire knowledge of them. Traditionally, logicism in the philosophy of mathematics has been developed within the setting of second-order logic. But second-order logic has come under fire from different charges, both formal—regarding, for instance, its incompleteness entailed by Gödel's 1931 theorems—and philosophical—such as that leveled by Quine (1970), who suggested that second-order logic is not logic but rather set theory in disguise. Switching to first-order logic may provide some reassurance but will prevent recovering not just, first and foremost, arithmetic in the way Frege envisaged but also larger parts of mathematics—whereas ideally, logicism is meant to extend to all branches of mathematics. One can then either try to defend second-order logic from various accusations (Shapiro, 1991; Wright, 2007) or even resort to some alternative interpretations—like those provided by plural logic (Boolos, 1984, 1985)—or even to non-classical logics (cf. e.g., Tennant, 1987). Needless to say, the same concerns apply when one considers the variety of accounts of logical validity that will have to underpin the inferences to be allowed in a given formal system, as well as some of the principles to be endorsed (think, for a case in point, about the Axiom of Choice or various kinds of comprehension principles).

Even once such formal problems are tackled, many remain open. Logical (or, in different guises, set-theoretical) platonism, that is, the identification of mathematical objects with logical objects, raises several challenges. First of all, many would conceive of logical discourse as entirely topic neutral, and this would boost skepticism against the idea that the subject-matter of logic involves *sui generis* objects and against the idea (see Dummett 1973, ch. 4; Hale 1994, 1996) that numerical expressions are genuine singular terms. One can thus either try to account for the logical form of mathematical statements so that such expressions are interpreted differently—for example, as complex predicates (per some of Russell's views)—or even adopt a background conception of logic as conventional in nature (as was the case for Carnap), with the side effect of threatening not just platonism, as intended, but also the objectivity of mathematics. More generally, the idea that the logical analysis of mathematical statements is to mirror faithfully their surface-grammar may be questioned and the identification of the grammatical and semantical roles of some mathematical expressions with those of singular terms jeopardized.[1] Also, the claim itself that the surface-grammar of those statements as it is adopted by standard platonist views is the correct one can be disputed (cf. e.g., Hofweber, 2005; Moltmann, 2013a, b).

---

1 This, in turn, would also jeopardize the semantical and theoretical import of Frege's famous and controversial Context Principle (see Cariani, 2018; Dummett, 1995; Linnebo, 2009b; Picardi, 2009, 2017; Resnik, 1967), as well as the particular interpretation it receives in neo-logicists' so-called Syntactic Priority Thesis (see Wright, 1983; Hale, 1987; MacBride, 2003).

Even if one accommodates logical objects, metaphysical questions about their nature come to the fore. The Fregean platonist view—at least as standardly presented—which locates such objects in a third realm additional to those of material objects and subjective representations and describes them as self-subsistent, mind-independent, abstract (aspatial, atemporal, acausal, possibly necessary existent) objects runs the risk of begging for more answers than it provides (cf. Dummett, 1991, ch. 18; Hale, 1988; Rosen, 2012). Qualms over our epistemic access to such objects have abounded, raising the suspicion that the so-called Integration Challenge (Peacocke, 1998) between semantics, ontology, and epistemology is hard to attain in mathematics, as paradigmatically emphasized by Benacerraf (1973).[2]

Concerns over the admissibility of abstract logical objects could be assuaged if one were able to provide some intuitive, immediate source of epistemic access to them, via clear, or unproblematic, or self-evident basic concepts. This points to another cradle of problems: definitions, which are the backbone of the conceptual reduction.

For one thing, as history has painfully taught us, naïve notions can be given apparently natural definitions and still engender paradoxes; Frege's Basic Law V as a definition of the (supposedly) logical notion of extension or—equivalently—Cantor's naïve notion of set or the notion of ordinal number all posed hardly surmountable challenges in this regard. Consistency clearly is a minimal requirement for a classical foundation of any branch of mathematics; how to ensure it is a complex matter. Even the discussion of where exactly the contradiction in Frege's system should be located—as emphasized by, among other things, debates on the cause of such inconsistency (Boolos, 1993; Dummett, 1991, ch. 17, 1994), on the tenability of impredicative vs. predicative definitions (cf. e.g., Feferman, 2005), and on Russell's Vicious Circle Principle (Chihara, 1973; Gödel, 1944; Jung, 1999)—opens up a range of diagnoses and solutions, including the possibility of modifying Frege's formal system enough to individuate consistent sub-systems in which his own definitions can be redeemed (cf. e.g., Ferreira, 2018; Ferreira & Wehmeier, 2002; Heck, 1996).[3]

Even when consistency is safeguarded, however, the nature and import of mathematical definitions, and definitions by abstraction in particular, maintain center stage. The Euclidean ideal of self-evidence is not easy to obtain, as the latter remains a highly problematic notion (see e.g., Jeshion, 2001, 2004; Shapiro, 2009). Nor it is uncontroversial, as many logicists have maintained (see e.g., Hale & Wright, 2000), to claim that

---

2 For the vast discussion on Benacerraf's challenge and its relevance for platonism, see, for example, Field (1989), Introduction, Hale and Wright (2002), Liggins (2010), Linnebo (2006), Panza and Sereni (2013).
3 For a survey, see Burgess (2005).

definitions are able to afford us *a priori* mathematical knowledge. This may lead to seeking alternative ways of characterizing basic mathematical notions. One major example is given by the stipulation of sets of axioms, possibly themselves conceived of as implicit definitions of the primitive terms occurring in them. While such a move is often associated with other traditions, such as the one inaugurated by Hilbert in the context of the development of the axiomatic method, it is well possible to merge logicist claims with similar, axiomatic characterizations of mathematical notions. If one also takes such characterizations to introduce not just individual mathematical objects as standardly conceived but either mathematical objects characterized by purely relational properties or even directly abstract mathematical structures (those ideally described categorically by given sets of axioms), a variety of logicism remarkably different from the one sketched previously may ensue (such as the one arguably retrievable in Dedekind [WSWSZ], despite the many possible interpretations of his views).

More generally, understanding the role and import of definitions one way or another is tantamount to characterizing the sought-after foundation in alternative ways. It is undeniable that definitions in a given branch of mathematics have a clear intra-theoretical and systematic function: they allow locating a given theory in the overall architecture of mathematics by establishing reduction, or at least translation procedures, from one theory to another (e.g., real analysis to arithmetic, arithmetic to set theory, etc.). But equally undeniable is that the role definitions play in the logicist enterprise, however characterized, is also exquisitely epistemological: they are meant to afford us *a priori* knowledge of basic mathematical concepts and laws and to provide the grounds upon which the entirety of the mathematical edifice is supposed to rest. Whether the structure of mathematical knowledge should mirror some metaphysical architecture of the mathematical universe is an open question,[4] as is the question of whether such knowledge, and thus the logicist target discourse, should be confined to pure mathematics or rather duly take into account from the start an explanation of its possible applications.[5]

LOGICISM AND ITS ORIGINS

Things are complicated, then, but rather than staring into the abyss of apparently endless possibilities, guidance into the many ways of logicism can be offered by a closer look at its origins and at the historical and conceptual development it underwent.

---

4 See, for example, Donaldson (2016), Rosen (2010), Schwartzkopff (2011).
5 See, for example, Hale (2002, 2016), Panza and Sereni (2019), Sereni (2019), Shapiro (2000), Snyder, Samuels and Shapiro (2018), Snyder and Shapiro (2019), Wright (2000).

Gottlob Frege is quite unanimously seen as the originator of logicism in the philosophy of mathematics. This is hardly questionable. Even though some ideas pulling towards what we now conceive of as logicism may be recovered in earlier authors, it is undeniable that the richness and formal and theoretical maturity of Frege's views make his position the first to earn the qualification of "logicism" in full right (even though the label became established only later).[6] The revolution Frege brought about in logic with his *Begriffsschrift* ([Beg]) provided a solid formal and conceptual basis on which the reduction of arithmetic to (the new) logic could be effected. The *Grundlagen der Arithmetik* ([GLA]) set the project in a wide philosophical and mathematical context and—beyond advancing some major conceptual changes that were to inaugurate the so-called "linguistic turn" and the analytic tradition in philosophy—helped sketch the essential features of the expected reduction and contrast the logicist proposal with other major philosophical traditions, the Kantian and empiricist ones (also including some brands of formalism) being the most relevant. The *Grundgesetze der Arithmetik* ([GGA]) finally offered the fully developed reduction in its proper formal setting and gestured at further extensions of the logicist view to real numbers. Frege's *opus magnum* inherited—something already being appealed to in the previous two works—the request of rigorous and gapless derivation of arithmetical truths that was consonant with the mathematical spirit of the time.

As the story goes, Frege's original project—the idea that all arithmetical truths are nothing but notational variants of logical statements, derivable from a restricted number of basic logical laws—collapsed, even before the second volume of the *Grundgesetze* was to appear, under the blow delivered by Russell's Paradox, affecting the very definition of extension, or better of value-range, as exposed in *Grundgesetze*'s Basic Law V, on which most of Frege's edifice was founded.

Russell and Whitehead undertook the hard task of salvaging the logicist enterprise from paradoxes (including Cantor's, Burali-Forti's, and others'). Their monumental work in *Principia Mathematica* ([PM])—as well as Russell's other works in this area (like [PoM] and [IMP])—testifies to the effort involved in their endeavor (see Giaquinto, 2002). In the end, however, their theory of types—in either the simple or ramified version—allowed preserving the technical stability of the expected foundation at high costs. Not only was the complexity of their formal setting to be superseded by the emerging Zermelo-Fraenkel set theory (ZFC), also, and

---

6 As Grattan-Guinness (2000), p. 501, recalls, the word "Logistique" was first adopted by Couturat during the 1904 International Congress of Philosophy. It was later used by Russell and others but also to cover views other than the one now usually intended. The real diffusion of the word is then mainly due to Carnap, who adopted it in his *Abriss der Logistik* (1929) in order to refer to views such as those presented in *Principia Mathematica*.

above all, some of the principles they had to assume—most notably, apart from the Axiom of Reducibility and the Multiplicative Axiom, the Axiom of Infinity, which was needed to grant the existence of as many classes as required by the envisioned foundation—jeopardized the expected logicist nature of their reconstruction.

Logicist ideas were not to be abandoned, nonetheless. Through the works of Wittgenstein and Ramsey,[7] they were inherited and espoused by Carnap (who also attended Frege's lectures in Jena between 1910 and 1914; see Carnap [FLL]) in the much-changed climate of Viennese neo-positivism. Here logicism preserved some of its original traits but was to encounter the needs of logical empiricism, the requirement to secure knowledge of non-empirical statements within a verificationist conception of meaning, and a revised understanding of logic in the framework of Carnap's conventionalism.

After the substantial demise of logical empiricism, logicism ended up, if not in a state of oblivion, at least in a mild hibernation. Even a cursory look at Benacerraf and Putnam's momentous collection (1964/1983)—which can reasonably be taken as a blueprint of the major threads in the debate in the philosophy of mathematics up to the sixties and then, with the second edition, up to the early eighties of the past century—one immediately realizes that the only positive logicist contributions are, indeed, those by Frege (from [GLA]), Russell (from [IMP]), and Carnap ([LFM]). Leaving aside the consideration it receives in the editors' introduction, logicism remains in the debate mainly as a term of opposition. In association with Frege, it is objected to due to semantic concerns vis à vis structuralism (Benacerraf, 1965) or for epistemological qualms concerning platonism (Benacerraf, 1973); in association with Russell, it is mentioned to emphasize the limits of the conception of mathematical logic underlying *Principia Mathematica* (Gödel, 1944); in association with Carnap, it is discussed within a broader charge of untenability of truth by convention (Quine, 1964). No other positive logicist proposal occurs in the volume.[8]

From this perspective, it is ironic that the second edition of Benacerraf and Putnam's selected readings was published in the same year in which Crispin Wright first attempted to resuscitate Frege's project (Wright, 1983). Neo-logicism—also known as Scottish neo-logicism, neo-Fregeanism, or abstractionism—as developed by Wright and Bob Hale (cf. Hale & Wright, 2001) and sustained by formal results mainly due to George Boolos and Richard Heck (cf. Boolos, 1987a, and Heck, 1993; for a general

---

7 For an overview of the contributions by Wittgenstein and Ramsey on this score, see Potter (2000), chs. 6, 8; Potter (2019), Parts III, IV.

8 It goes without saying, however, that since the early seventies, Frege's ideas, including those underlying his logicism, were to be considerably explored, disseminated, and developed through the works of Michael Dummett.

discussion, see Tennant, 2017, and for a survey Zalta, 2018)—was to mark a new phase in the consideration of logicist ideas. The possibility of reviving a logicist foundation of arithmetic—as weak as it was with respect to Frege's outright identification of arithmetical statements with logical statements—by abandoning the ill-fated Basic Law V and resorting to what is today known as Hume's Principle as a suitable definition of the concept of cardinal (and hence finite cardinal, or natural) numbers[9] initiated a logicist renaissance which resulted in an impressive wealth of new lines of research and stands out as a major contribution in contemporary times to the history and development of logicism.[10]

EXPANDING THE LANDSCAPE

This short story has several gaps. Obviously, one first missing element consists of a full contextualization of early logicist ideas in the mathematical and philosophical milieu of the preceding and contemporary times. As for the former, one should look at the legacy of possible predecessors stemming from Plato to Euclid to Leibniz, to mention just a few cases in point. As for the latter, it impossible to properly understand the emergence of Frege's, as well as Dedekind's, thought without an extensive contrast with Kant's views on the nature of logic, mathematical judgments, analyticity, and the intuition of space and time (see e.g., Coffa, 1991; MacFarlane, 2002)—nor, on the positive side, without placing their mathematical enterprise in the context of the rigorization and arithmetization of analysis permeating the mathematical community of their times (see e.g., Benacerraf, 1981; Demopoulos, 1994; Ferreirós, 2007, chs. 3, 4, 5, 7; Reck, 2020). Above all, a full appreciation of the early stages of logicist ideas cannot be obtained without realizing that even authors—be they philosophers or mathematicians or both—commonly placed outside, or in

---

9 Both Basic Law V (BLV) and Hume's Principle (HP) are instances of abstraction principles, that is, principles of the form $\S\varphi=\S\psi\leftrightarrow R_E(\varphi, \psi)$, where $\S$ is an abstraction function mapping the φs and ψs into a domain of entities of a possibly different sort, and R_E is an equivalence relation holding between the φs and ψs. BLV (in one of its readings) states that: $\{F\}=\{G\}\leftrightarrow F\equiv G$ (the extension of the concept F is identical to the extension of the concept G iff F and G are equipollent, that is, the same elements fall under both of them); HP states that $\#F=\#G\leftrightarrow F\approx G$ (the number of the concept F is identical to the number of the concept G iff F and G are equinumerous, that is, one has as many elements falling under it as the other); another notable abstraction is the Direction Principle, which states that: $d(a)=d(b)\leftrightarrow a//b$ (the direction of line a is identical to the direction of line b iff a and b are parallel). The discussion on the role and import of abstraction principles, both in traditional and contemporary varieties of logicist projects, can hardly be overestimated.
10 Beyond Hale's and Wright's own works, one can see the collections Cook (2007), Ebert and Rossberg (2016a), and Heck (2011). On the same track, the new translation of Frege's *Grundgesetze* has opened new studies on Frege's major work, such as Ebert and Rossberg (2019) and Heck (2012).

open contrast with, the logicist tradition advocated varieties of logicism. Most notable among these is Dedekind.[11] Despite his role as the initiator, or forefather, of mathematical structuralism, Dedekind explicitly qualifies himself in logicist terms, and only uncompromising assumptions on what logicism entails—like the anti-psychologistic and anti-creationistic ones famously advanced by Russell in his objections—can prevent acknowledging him such qualification. Assessing the origins of logicism without considering Dedekind's view would be at best partial.[12]

Second, even the most representative views of Frege and Russell unfolded via a number of discussions and exchanges with philosophers and mathematicians of the times. Appreciating their proposals and their legacy in philosophy and mathematics cannot overlook Frege's reactions to Boole and Peano; the controversy between Frege and Hilbert on axioms and definitions; the unequal treatment Frege provided of mathematics and geometry and his reaction to the development of non-Euclidean geometries; the heated contrast between Frege and Husserl (basic references on the previous issues are to be found in Frege's own works, especially [PMC] and [PW]); the philosophical connections that could be established between Frege's and Bolzano's views, even beyond the shared conviction in a rigorous and gapless presentation of mathematical theories (e.g., Chihara, 1999; Sundholm, 2000); Frege's (and Russell's) reception of the mathematical advancements of their time as well as the development of Cantorian set theory and, later, of axiomatic set theory;[13] the comparison between logicism and not only rival views such as those of Hilbert and Brouwer but also stretching to critics such as Poincaré;[14] and a closer consideration of the role of Wittgenstein and Ramsey in conveying logicist ideas to Carnap, as well as the reception of those ideas from other leading figures of logical empiricism.[15]

When it comes to the present day, moreover, that story should be complemented by an impressive number of factors. To mention a few major

11 Logicist ideas can be detected in Hilbert, too, according to Ferreirós (2009).
12 The rightful insertion of Dedekind among early logicists has recently and luckily become more and more explicit. For some examples, beyond those included in this volume, see Benis-Sinaceur, Panza and Sandu (2015), Demopoulos and Clark (2005), Detlefsen (2011), Ferreirós (forthcoming), Hellman and Shapiro (2018), Klev (2017), Reck (2013a, b, 2019), Reck and Schiemer (2020), Tennant (2017).
13 Regarding Frege, as far as the mathematical role of abstractive definitions is concerned, see, for example, Mancosu (2016); for a contextualization of Frege's works in the mathematical setting of his time, see Wilson (2010). On a more general exploration of the development of set theory, see Ferreirós (2007), Hallett (1984).
14 For some basic references on Hilbert's formalism and Brouwer's intuitionism, see, for example, the essays in Benacerraf and Putnam (1964/83), in Mancosu (1998), and, for their legacies (beyond other references given previously) in Lindström, Palmgren, Segerberg and Stoltenberg-Hansen (2009); on Poincaré, see, for example, Detlefsen (1992).
15 See, for example, Potter (2019).

elements, one can think of the extended debate on the nature of logic due to the extensive development of mathematical and philosophical logic (think, for example, of the already mentioned discussions on the status of second-order logic—on which see, regarding neo-logicism, for example, Heck, 2018; Shapiro & Weir, 2000—and appeals to alternative logics as the background for neo-logicism—see, for example, Boccuni, 2013); the presence of alternative varieties of neo-logicism, such as those by Tennant (1987) or Zalta (1999); and the emergence of non-logicist abstractionist views, or anyway views that acknowledge the relevance, albeit in different interpretations, of abstraction principles (e.g., Linnebo, 2018; Rayo, 2013), as well as a significantly more nuanced exploration of forms of analiticity on the one hand and aprioricity on the other, including the discussion on non-propositional forms of *a priori* justification (like entitlement; see Wright, 2016), which may underlie the justification of basic mathematical and logical principles. But in order to be completed, that story could lead us even further, to a number of issues to which it is impossible to give due justice here: the possibility of recovering viable versions not only of Frege's logicism but also of other varieties, starting from Russell's; a careful disentanglement of similarities and dissimilarities between logicist and structuralist views, as made possible by the contemporary evolutions of both traditions; the suggestions stemming from a renovated interested in the interaction between traditional philosophy of mathematics and the philosophy of mathematical practice; an improved examination of the relative weight of pure vs. applied mathematics in foundational programs; the relation between epistemological and metaphysical foundations, also in light of recent discussion about conceptual and metaphysical grounding, as applied to mathematics; the exploration of non-monist accounts of foundations in the philosophy of logic and mathematics, including both logico-mathematical pluralism and the discussion of abductive (or "anti-exceptional") methodologies in the choice of logical and mathematical theories; and, last, a discussion of the relation between traditional epistemological concerns on the one side and the empirical adequacy of foundational programs on the other, in light of conceptions of developmental psychology and cognitive sciences that have by far surpassed the theories that Frege's and Russell's anti-psychologism targeted.

There are surely other gaps that we are omitting. Only some of these topics will be covered in the chapters collected here. Many others amount, as far as the present volume is concerned, to work undone. Additional contributions on some of those items were planned but could not be included in the end; some were wished for but could not be considered for lack of time and space. If the number of possible additions and ramifications is so conspicuous, this is because logicism stands, then, as today, at the crossroads of an inextricable maze of connections with some of the most pressing questions in both philosophy and mathematics.

* * *

The present volume is motivated—as were the workshops behind it—by four convictions.

The first—quite unsurprisingly for anyone familiar with either the philosophy of mathematics or Fregean exegesis itself—is that even in its most paradigmatic version, that is, Frege's, logicism still leaves many open issues to be explored at both the exegetical, philosophical, and formal levels.

The second is that a full appreciation of the value of logicism, and the extent to which it can be pursued in one version or another, cannot but be obtained by looking at a network of comparisons with both cognate and rival views, for only this allows identifying its essential traits in an unprejudiced manner, assessing the ways in which common tenets can be differently refined, and locating logicist views within a wider philosophical and mathematical context.

A third motivation is that logicism should not be seen, today less than ever, as a monolithic standpoint. A richness in varieties of logicist views is not only intrinsic to its early stages but has reverberated up to the present day and still yields a plurality of paths that can be taken in reviving its original suggestions or some reevaluation of them.

Finally, a last (but not least) motivation is the belief that logicism still has a lot to say and many viable options to offer and that only a careful combination of historical, exegetical, mathematical, and philosophical explorations can provide the proper setting within which its prospects can be valued.

The themes running through the chapters collected here, to which we now turn, are meant to exemplify and substantiate these guiding thoughts.

THEMES AND CHAPTERS

Part I: Origins

Any exhaustive discussion of logicism must commence by looking closely at the authors with which the view, even possibly in different concurrent varieties, originated. As is predictable, then, the volume opens with a section devoted to the major historical figures in this tradition. It will not come as a surprise to find chapters on Frege and Russell in this section, but there are two slightly more controversial features. One is the absence of Carnap, or anyway of the conception(s) of logicism that pervaded logical empiricism. Surely Carnap is often recalled—especially thanks to some well-known views such as those presented in Carnap [LFM]—as an influential proponent of (some form of) logicism, but it is clear that his views (like those of his fellow neo-positivists) were influenced by a much more variegated philosophical, scientific, and cultural context than the one in which Frege's and Russell's accounts originated (cf. Coffa, 1991). For this reason, the reception and evolution of logicism in Carnap and logical empiricism are here considered as parts of a later development and are treated in Part II of the volume. Equally controversial to some may be to find Dedekind among early logicist thinkers. Some theoretical aspects of Dedekind's works were subjected to fierce objections by Frege and (above all) Russell; also, the structuralist tradition Dedekind is often thought to inaugurate stood, and still stands, in stark contrast with the Fregean one. But it is undeniable that many of Dedekind's remarks on the relation between mathematics and logic are written in a spirit that could easily be shared by Frege and Russell. As this has become more and more appreciated in recent literature, we think it is rightful to fully count Dedekind among early logicists.

Chapters in Part I unfold in an approximate chronological order, which is also meant to reflect conceptual developments. In considering Frege's logicist project, it is both natural and recommended to move from earlier conceptions in *Grundlagen* to more mature evolutions in *Grundgesetze*. To this aim, in her chapter "Frege on Caesar and Hume's Principle", Patricia Blanchette opens the volume by discussing one of the crucial issues of Frege's project as exposed in *Grundlagen*, namely the reasons Frege abandoned Hume's Principle as a good definition of the concept of cardinal number capable of providing truth-conditions to a sufficiently wide array of arithmetical identity statements (an issue later to take central stage in neo-logicist views, discussed in Part III). Despite a common belief that those reasons are to be found in the so-called Caesar Problem, Blanchette advances alternative readings of the relevant texts concluding that the limitations imposed by such problem are in fact less pressing than a broader impossibility of introducing object-names via abstraction principles, as a retrospective analysis of Russell's paradox allows appreciating. A rather different conception of Frege's realism, usually associated with his logicism, then ensues.

Such reflections naturally lead one to consider how Frege conceived of truth and truth-values in providing a semantics for arithmetical language. Robert May's chapter "The Role of Truth" focuses directly on this, especially on Frege's conception of truth-values as objects. While placing the most fundamental question for Frege not in the nature of truth (given how truth-values can be stipulated to be appropriate but still arbitrary value-ranges) but in the role truth plays in semantics and logic, May investigates how the identification of truth-values with objects is supposed to discharge the required roles. He does this by focusing both on Frege's characterization of his formal language as a referential language (in [GGA], I.§§29–32) and on his identification of truth-values as 0-level entities (in [GGA], I.§10). May considers how the role of truth is tied to Frege's conceptual hierarchy and ends up emphasizing a significant tension in Frege's framework, brought about by the indefinability of truth, on the one hand, and the necessity, on the other hand, of precisely identifying truth-values as objects of some sort in order to provide an adequate account of the language of logic. Such an identification overtly clashes with Frege's explicit rejection of creative definitions, which nevertheless are clearly playing a role in Frege's stipulation that truth-values are value-ranges of some kind.

Further related tensions, or gaps, in Frege's systems emerge when one considers the nature of value-ranges themselves. In his chapter *"Anzahlen and Werthferläufe. Are There Objects Other Than Value-Ranges in Frege's Universe? An Argument From Grundgesetze*, §§I.34–40", Marco Panza explores Frege's formal system in this regard, starting from the widespread conception that such a system assigns invariant meanings to its primitive functional constants and fixed quantifier domains and focusing on the chain of definitions leading to the definition of *Anzahlen* in *Grundgesetze*. Such invariance and fixity depend on how value-ranges are conceived of, but Frege never explicitly clarifies their nature. Moreover, he never explicitly rules out whether his system allows for objects different from value-ranges. Lacking these clarifications, the sought-for precision for mathematical definitions and theorems seems threatened.

Together, the first three chapters depict an inherently nuanced account of some of the most basic elements underlying Frege's logicism: his conception of definitions and reference, his understanding of truth and truth-values, and the role of logical (or other) objects in delivering (paradoxes apart) a stable logicist foundation for arithmetic. One is not bound, however, to take all aspects of Frege's philosophy on board in order to argue for logicism. As Erich Reck shows in his chapter, "Dedekind's Logicism: A Reconsideration and Contextualization", Dedekind should be rightfully held to be a proponent of logicism. Even independently of Dedekind's own declarations on this score, his works clearly deliver a variety of logicism. The chapter defends this reading against previous criticism and shows how Dedekind's logicism is deeply entrenched in the mathematical developments of the time. Finally, the chapter advances an interpretive and methodological suggestion which is especially relevant for the purposes of this volume: namely that we should give a sufficiently general explanation of what can count as logicism and avoid identifying its essential features by bringing in specific traits and philosophical assumptions relating to only some historically major figures like Frege and Russell.

A closer look at the latter is then needed to complete a survey on the original varieties of logicism, and Kevin C. Klement provides one in his chapter "Logical Form and the Development of Russell's Logicism". An underlying theme in Fregean and Dedekindian conceptions of logicism is the role individual abstract mathematical objects, if there are any, are to play to sustain a logical analysis of mathematical statements and at the same time provide such analysis with the expected general validity. Russell, also in response to the paradoxes affecting set theory and mathematical logic, moved progressively towards a conception according to which arithmetical statements are indeed not about objects at all but about "fragments of logical forms". Surface syntax—contrary to a standard reading of Frege's position—thus appears misleadingly to be about objects. However, once unveiled, the proper logical form of mathematical statements is able to guarantee the general validity expected of logical claims. This reading helps both to individuate what is peculiar to Russell's brand of logicism as opposed to other varieties and also to draw comparisons with some prominent contemporary readings of definitions by abstraction, to be explored later in the volume.

Part II: Dangerous Liaisons

Logicism was not born in either a philosophical or mathematical vacuum. Inevitably, the views of its founding fathers overhung their critical contemporaries, as well as their disciples and successors, but also inspired positions that are unexpectedly connected with theirs. The chapters in Part II of this volume shed light on some of those liaisons.

When considering the interconnections between different forms of logicism and other views on mathematics, typically, one is reminded of Frege's correspondence with, among others, Couturat, Jourdain, Vailati, likely Wittgenstein, and, of course, Russell. Among the scholars Frege entertained intellectual relations with, authors such as Hilbert and Husserl easily come to mind, especially for the views they held in opposition to Frege's (cf. [PMC]). Famously, their disagreement with Frege concerned respectively the understanding of the role of axioms in mathematical theories and the relation between existence and consistency and the role of psychology (or more generally psychologism) in the logic and philosophy of mathematics. Traditionally, also other authors of that time, such as Peano and his mathematical school, are considered to maintain views incompatible with logicism. Nevertheless, at a closer look, this incompatibility may turn out to be not as divisive as it is usually reckoned. Finally, the work of subsequent authors, above all Carnap, is justly considered, not only as a matter of scholarship, as continuing the legacy of the forefathers, but at the same time as inaugurating a more or less radical departure from early logicism.

Given the role Frege played in establishing logicism as a foundational view, the crucial exchanges he had with contemporary authors on the philosophical core issues of his program, and the legacy he himself passed on, Part II opens with two chapters on renowned debates those exchanges prompted.

The Frege-Hilbert controversy is famous, particularly for the different views those two authors held on existence and consistency (cf. e.g., Blanchette, 2018) and the role they ascribed to axioms in mathematical theories. Consistently with this latter debate, in "Frege and Hilbert on Conceptual Analysis and Foundations: Some Remarks", Michael Hallett deepens the analysis of the Frege-Hilbert controversy by highlighting

Frege's and Hilbert's different positions on the conceptual analysis of basic theoretical concepts, also with respect to Euclidean geometry. Where Frege held the view that the investigation of the foundations of mathematics amounted to a quest for the ultimate logical primitives, which was so clearly attuned with his logicist view, Hilbert rejected the very idea of ultimate primitives and ascribed to axioms the task of providing meaning to basic concepts, thus paving the way to the possibility of further analysis and to an overall structural conception of mathematics.

In "The Chimera of Logicism: Husserl's Criticism of Frege", Mirja Hartimo focuses on another well-known debate of that time: the one that placed Frege and Husserl in opposition. Famously, the two mutually criticized their respective views on arithmetic and therefore on the core target of logicism. In particular, Frege harshly criticized Husserl's *Philosophy of Arithmetic* on the basis of its psychological gist. Hartimo reads Husserl's reaction to Frege's criticism as the motivation that led him to his mature understanding of the foundations of mathematics. According to Husserl, Frege's methods surely provided rigorous definitions—for instance, of the notions of number and quantity—but were in the end unable to deliver a proper philosophical analysis of what is essential to mathematics, namely all those notions that are central to abstract mathematical structures and are commonly appealed to in actual mathematical practice.

While the two previous chapters concern more classical divides logicism inspired, in "Peano's Philosophical Views Between Structuralism and Logicism", Paola Cantù sheds light on yet another traditional dissonance, which nevertheless turns out to be not as radical as it is usually reckoned: the one between logicism and Peano's mathematical school. Especially in the *Formulario*, Peano's mathematical efforts amount to understanding the roots of mathematics rather than providing a foundation for it. Still, in his framework, the role played by logical symbols is such as to grasp the invariant features of mathematical functions and operations, which to some extent is more akin to a logicist view than is commonly acknowledged, largely depending on what the main aim of logicism is taken to be.

If Peano's school can be intended as having a logico-philosophical and mathematical aim that, to some extent, has more in common with logicism than expected, indubitably the torch of a direct legacy of early logicist views is held by Rudolph Carnap. In "Logicism in Logical Empiricism", Georg Schiemer clearly states the extent to which Carnap's logicism owes early logicist positions a philosophical debt but also highlights the varied kind of logicism Carnap championed in logical positivism through the influence that the reading of Wittgenstein's *Tractatus* exerted upon the Vienna Circle. In particular, Schiemer focuses on the development of Carnap's views on logic and logicism in the years between 1920 and 1940. In this respect, Carnap revised the early logicists' conception of logic in the light of Wittgenstein's view of the tautological status of logical truths, thus yielding, via Ramsification, an if-thenist conception of the logicist endeavor that he also aimed at extending to branches of mathematics other than arithmetic.

Part III: The Roads Ahead

The legacy of a foundational view can be measured, among other things, by its ongoing influence. In the past decades, the interest in classical logicism in its varied forms has been revived, and nowadays works in and around logicist views in the philosophy of mathematics are indeed multifaceted. Still, a place of honor is clearly occupied by neo-logicism. Part III of this volume is then to some extent naturally devoted to it. At the same time, though, this part includes contributions that highlight directions of future research either by interconnecting early logicist views and more recent approaches in a novel way or by building bridges between traditional logicist themes and other current areas of research.

A rather crucial issue with any logicist framework concerns how much mathematics it can recover—while consistent. Famously, arithmetic and real analysis can indeed be interpreted in more or less appropriate logicist axiomatizations quite easily.[16] This might seem fair enough, if indeed the main goal of a logicist foundation of mathematics was Frege's. Still, a more contemporary worry concerns whether a logicist foundation is also available for the theory of ordinals and set theory more generally.

While the literature on this latter topic is by now quite extensive,[17] works on the theory of ordinals in a plainly logicist framework are lacking. Bob Hale's "Ordinals by Abstraction" aims at filling this gap. Clearly, the main difficulties here concern first and foremost the consistency of the theory (Burali-Forti's paradox looms large) but also whether such a project is indeed viable without recurring to the notion of ordinality embedded in ZFC set theory. Bob Hale's contribution tackles these two issues by providing a truly neo-logicist proposal, appealing to a consistent abstraction principle for ordinals that relies on Hume's Principle.

The divide between a neo-logicist set theory and ZFC's conception of sets can be seen in two different ways. First of all, one might argue that, if a logicist foundation of mathematics is to be feasible at all, it has to interpret all the axioms of ZFC. This demand might sound at best unjust to some: all in all, Frege's conception of value-ranges can be conceived of as an alternative to ZFC's iterative conception of sets, and it would be unnatural to attempt to collapse the former to the latter or the other

---

16 See, for example, Boccuni and Panza (2021), Dummett (1991), Hale (2000b), Schirn (2013), Shapiro (2000), Simons (1987), Snyder and Shapiro (2019), Wright (1983, 2000).

17 See, for example, Boolos (1987b, 1989), Cook (2003), Hale (2000a), Jané and Uzquiano (2004), Shapiro (2003).

way around. In fact, a second way to look at this divide focuses on the incommensurability of these two conceptions.[18] Roy Cook's "Logicism, Separation, and Complement" takes this latter standpoint. Starting from one of the traditional issues for (neo-)logicism, that is, the so-called *Bad Company Objection*,[19] Cook argues that any neo-logicist foundation of set theory, when combined with the criteria formulated for solving that objection, implies the axiom of complement, in contradiction with ZFC's axiom of separation. In this respect, a viable neo-logicist conception of sets does not plainly overlap with the iterative conception embodied in ZFC.

In "*Principia Mathematica* Redux", Gregory Landini highlights the structural nature of Whitehead and Russell's type theory in *Principia Mathematica*. This feature is not only somewhat opposed to Frege's view concerning the crucial role of logical objects in the reconstruction of arithmetic but is also a rather significant contribution to the debate on (Hilbertian) structural axiomatic conceptions of mathematics versus Fregean object-based approaches, also investigated in Hallett's contribution to this volume. A re-evaluation of Whitehead and Russell's core conception in *Principia* is therefore in order.

The renewal of the philosophical interest in Frege's logicism, in particular as represented by its main successor, namely Scottish neo-logicism, and in Dedekind's seminal contribution to that view is no surprise anymore. It is therefore only natural to bring the two together. Fiona Doherty's "The Ontology of Abstraction, From Neo-Fregean to Neo-Dedekindian Logicism" investigates the sharp ontological divide between Frege and neo-logicism. On the one hand, this divide leads neo-logicism to a thin conception of mathematical objects, in opposition to Frege's original view; on the other, it prompts a reconsideration of Hale and Wright's neo-logicism, yielding a neo-Dedekindian form of logicism.

Last but not at all least, a relatively recent area of research has questioned the rather entrenched divide between logicism and psychologism. Traditionally, those views are taken to contrast each other fiercely—if only on the basis of Frege's and Russell's scorn for the latter. Recently, though, a revived attention on this divide due, among other things, to studies on Frege's arithmetic and ordinary arithmetic (cf. e.g., Heck, 2000), on the one hand, and the development of numerical cognition, on the other,[20] has brought to light that such an opposition is less radical than the traditional debate might have foreseen. Lieven Decock's "Frege's Theorem

---

18 To some extent, this view is also explored in Jané and Uzquiano (2004).

19 For a survey, see, for example, Ebert and Rossberg (2016b), Linnebo (2009a), Tennant (2017).

20 For classic references, see Carey (2009) and Dehaene (2011); for some recent developments, see the essays in Bangu (2018).

and Mathematical Cognition" takes stock of this debate by underscoring important connections between a Fregean conception of cardinality and results from developmental psychology and linguistic anthropology. In this respect, both equinumerosity and the successor relation seem to play independent roles in the development of number concepts. However, methodological problems with cognitive explanations of the acquisition of numerical concepts suggest adopting a form of apsychologism in the analysis of such concepts as used in mathematical practice.

* * *

As these final remarks, and more generally those in the chapters of this last Part, suggest, logicism is far from being a view of purely historical interest. Revised interpretations, novel proposals, and unexpected connections between traditionally different stances all suggest that the prospects for the study of logicism and its legacy are more open than ever and that many roads lie ahead.

References

Bangu, S. (Ed.) (2018). *Naturalizing Logico-Mathematical Knowledge: Approaches From Psychology and Cognitive Science*. New York and Abingdon: Routledge.

Benacerraf, P. (1973). "Mathematical truth". *Journal of Philosophy* 70(19), 661–679. Reprinted in Benacerraf & Putnam (1964/1983), pp. 403–420. Unabridged 1968 edition published in F. Pataut (Ed.), *Truth, Objects, Infinity: New Perspectives on the Philosophy of Paul Benacerraf*, Springer, 2016, pp. 263–287.

Benacerraf, P. (1981). "Frege: The last logicist". *Midwest Studies in Philosophy* 6(1), 17–36.

Benacerraf, P. (1965). "What numbers could not be". *The Philosophical Review* 74(1), 47–73.

Benacerraf, P. & Putnam, H. (1964/1983). *Philosophy of Mathematics, Selected Readings*. Cambridge: Cambridge University Press, 1st edition 1964, 2nd edition 1983.

Benis-Sinaceur, H., Panza, M. & Sandu G. (Eds.) (2015). *Functions and Generality of Logic: Reflections on Dedekind's and Frege's Logicisms*. Berlin: Springer.

Blanchette, P. (2018). "The Frege-Hilbert controversy". In E.N. Zalta (Ed.). *The Stanford Encyclopedia of Philosophy* (Fall 2018 edition). Retrieved from https://plato.stanford.edu/archives/fall2018/entries/frege-hilbert/.

Boccuni, F. (2013). "Plural logicism". *Erkenntnis* 78(5), 1051–1067.

Boccuni, F. & Panza, M. (2021). "Frege's theory of real numbers: A consistent rendering". *Review of Symbolic Logic*, 1–44.

Boolos, G. (1984). "To be is to be a value of a variable (or to be some values of some variables)". *The Journal of Philosophy* 81(8), 430–449.

Boolos, G. (1985). "Nominalist platonism". *Philosophical Review* 94(3), 327–344.

Boolos, G. (1987a). "Saving Frege from contradiction". *Proceedings of the Aristotelian Society 87*, 137–151.

Boolos, G. (1987b). "The consistency of Frege's *Foundations of Arithmetic*". In J. Thomson (Ed.). *On Being and Saying: Essays in Honor of Richard Cartwright* (pp. 3–20). Cambridge, MA: MIT Press.

Boolos, G. (1989). "Iteration again". *Philosophical Topics 17*(2), 5–21.

Boolos, G. (1993). "Whence the contradiction?". *Aristotelian Society Supplementary Volume 67*, 211–233.

Burgess, J.P. (2005). *Fixing Frege*. Princeton: Princeton University Press.

Carey, S. (2009), *The Origin of Concepts*. Oxford: Oxford University Press.

Cariani, F. (2018). "A context principle for the twenty-first century". In A. Coliva, P. Leonardi & S. Moruzzi (Eds.). *Eva Picardi on Language, Analysis and History* (pp. 183–203). Basingstoke, UK: Palgrave Macmillan.

Chihara, C. (1973). *Ontology and the Vicious Circle Principle*. Ithaca, NY: Cornell University Press.

Chihara, C. (1999). "Frege's and Bolzano's rationalist conceptions of arithmetic". In Sinaceur H. (Ed.). *Mathématique et logique chez Bolzano*. Special issue of *Revue d'histoire des sciences 52*(3–4), 343–361.

Coffa, A. (1991). *The Semantic Tradition from Kant to Carnap: To the Vienna Station*. Cambridge: Cambridge University Press.

Cook, R. (2003). "Iteration one more time". *Notre Dame Journal of Formal Logic 44*(2), 63–92.

Cook, R. (Ed.) (2007). *The Arché Papers on the Mathematics of Abstraction*. Dordrecht: Springer.

Dehaene, S. (2011). *The Number Sense: How the Mind Creates Mathematics*. Oxford: Oxford University Press.

Demopoulos, W. (1994). "Frege and the rigorization of analysis". *Journal of Philosophical Logic 23*(3), 225–245.

Demopoulos, W. & Clark, P. (2005). "The logicism of Frege, Dedekind, and Russell". In S. Shapiro (Ed.). *The Oxford Handbook of Philosophy of Mathematics and Logic* (pp. 166–202). Oxford: Oxford University Press.

Detlefsen, M. (1992). "Poincaré against the logicians". *Synthese 90*, 349–378.

Detlefsen, M. (2011). "Dedekind against intuition: Rigor, scope, and the motives of his logicism". In C. Cellucci et al. (Eds.). *Logic and Knowledge* (pp. 205–217). Cambridge: Cambridge Scholars Publishing.

Donaldson, T. (2016). "The (metaphysical) foundations of arithmetic?". *Noûs 51*(2), 775–801.

Dummett, M. (1973). *Frege: Philosophy of Language*. London: Duckworth; 2nd edition 1981.

Dummett, M. (1991). *Frege: Philosophy of Mathematics*. Cambridge, MA: Harvard University Press.

Dummett, M. (1994). "Chairman's address: Basic Law V". *Proceedings of the Aristotelian Society 94*, 243–251.

Dummett, M. (1995). "The context principle: Centre of Frege's philosophy". In I. Max & W. Stelzner (Eds.). *Logik und Mathematik: Frege-Kolloquium Jena 1993* (pp. 3–19). Berlin: de Gruyter.

Ebert, P.A. & Rossberg, M. (Eds.) (2016a). *Abstractionism: Essays in Philosophy of Mathematics*. Oxford: Oxford University Press.

Ebert, P.A. & Rossberg, M. (2016b). "Introduction to abstractionism". In Ebert & Rossberg (Eds.) (2016a) (pp. 3–33).

Ebert, P.A. & Rossberg, M. (Eds.) (2019). *Essays on Frege's Basic Laws of Arithmetic*. Oxford: Oxford University Press.

Feferman, S. (2005). "Predicativity". In S. Shapiro (Ed.). *Oxford Handbook of Philosophy of Mathematics and Logic* (pp. 590–624). Oxford: Oxford University Press.

Ferreira, F. (2018). "Zigzag and Fregean arithmetic". In H. Tahiri (Ed.). *The Philosophers and Mathematics: Logic, Epistemology, and the Unity of Science* (Vol. 43, pp. 81–100). Dordrecht: Springer.

Ferreira, F. & Wehmeier, K. (2002). "On the consistency of the Δ^1_1-CA fragment of Frege's *Grundgesetze*". *Journal of Philosophical Logic 31*, 301–311.

Ferreirós, J. (2007). *Labyrinth of Thought. A History of Set Theory and Its Role in Modern Mathematics*. Basel & Boston: Birkhäuser Verlag (2nd Edition; 1st Edition *Science Networks – Historical Studies*, Vol. 23, 1999).

Ferreirós, J. (2009). "Hilbert, logicism, and mathematical existence". *Synthese 170*(1), 33–70.

Ferreirós, J. (forthcoming). "On Dedekind's logicism". In A. Arana & C. Alvarez (Eds.). *Analytic Philosophy and the Foundations of Mathematics*. London: Palgrave.

Field, H. (1989). *Realism, Mathematics, and Modality*. Oxford: Basil Blackwell.

Giaquinto, M. (2002). *The Search for Certainty: A Philosophical Account of Foundations of Mathematics*. Oxford: Oxford University Press.

Gödel, K. (1944). "Russell's mathematical logic". In P. Schilpp (Ed.). *The Philosophy of Bertrand Russell* (Library of Living Philosophers) (pp. 123–153). New York: Tudor.

Grattan-Guinness, I. (2000). *The Search for Mathematical Roots: Logics, Set Theories and the Foundations of Mathematics from Cantor Through Russell and Goedel*. Princeton: Princeton University Press.

Hale, B. (1988). *Abstract Objects*. Oxford: Blackwell.

Hale, B. (1994). "Singular terms (2)". In B. McGuinness & G. Oliveri (Eds.). *The Philosophy of Michael Dummett* (pp. 17–44). Dordrecht, Boston & London: Kluwer. Reprinted in Hale & Wright (2001), Essay 2.

Hale, B. (1996). "Singular terms (1)". In M. Schirn (Ed.). *Frege: Importance and Legacy* (pp. 438–457). Berlin & New York: Walter de Gruyter. Reprinted in Hale & Wright (2001), Essay 1.

Hale, B. (2000a). "Abstraction and set theory". *Notre Dame Journal of Formal Logic 41*(4), 379–398.

Hale, B. (2000b). "Reals by abstraction". *Philosophia Mathematica 8*(2), 100–123.

Hale, B. (2002). "Real numbers, quantities, and measurement". *Philosophia Mathematica 10*(3), 304–323.

Hale, B. (2016). "Definitions of numbers and their applications". In Ebert & Rossberg (2016) (pp. 332–347).

Hale, B. & Wright, C. (2000). "Implicit definitions and the *a priori*". In P. Boghossian & C. Peacocke (Eds.). *New Essays on the A Priori* (pp. 286–319). Oxford: Oxford University Press. Reprinted in Hale & Wright (2001), Essay 5.

Hale, B. & Wright, C. (2001). *The Reason's Proper Study: Essays Towards a Neo-Fregean Philosophy of Mathematics*. Oxford: Clarendon Press.

Hale, B. & Wright, C. (2002). "Benacerraf's dilemma revisited". *European Journal of Philosophy 10*(1), 101–129.

Hallett, M. (1984). *Cantorian Set Theory and Limitation of Size*. Oxford: Clarendon Press.

Heck, R.K. (1993). "The development of arithmetic in Frege's *Grundgesetze der Arithmetik*". *Journal of Symbolic Logic 58*, 579–601 (originally published under the name "Richard G. Heck, Jr").

Heck, R.K. (1996). "The consistency of predicative fragments of Frege's *Grundgesetze der Arithmetik*". *History and Philosophy of Logic 17*(1), 209–220 (originally published under the name "Richard G. Heck, Jr").

Heck, R.K. (2000). "Cardinality, counting, and equinumerosity". *Notre Dame Journal of Formal Logic 41*(3), 187–209 (originally published under the name "Richard G. Heck, Jr").

Heck, R.K. (2011). *Frege's Theorem*. Oxford: Oxford University Press (originally published under the name "Richard G. Heck, Jr").

Heck, R.K. (2012). *Reading Frege's Grundgesetze*. Oxford: Oxford University Press (originally published under the name "Richard G. Heck, Jr").

Heck, R.K. (2018). "Logicism, ontology, and the epistemology of second-order logic". In I. Fred-Rivera & J. Leach (Eds.). *Being Necessary: Themes of Ontology and Modality From the Work of Bob Hale* (pp. 140–169). Oxford: Oxford University Press.

Hellman, G. & Shapiro, S. (2018). *Mathematical Structuralism*. Cambridge: Cambridge University Press.

Hofweber, T. (2005). "Number determiners, numbers, and arithmetic". *Philosophical Review 114*(2), 179–225.

Jané, I. & Uzquiano, G. (2004). "Well- and non-well founded Fregean extensions". *Journal of Philosophical Logic 33*(5), 437–465.

Jeshion, R. (2001). "Frege's notions of self-evidence". *Mind 110*(440), 937–976.

Jeshion, R. (2004). "Frege: Evidence for self-evidence". *Mind 113*(449), 131–138.

Jung, D. (1999). "Russell, presupposition, and the vicious-circle principle". *Notre Dame Journal of Formal Logic 40*(1), 55–80.

Klev, A. (2017). "Dedekind's logicism". *Philosophia Mathematica 25*, 341–368.

Liggins, D. (2010). "Epistemological objections to platonism". *Philosophy Compass 5*(1), 67–77.

Lindström, S., Palmgren, E., Segerberg, K. & Stoltenberg-Hansen, V. (Eds.) (2009). *Logicism, Intuitionism, and Formalism: What Has Become of Them?* Dordrecht: Springer.

Linnebo, Ø. (2006). "Epistemological challenges to mathematical platonism". *Philosophical Studies 129*(3), 545–574.

Linnebo, Ø. (Ed.) (2009a). "The bad company problem". *Synthese* special issue *170*(3).

Linnebo, Ø. (2009b). "Frege's Context Principle and reference to natural numbers". In Lindström, Palmgren, Segerberg & Stoltenberg-Hansen (2009) (pp. 47–68).

Linnebo, Ø. (2018). *Thin Objects*. Oxford: Oxford University Press.

MacBride, F. (2003). "Speaking with shadows: A study of neo-logicism". *British Journal for the Philosophy of Science 54*(1), 103–163.

MacFarlane, J. (2002). "Frege, Kant, and the logic in logicism". *Philosophical Review 111*(1), 25–65.

Mancosu, P. (Ed.) (1998). *From Brouwer to Hilbert: The Debate on the Foundations of Mathematics in the 1920s*. Oxford: Oxford University Press.

Mancosu, P. (2016). *Abstraction and Infinity*. Oxford: Oxford University Press.

Moltmann, F. (2013a). "Reference to numbers in natural language". *Philosophical Studies 162*(3), 499–536.

Moltmann, F. (2013b). *Abstract Objects and the Semantics of Natural Language*. Oxford: Oxford University Press.

Panza, M. & Sereni, A. (2013). *Plato's Problem. An Introduction to Mathematical Platonism*. Basingstoke: Palgrave Macmillan.

Panza, M. & Sereni, A. (2019). "Frege's Constraint and the nature of Frege's foundational program". *Review of Symbolic Logic 12*, 97–143.

Peacocke, C. (1998). *Being Known*. Oxford: Oxford University Press.

Picardi, E. (2009). "Wittgenstein and Frege on proper names and the Context Principle". In P. Frascolla, D. Marconi & A. Voltolini (Eds.). *Wittgenstein, Mind, Meaning and Metaphilosophy* (pp. 167–188). Basingstoke, UK: Palgrave Macmillan.

Picardi, E. (2017). "Michael Dummett's interpretation of Frege's Context Principle. Some reflections". In M. Frauchinger (Ed.). *Justification, Understanding, Truth, and Reality* (pp. 29–62). Berlin: De Gruyter.

Potter, M. (2000). *The Reason's Nearest Kin*. Oxford: Oxford University Press.

Potter, M. (2019). *The Rise of Analytic Philosophy, 1879–1930. From Frege to Ramsey*. New York and Abingdon: Routledge.

Quine, W.V.O. (1964). "Truth by convention". In P. Benacerraf & H. Putnam (Eds.). *Philosophy of Mathematics: Selected Readings* (pp. 329–354). Cambridge: Cambridge University Press.

Quine, W.V.O. (1969). "Epistemology naturalized". In *Ontological Relativity and Other Essays*. New York: Columbia University Press.

Quine, W.V.O. (1970). *Philosophy of Logic*. Cambridge, MA: Harvard University Press, 2nd edition 1986.

Rayo, A. (2013). *The Construction of Logical Space*. Oxford: Oxford University Press.

Reck, E. (2013a). "Frege or Dedekind? Towards a reevalaution of their legacies". In *The Historical Turn in Analytic Philosophy*. Basingstoke: Palgrave-Macmillan.

Reck, E. (2013b). "Frege, Dedekind, and the origins of logicism". *History and Philosophy of Logic 34*, 242–265.

Reck, E. (2019). "The logic in Dedekind's logicism". In S. Lapointe (Ed.). *Logic From Kant to Russell: Laying the Foundations for Analytic Philosophy* (pp. 171–188). London: Routledge.

Reck, E. (2020). "Dedekind's contributions to the foundations of mathematics". In E.N. Zalta (Ed.). *The Stanford Encyclopedia of Philosophy* (Winter 2020 edition). Retrieved from https://plato.stanford.edu/archives/win2020/entries/dedekind-foundations/.

Reck, E. & Schiemer, G. (Eds.) (2020). *The Prehistory of Mathematical Structuralism*. Oxford: Oxford University Press.

Resnik, M. (1967). "The context principle in Frege's philosophy". *Philosophy and Phenomenological Research 27*, 356–365.

Rosen, G. (2010). "Metaphysical dependence: Grounding and reduction". In B. Hale & A. Hoffmann (Eds.). *Modality: Metaphysics, Logic, and Epistemology* (pp. 109–136). Oxford: Oxford University Press.

Rosen, G. (2012). "Abstract objects". In E.N. Zalta (Ed.). *The Stanford Encyclopedia of Philosophy*. Retrieved from https://plato.stanford.edu/archives/spr2020/entries/abstract-objects/.

Schirn, M. (2013). "Frege's approach to the foundations of analysis (1874–1903)". *History and Philosophy of Logic 34*(3), 266–292.

Schwartzkopff, R. (2011). "Numbers as ontologically dependent objects—Hume's Principle revisited". *Grazer Philosophische Studien 82*, 353–373.

Sereni, A. (2019). "On the philosophical significance of Frege's Constraint". *Philosophia Mathematica 27* (2), 244–275.

Shapiro, S. (1991). *Foundations Without Foundationalism: A Case for Second-Order Logic*. Oxford: Oxford University Press.

Shapiro, S. (2000). "Frege meets Dedekind: A neologicist treatment of real analysis". *Notre Dame Journal of Formal Logic 41*, 335–364.

Shapiro, S. (2003). "Prolegomenon to any future neo-logicist set theory: Abstraction and indefinite extensibility". *British Journal for the Philosophy of Science 54*(1), 59–91.

Shapiro, S. (2009). "We hold these truths to be self-evident: But what do we mean by that?". *Review of Symbolic Logic 2*(1), 175–207.

Shapiro, S. & Weir, A. (2000). "Neo-logicist logic is not epistemically innocent". *Philosophia Mathematica 8* (2), 160–189.

Simons, P. (1987). "Frege's theory of real numbers". *History and Philosophy of Logic 8*, 25–44.

Snyder, E., Samuels, R. & Shapiro, S. (2018). "Neologicism, Frege's Constraint, and the Frege-Heck condition". *Noûs*. https://doi.org/10.1111/nous.12249.

Snyder, E. & Shapiro, S. (2019). "Frege on the real numbers". In Ebert & Rossberg (2019) (pp. 343–383).

Sundholm, G. (2000). "When and why, did Frege read Bolzano?". In *LOGICA Yearbook 1999* (pp. 164–174). Prague: Filosofia.

Tennant, N. (1987). *Anti-Realism and Logic: Truth as Eternal*. Clarendon Library of Logic and Philosophy. Oxford: Oxford University Press, June.

Tennant, N. (2017). "Logicism and neologicism". In E.N. Zalta (Ed.). *The Stanford Encyclopedia of Philosophy* (Winter 2017 edition). Retrieved from https://plato.stanford.edu/archives/win2017/entries/logicism/.

Wilson, M. (2010). "Frege's mathematical setting". In T. Ricketts & M. Potter (Eds.). *The Cambridge Companion to Frege* (pp. 379–412). Cambridge: Cambridge University Press.

Wright, C. (1983). *Frege's Conception of Numbers as Objects*. Aberdeen: Aberdeen University Press.

Wright, C. (2000). "Neo-Fregean foundations for real analysis: Some reflections on Frege's Constraint". *Notre Dame Journal of Formal Logic 41*, 317–334.

Wright, C. (2007). "On quantifying into predicate position: Steps towards a new(tralist) perspective". In M. Leng, A. Paseau & M. Potter (Eds.). *Mathematical Knowledge* (pp. 150–174). Oxford: Oxford University Press.

Wright, C. (2016). "Abstraction and epistemic entitlement: On the epistemological status of Hume's Principle". In Ebert, P.A. & Rossberg, M. (Eds.) (2016a). *Abstractionism: Essays in Philosophy of Mathematics* (pp. 161–202). Oxford: Oxford University Press.

Zalta, E.N. (1999). "Natural numbers and natural cardinals as abstract objects: A partial reconstruction of Frege's *Grundgesetze* in object theory". *Journal of Philosophical Logic 28*(6), 619–660.

Zalta, E.N. (2018). "Frege's Theorem and foundations for arithmetic". In E.N. Zalta (Ed.). *The Stanford Encyclopedia of Philosophy* (Winter 2018 edition). Retrieved from https://plato.stanford.edu/archives/win2018/entries/frege-theorem.

* * * * * * * *

Selected Primary References

[LFM] Carnap, R. (1983). "The logicist foundation of mathematics". In P. Benacerraf & H. Putnam (Eds.). *Philosophy of Mathematics: Selected Readings* (2nd edition, pp. 41–52). Cambridge: Cambridge University Press.

[FLL] Carnap, R. (2004). *Frege's Lectures on Logic: Carnap's Student Notes, 1910–1914.* Translated and ed. with introductory essay by Erich H. Reck & Steve Awodey; based on the German text, edited with introduction and annotations by Gottfried Gabriel. Chicago, IL: Open Court, 2004.

[WSWSZ] Dedekind, R. (1888). *Was sind und was sollen die Zahlen?*, Braunschweig: Vieweg (originally published as a separate booklet); reprinted in Dedekind (1930–32), Vol. 3, pp. 335–391, in Dedekind (1965), pp. III–XI and 1–47, and in Dedekind (2017), pp. 53–109; English trans., (Dedekind 1901c) and (revised) Dedekind (1995).

[Beg] Frege, G. (1879). *Begriffsschrift. Eine der arithmetischen nachgebildete Formelsprache des reinen Denkens.* L. Nebert, Halle, 1879. Translated as *Concept Script, a Formal Language of Pure Thought Modelled Upon That of Arithmetic*, by S. Bauer-Mengelberg in J. van Heijenoort (Ed.). *From Frege to Gödel: A Source Book in Mathematical Logic, 1879–1931.* Cambridge, MA: Harvard University Press, 1967.

[GLA] Frege, G. (1884). *Die Grundlagen der Arithmetik. Eine logisch mathematische Untersuchung über den Begriff der Zahl.* Wilhelm Koebner, Breslau, 1884. *The Foundations of Arithmetic.* Translated by J.L. Austin. Oxford: Basil Blackwell, 1950.

[GGA] Frege, G. (1893/1903). *Grundgesetze der Arithmetik. Begriffsschriftlich abgeleitet.* Hermann Pohle, Jena, 1893/1903. *Basic Laws of Arithmetic. Derived Using Concept-Script.* Translated and edited by Philip A. Ebert & Marcus Rossberg. Oxford: Oxford University Press, 2013.

[PW] Frege, G. (1979). *Posthumous Writings.* Translated by Peter Long & Roger White, with the assistance of Raymond Hargreaves. Chicago: University of Chicago Press.

[PMC] Frege, G. (1980). *Philosophical and Mathematical Correspondence.* Gottfried Gabriel, Hans Hermes, Friedrich Kambartel, Christian Thiel, Albert Veraart, Brian McGuinness, & Hans Kaal (Eds.). Oxford: Blackwell Publishers.

[PoM] Russell, B. (1903). *The Principles of Mathematics.* Cambridge: Cambridge University Press.

[IMP] Russell, B. (1919). *Introduction to Mathematical Philosophy.* London: Allen & Unwin.

[PM] Russell, B. & Whitehead, A.N. (1910–1913). *Principia Mathematica*, 3 Vols. Cambridge: Cambridge University Press.

Part I
Origins

2 Frege on Caesar and Hume's Principle

Patricia Blanchette

1. Introduction

Frege's reaction to Russell's Paradox was, in the end, to abandon his logicist project. That he did so without seriously considering a now widely discussed alternative, the alternative of grounding arithmetic in pure logic together with Hume's Principle, is significant. Clearly there is something in such a grounding that Frege took to be incompatible with either his view of the nature of logic or of arithmetic, or his view of acceptable ways to pursue a reduction of one to the other. The purpose of this chapter is to investigate what, exactly, Frege took the difficulty to be, and what we can learn from this about his view of the nature of mathematics and of logic. I will be concerned to argue that there are some important difficulties with what might be called the "standard" answer to these questions, an answer according to which the central difficulty with the Hume's Principle strategy is to be found in the so-called "Caesar passages" of *Grundlagen*. The alternative answer offered here involves a different account of the importance of the Caesar passages and a revisionary, though I think better supported, account of Frege's view of the role of identity-statements in the reduction. Finally, I'll suggest that the alternative reading attributes to Frege a more plausible view of mathematical discourse than does the standard reading.

2. Background

Frege's attempt to prove the fundamental truths of arithmetic from purely logical principles involved a crucial appeal to the *extensions* of concepts. Both the pre-formal account in *Grundlagen* and the formal account in *Grundgesetze* employ a term-forming operator whose output is a collection of singular terms intended to refer to extensions, and a principle of

DOI: 10.4324/9780429277894-3

extensionality for those extensions.[1] The most important role of extensions in the project is Frege's identification of numbers as the extensions of concepts.

What Frege learned from Russell in 1902 was that his conception of extensions was incoherent: there cannot be objects which both answer to the singular terms just mentioned and obey the principle of extensionality. Frege's reaction to the discovery was that it entirely undermined his attempt to provide a logical foundation for arithmetic. He abandoned the project for this reason, holding in the end that logic is insufficient to furnish us with the objects needed for arithmetic.[2]

The neo-logicist project, following on from Crispin Wright's [(1983)], seeks to establish that Frege's abandonment of logicism was hasty, in the sense that a logicism very much in the spirit of Frege's project can be successfully pursued by making a straightforward adjustment to the foundation Frege attempted to give. That adjustment is to discard the notion of "extension" altogether and to provide reference to numerical singular terms not by stipulating that those terms refer to extensions, but by characterizing the nature of those references directly by appeal to an arguably analytic principle governing numbers. That principle, now known as "Hume's Principle," is the principle that the number that belongs to a concept F is the same as the number that belongs to a concept G iff the items falling under those concepts can be correlated one to one:[3]

(HP) NxFx = NxGx iff there is a bijection from the Fs to the Gs.

The crucial fact about HP is that, as noted by Charles Parsons and demonstrated in detail by Crispin Wright, the addition of HP to an unproblematic system of second-order logic provides resources sufficient for the proof of those fundamental principles of natural-number arithmetic that Frege himself sought to prove.[4] Indeed, Frege's own strategy turns on just this point: Frege's proof-sketches in *Grundlagen* and rigorous proofs in *Grundgesetze* can be understood as comprising two separable steps: first,

---

1 The extensions of *Grundgesetze* are known as "value ranges" (*Werthverläufe*). The principle of extensionality (that the extension of concept F is identical to the extension of concept G iff exactly the same things fall under those concepts, or more generally that the extension of F = the extension of G iff F and G give the same values for every argument) is entailed by Basic Law V of *Grundgesetze*; it is not stated explicitly in *Grundlagen* but is assumed throughout the proof-sketches given there.

2 See the unpublished fragment entitled "A new attempt at a foundation for arithmetic," [Frege (1979)] 278–281.

3 The name, introduced by George Boolos, is prompted by Frege's mention in *Grundlagen* §63 of Hume's recognition that equinumerosity can be characterized in terms of one-one correlation. See [Boolos (1990)]. On the infelicity of that name, see [Tait (1996)].

4 [Parsons (1965)], [Wright (1983)]. For helpful discussion of the history here, see [Heck (2011)], [Heck (2012)], and [Ebert and Rossberg (2016)].

the proof of HP from principles (including extensionality) that Frege took to be purely logical and, second, the proof from HP of the fundamental truths of arithmetic. As Richard Kimberly Heck has demonstrated, not only is it the case that Frege's proof of those fundamental truths relies on no further essential appeal to extensions once HP has been established, but importantly that Frege's construction shows that he himself knew this. In short, as Heck has put it, "there was no *formal* obstacle to the logicist program, even after Russell's discovery of the contradiction, and Frege knew it."[5]

This raises the following important question: Why, in the aftermath of the paradox, didn't Frege do as just suggested: why didn't he excise appeal to the problematic extensions, adopt HP as a fundamental principle, and proceed as before? Without an understanding of this, we certainly lack an understanding of Frege's conception of his own project and of his view of the constraints on its successful completion.

The standard answer to this question is essentially that Frege felt that such a strategy would fail to determine the references of numerical terms. That it would fail to do so is due, according to this account, to the fact that the HP strategy provides truth-conditions for insufficiently many kinds of identity sentences involving terms of the form "the number of Fs."[6] Appeal to HP will determine that each sentence of the form

"The number of Fs = the number of Gs"

shares the truth-value of the corresponding sentence

"There is a bijection mapping the Fs to the Gs"

but will not provide truth-conditions for sentences of the form

"The number of Fs = q"

for q not of the form "the number of Gs." In failing to do this, it fails, on this account, to determine the reference of terms of the form "the number of Fs." Finally, because numeral-reference is stipulated by appeal to terms of the form "the number of Fs," the HP strategy would, on this account, fail to determine the reference of the numerals. As Dummett puts it, the difficulty Frege saw with grounding arithmetic in HP was that "it affords us no possibility of determining the truth or falsity of a sentence like 'the

---

5 [Heck (2011)] 92.
6 This, as I understand it, is the view of Michael Dummett (see [Dummett (1991)] especially ch. 13), Richard Kimberly Heck (see [Heck (2005)] and [Heck (1997)]), Crispin Wright ([Wright (1983)] especially ch. 3), and Bob Hale ([Hale and Wright (2001b)]).

number of planets is Julius Caesar,' and thereby fails to determine the references of numerical terms."[7]

In what follows, it will be argued (i) that the texts usually thought to support this understanding of Frege's failure to adopt the HP strategy do not in fact do so and (ii) that by the time he learned of Russell's paradox, Frege did not hold that a successful account of arithmetical discourse must provide a semantics or truth-conditions for such non-arithmetical sentences as "the number of planets is Julius Caesar." An alternative account of Frege's failure to adopt the HP strategy will be proposed.

3.   *Grundlagen* §§55–56: The Recursive Analysis

Frege sets the groundwork for his eventual analysis of numerical discourse by first presenting two inadequate analyses and explaining their inadequacies. All of the analyses, both rejected and successful, focus on what Frege calls "assignments of number," that is, on statements like "There are nine planets," which he understands as involving the claim that a particular object (here, the number nine) bears a particular relationship (the relation of numbering) to a given concept (in this case, the concept *planet*). In his canonical mode of expression, each such statement is equivalent to an instance of the form

(*An*) "The number *n* belongs to the concept *F*."

It is crucial to Frege's overall strategy to treat the place occupied by "*n*" in this schema as open to quantification, that is, to treat the position marked by the ellipsis in

(*A*) The number . . . belongs to the concept F

as one into which it is possible to quantify with an object-level quantifier. Accordingly, his analyses of, for example,

(*A*0) The number 0 belongs to the concept F

and

(*A*1) The number 1 belongs to the concept F

must treat these sentences as predicating the content of the open sentence (*A*) of the content of "0" and of "1," respectively. This is not a trivial requirement: the analysis must provide an account of the content of (*A*0),

---

7 [Dummett (1991)] 156–157.

(A1), and so on that preserves the function-argument structure of the originals: each of the analysans-sentences must predicate a first-level concept of an object, the number in question.

The first rejected analysis, which we'll call "the recursive analysis," appears in §55 of *Grundlagen*, and the discussion of its failings in §56. Its core is the proposal to define each (finite) instance of (An) by stipulating that:

- The number 0 belongs to F iff for all x ¬Fx
- The number 1 belongs to F iff $\exists xFx$ & $\forall y\forall z(Fy$ & $Fz \rightarrow y = z)$
- The number n+1 belongs to F iff $\exists y(Fy$ & the number n belongs to $(Fx$ & $y \neq x))$

Frege's objection to the recursive analysis is as follows:

§56 These definitions suggest themselves so spontaneously in the light of our previous results, that we shall have to go into the reasons why they cannot be reckoned satisfactory.

The most likely to cause misgivings is the last; for strictly speaking we do not know the sense of the expression "the number *n* belongs to the concept *G*" any more than we do that of the expression "the number (*n* + 1) belongs to the concept F." We can, of course, by using the last two definitions together, say what is meant by "the number 1+1 belongs to the concept F" and then, using this, give the sense of the expression "the number 1+1+1 belongs to the concept F" and so on; but we can never—to take a crude example—decide by means of our definitions whether any concept has the number Julius Caesar belonging to it, or whether that same familiar conqueror of Gaul is a number or is not. Moreover we cannot by the aid of our suggested definitions prove that, if the number *a* belongs to the concept *F* and the number *b* belongs to the same concept, then necessarily *a = b*. Thus we should be unable to justify the expression "*the* number which belongs to the concept F," and therefore should find it impossible in general to prove a numerical identity, since we should be quite unable to achieve a determinate number. It is only an illusion that we have defined 0 and 1; in reality we have only fixed the sense of the phrases

"the number 0 belongs to"
"the number 1 belongs to";

but we have no authority to pick out the 0 and 1 here as self-subsistent objects that can be recognized as the same again.

The fundamental failing of the recursive analysis is that it fails to treat numerical statements like (A1) as each affirming of an object (e.g., the content of "1") that it falls under a concept [the content of (A)]. In providing meaning only to a series of semantically simple second-level predicates ("the number 0 belongs to . . .," "the number 1 belongs to . . .," etc.) and not to the two parts that Frege needs to treat as semantically significant units (the numerical singular terms "0," "1," etc. and the relation-term "The number . . . belongs to . . ."), the recursive analysis gives, as Frege says, "only an illusion" of having defined the numbers.[8]

The failure of the recursive analysis to treat the numerals in instances of (An) as singular terms blocks Frege's route to an account of the content of the concept-phrase

(**Num**) ". . . is a number."

That route (§72) is to stipulate that

(**Num**$_n$) "n is a number"

is to be understood as meaning

(**Num**$_n^*$) "there exists a concept such that n is the number which belongs to it,"

and again requires that the syntactic position marked by "n" be of the kind that can be filled by arbitrary singular terms, and hence of the kind into which one can place a first-level bound variable. That the recursive analysis fails to do this is made plain by noting that neither (Num$_n^*$) nor, consequently, (Num$_n$), is treated by the recursive analysis in a way that yields a syntactically well-formed sentence on the insertion of an arbitrary singular term at the position marked by "n."

If numerical discourse is understood in the way the recursive analysis proposes, then we cannot form the claim that a concept F has the number Julius Caesar belonging to it, and we cannot form the claim that Julius Caesar, or any arbitrary object, is a number. It will be impossible, that is, to understand assignments of number as affirming a relationship between an object and a concept, and impossible to understand (Num) as expressing a concept under which objects fall.

At least part of the role of the example "Julius Caesar," then, is to stand as an uncontroversial singular term used to illustrate the failure of the recursive analysis to treat numerals as singular terms. It is sometimes supposed that the example is used to point out a further failing of the recursive analysis, that is, a failure to provide criteria by means of which

---

8 For a clear discussion of this point, see [Parsons (1965)].

to determine that Julius Caesar is not a number. There is a sense in which the recursive analysis is guilty of this failing: in failing to treat "Julius Caesar is a number" as well formed, it stands in the way even of asking the question whether Caesar is a number. It is important, though, to note that the failure here is no indication that an adequate account of the semantics of (Num) must come along with criteria distinguishing numbers from, for example, people. Keeping in mind Frege's central goal, an accurate analysis of the syntax and semantics of arithmetical discourse—which itself has nothing to say about the distinction between numbers and people—we should note that such a requirement is considerably stronger than is required by the solution of the more general problem on display in §§55–56.

4.   *Grundlagen* §62: Identity Sentences

In the next five sections (§§57–61), Frege clarifies what he means by the term "object," and notes again the importance of identity-sentences, like "1 + 1 = 2," in understanding the role of the numerals.[9] The crucial points here regarding objects are (i) that it makes sense to affirm identity-statements about them and (ii) that some of these identity-statements are true even when the terms flanking the identity-sign present their objects in different ways.[10] Finally, in §62, Frege makes these remarks about identity and objects:

> [W]e have already settled that number words are to be understood as standing for self-subsistent objects. And that is enough to give us a class of propositions which must have a sense, namely those which express our recognition of a number as the same again. If we are to use the symbol *a* to signify an object, we must have a criterion for deciding in all cases whether *b* is the same as *a*, even if it is not always in our power to apply this criterion. In our present case, we have to define the sense of the proposition
>
> > "the number which belongs to the concept *F* is the same as that which belongs to the concept *G*";

---

9 §57.

10 This passage is independently interesting for its hint of what will soon become Frege's idea that significant pieces of language carry a sense as well as a reference: ". . . what we have here is an identity, stating that the expression "the number of Jupiter's moons" signifies the same object as the word "four." And identities are, of all forms of proposition, the most typical of arithmetic. It is no objection to this account that the word "four" contains nothing about Jupiter or moons. No more is there in the name "Columbus" anything about discovery or about America, yet for all that it is the same man that we call Columbus and the discoverer of America." [§57]

that is to say, we must reproduce the content of this proposition in other terms, avoiding the use of the expression

"the Number which belongs to the concept *F*".

In doing this, we shall be giving a general criterion for the identity of numbers. When we have thus acquired a means of arriving at a determinate number and of recognizing it again as the same, we can assign it a number word as its proper name.[11]

Parts of this passage are exactly as one would expect, given what has gone before. The requirement that we treat numerical terms as genuine singular terms, that is, as referring to "self-subsistent objects," involves the requirement that these terms be able to stand in the appropriate place in sentences. And a particularly important such place is the whole of the position to either side of the identity sign. The requirement is not merely that the analysis must assign a sense to sentences that appear superficially to be identity-sentences linking singular terms, since this weak criterion is satisfiable by an analysis that fails to preserve the singular-term status of the numerals. What's required, if Frege is to be able to put the analysis to work, is that the resulting analysans-sentences themselves have the structure of identity-sentences. Just as the recursive analysis provides only "an illusion" of defining the singular terms "0" and "1" etc., an analysis of

"The number which belongs to the concept *F* is the same as that which belongs to the concept *G*"

that does not contain two places into which uncontroversial singular terms can be meaningfully substituted, linking these places by the identity-sign, will fail to satisfy Frege's crucial requirement that arithmetical sentences treat numbers as objects.

All of this is familiar, given the discussion in §§55–61. But there is one odd line in this §62 passage, a line which has received a lot of attention, and which I will argue below has been the source of much confusion. This is Frege's claim that the use of a term *a* as a genuine singular term—as he puts it, its use "to signify an object"—requires that we "must have a criterion for deciding in all cases whether *b* is the same as *a*."[12] At first glance, this might seem uncontroversial: after all, if *a*

---

11 *Grundlagen* §62.
12 The German here reads as follows: "Wenn uns das Zeichen a einen Gegenstand bezeichenen soll, so müssen wir ein Kennzeichen haben, welches überall entscheidet, ob b dasselbe sei wie a, wenn es auch nicht immer in unserer Macht steht, dies Kennzeichen

and *b* are singular terms, then either they refer to the same object or they don't, and a failure to make sense of the apparent identity-claim "*a* = *b*" might be thought an indication that we do not, after all, have a genuine identity claim on our hands, or that we haven't determined what (at least) one of the apparent singular terms stands for. But while this thought is indeed natural when *a* and *b* are terms for ordinary household goods, it is not at all clear that the thought is plausible once we include under "singular term" and "object" all of the things that Frege includes. Mathematical singular terms appear in contexts: terms for real numbers appear in analysis, terms for rationals and terms for natural numbers in their own respective theories, and so on. And ordinary mathematical practice counts, for example, "the natural number 2" and "the real number 2" as genuine singular terms (for the reasons Frege has already given) without providing any means for determining the truth-value of the identity-sentence linking them. If one can make any sense of there being a "fact of the matter" whether the natural number 2 is identical to the real number 2, it is certainly one that has no bearing on mathematics. So if we take Frege's (apparent) requirement at face value, we will have to understand him as insisting on a very strong, and mathematically strange, requirement. The alternative is to understand Frege's use of "all cases" as restricted to a contextually relevant domain, for example, a particular mathematical theory.[13]

The two interpretive options regarding this passage are these: *First,* take Frege to have been speaking carefully and strictly in the target sentence of §62 and to have insisted on the following strong requirement:

> **Strong Identity Condition (SIC)** *t* is a genuine singular term only if for every genuine singular term *t'*, we have a criterion for determining whether the sentence *t* = *t'* is true or false.

Second, take Frege's prose here to be loose, with the intended requirement just the one we have already discussed above namely that

> **Weak Identity Condition (WIC)** *t* is a genuine singular term only if it is of the right syntactic kind to occur as a term in identity-sentences, which is to say that it can appear in the gap in true or false instances of the form "… = *t'*," where *t'* is a genuine singular term.

---

anzuwenden." A certain unclarity that isn't preserved in Austin's translation is what Frege means at the opening of the sentence: "If *for us* the sign a is to designate an object. . . ." An interesting question is whether this qualification is relevant to what's meant by "all cases" in this context. Thanks to Philip Ebert for pointing out the interest of this passage.

13 For further discussion of this remark about "all cases," and the interpretation of similar passages throughout Frege's work, see [Blanchette (2012b)].

The difference between the two is illustrated by the fact that the former, but not the latter, entails that at least one of the terms "the natural number 2" and "the real number 2" fails to count as a genuine singular term under ordinary conditions—that is, in the absence of some further stipulation that grounds the truth or the falsehood of the identity-sentence linking these terms.[14]

The "standard account" of Frege's later failure to adopt the HP strategy in light of Russell's paradox follows immediately from the idea that Frege holds (both here and in 1902) the stronger condition (SIC). The argument in favor of interpreting Frege as holding (SIC) is that he seems to say just this: the passage just quoted is perhaps most straightforwardly read as endorsing it. The argument against this interpretive option is that, as argued below (SIC) is inconsistent with Frege's favored treatment of arithmetical discourse (both in *Grundlagen* and *Grundgesetze*), and that there is a straightforward alternative account of the passage in question—that is, the contextually restricted reading just mentioned—which is consistent with (WIC) but not with (SIC). We return to this interpretive issue after laying out Frege's rejection of the second inadequate analysis of arithmetical discourse, and his own favored analysis.

5. *Grundlagen* §§63–68: The HP Strategy

Immediately after the passage just discussed, Frege presents the second inadequate proposal for the analysis of numerical terminology. The proposal is to define terms of the form "the number that belongs to the concept F" by stipulating that each instance of

(N=) "The number that belongs to the concept F = the number that belongs to the concept G"

is to mean:

"There is a 1-1 correlation between the objects falling under F and those falling under G."

That is to say, the proposal is just what we have called the "HP strategy" above and is essentially the proposal at the heart of neo-logicism. Frege's criticism of the strategy focuses not on the actual case of central concern, that is, on the analysis of terms of the form "the number that belongs to the concept F," but on a simpler analogical case having to do

---

14 Both conditions involve the usual Fregean "boot-strapping" in the sense that we gain from each a significant requirement on the genuine singular-term status of a term *t* only against the backdrop of some collection of other terms already recognized as genuine singular terms. For more on his understanding of such boot-strapping in a formal environment, see *Grundgesetze* I §30.

with directions. The proposal, outlined in §§64–65, is to give an analysis of direction-talk by stipulating that statements of the form

(**Dir=**) "the direction of line *a* = the direction of line *b*"

are to express what their counterpart sentences of the form

(**Par**) "line *a* is parallel to line *b*"

express. The criticism, in §66, is as follows:

> In the proposition
>
>> "the direction of *a* is identical with the direction of *b*"
>
> the direction of *a* plays the part of an object [*footnote*], and our definition affords us a means of recognizing this object as the same again, in case it should happen to crop up in some other guise, say as the direction of *b*. But this means does not provide for all cases. It will not, for instance, decide for us whether England is the same as the direction of the Earth's axis—if I may be forgiven an example which looks nonsensical. Naturally no one is going to confuse England with the direction of the Earth's axis; but that is no thanks to our definition of direction. That says nothing as to whether the proposition
>
>> "the direction of *a* is identical with *q*"
>
> should be affirmed or denied, except in the one case where *q* is given in the form of "the direction of *b*". *What we lack is the concept of direction;* for if we had that, then we could lay it down that, if *q* is not a direction, our proposition is to be denied, while if it is a direction, our original definition will decide whether it is to be denied or affirmed. So the temptation is to give as our definition:
>
>> *q* is a direction, if there is a line *b* whose direction is *q*.
>
> But then we have obviously come round in a circle. For in order to make use of this definition, we should have to know already in every case whether the proposition
>
>> "*q* is identical with the direction of *b*"
>
> was to be affirmed or denied.[15]

---

15 *Grundlagen* §66; emphasis added.

If we analyze (*Dir=*) as (Par), then, for the reasons given above in the rejection of the recursive analysis, we will have failed to treat direction-terms as genuine singular terms. We will have failed to provide a means of explaining direction-talk that makes sense of direction-terms as occupying a syntactic place suitable for substitution by arbitrary singular terms and therefore suitable for first-level generalization. Frege's means of demonstrating this failure is familiar: one demonstrates that the analysis provides no account of the meaning of instances of

(*Dir$_{arb}$*) ⟨*arbitrary singular term*⟩ = the direction of the Earth's axis

or of

(**Dir**) ⟨*arbitrary singular term*⟩ is a direction

by pointing out that the completion of these phrases by a token, uncontroversial singular term (here, "England") produces a sentence for which the HP analysis provides no counterpart analysans. We don't even get to the question of the semantic value of the sentences delivered by the analysis; the difficulty is that the analysis delivers nothing. In failing to provide an account that treats direction-terms as occupying a position into which uncontroversial genuine singular terms can be substituted, the "parallels analysis" fails to treat the position of these terms as susceptible to first-order generalization. And it therefore fails (i) to treat directions as objects and (ii) to treat the phrase ". . . is a direction" as referring to a concept. "What we lack," says Frege, "is the concept of direction." This is not because the analysis assigns to ". . . is a direction" a semantic value that provides no answer to whether England is a direction; it is because the analysis stands in the way of assigning to ". . . is a direction" any semantic value at all.

Similarly for the HP strategy and the concept of *number*. That strategy provides no account of instances of

(*N =$_{arb}$*) ⟨*arbitrary singular term*⟩ = the number of planets,

and so stands in the way of an account of

(*Num$_{arb}$*) ⟨*arbitrary singular term*⟩ is a number.

It fails, in short, to give us a concept of number.

Just as in the discussion of the recursive analysis, Frege's way of making vivid the fact that the HP analysis fails to reconstruct the apparent concept-phrase (". . . is the number that belongs to the concept *F*," or ". . . is the direction of the line *a*") as a genuine concept-phrase is to point out via the direction analogy that that reconstruction provides no argument-place suitable for a singular term to be inserted. There is no part of the recon-structed version of (N=) that can be swapped out for "England."

The interpretive question noted above can now be framed as follows. When using the unproblematic singular terms "Julius Caesar" and "England" to make his point, does Frege mean to convey the strong requirement that an analysis of the concept-phrases ". . . is the number that belongs to the concept *F*" and ". . . is the direction of line *a*" must come along with a criterion that distinguishes numbers from people and directions from countries? Or does he mean to convey just the weaker requirement that the analysans provided by an analysis of these concept-phrases must be one that has an argument place suited for a first-level bound variable and hence of the kind that takes genuine singular terms as arguments?

Focusing just on the text of *Grundlagen*, it is difficult to find a conclu-sive basis for preferring the strong or the weak reading of Frege's texts. The difficulties are these. In favor of the strong criterion is Frege's state-ment, quoted above that *a* is a genuine singular term only if it comes along with a criterion for deciding "in all cases" whether *b* is the same as *a*. Arguably weakening the force of this claim is that Frege makes it in service of his immediately preceding remark, for which he has clear reasons on either interpretive option, that genuine singular terms must be able to serve as terms in identity-sentences. Most importantly (or at least most influentially) in favor of the strong reading is Frege's choice of "Julius Caesar" and "England" as examples to illustrate the failure of the recur-sive and the Hume analyses to secure the singular-term status of numerical terms. Arguably weakening the force of these examples is the fact that in order to make his point, as sketched above Frege required terms that his audience would immediately grant as genuine singular terms. Since the status of numerical terms is at issue just here, Frege's point cannot be made via the use of numerical terms. Independently, then, of the question of whether he requires an adequate analysis of arithmetical discourse to settle such extra-mathematical issues as the identity of numbers and people, we should expect his rhetorical examples to be non-mathematical.

The strongest, though still inconclusive, argument for attributing to Frege merely the weaker condition is the argument that his own solution in *Grundlagen* meets the weak condition and fails to meet the strong. Frege's solution, presented in §68, is to understand "the number that belongs to the concept F" as referring to the extension of that second-level concept under which fall all and only the first-level concepts equinumer-ous with F. Extensions are clearly objects, according to Frege, and hence our core problem is solved: the weak requirement is satisfied. In analyzing instances of (N=) as instances of

(**Ext=**) "The extension of the concept '*equinumerous with the con-cept F*' = the extension of the concept '*equinumerous with the concept G*'," the favored analysis treats the numerical terms as occupying uncontroversially genuine singular-term positions. But it is far from clear that the analysis satisfies the strong criterion. The identity-statements involving extension-terms that are clearly taken in *Grundlagen* to be meaningful are those of the form

(**ID-ext1**) The extension of F = the extension of G

where F and G are first-level concepts,

(**ID-ext2**) The extension of Φ = the extension of Ψ

where Φ and Ψ are second-level concepts, and

(**ID-mix**) The extension of Φ = the number that belongs to the (first-level) concept F.

Instances of (ID-mix) are provided truth-conditions by reducing them to corresponding instances of (ID-ext2). Frege's proof-sketches indicate that he takes the truth-value of each instance of (ID-ext1) and of (ID-ext2), as he would have expected his audience to know, to be given by the principle of extensionality for concepts of that level. But Frege is silent, in *Grundlagen*, about truth-conditions for any other identity-statements involving extension-terms. As a consequence, the *Grundlagen* account does not provide truth-conditions for sentences of the form

(**ID-q**) The number that belongs to the concept F = q,

for q not of the form "the number that belongs to the concept G" or "the extension of the (second-level) concept Φ." This is at least odd, if it has been essential to Frege's line of reasoning all along that genuine singular terms must come along with criteria determining the truth-value of all such statements. If we take Frege to have insisted on what we have called the "strong condition," (SIC), above then his own analysis of arithmetical discourse fails, in a very straightforward way, to meet his requirement.

One *might* hold in response to this line of thought that Frege presumes his audience to know, for example, that extensions (and hence numbers) are not identical to people or to countries; this might be part of what he intends to convey in the footnote remark: "I assume it is known what the extension of a concept is."[16] But even this is not enough to satisfy the strong condition. Frege is well aware, by the time of writing *Grundlagen*, of the difference

---

16 *Grundlagen* §69, footnote.

between first-level and second-level concepts: quantification over the latter is essential to all of the interesting work in Part III of *Begriffsschrift*. *Grundlagen* treats directions as extensions of first-level concepts, and numbers as extensions of second-level concepts. Even granting Frege the assumption in *Grundlagen* of the extensionality condition for concept-extensions, we still get no truth-conditions for statements of the forms

The extension of (first-level) F = the extension of (second-level) Φ,
The extension of (second-level) Φ = the direction of a;

or

The extension of (first-level) F = the number that belongs to the (first-level) concept G.

For the extensionality principle provides truth-conditions only for identity-statements between extensions of concepts of the same level. We could of course make a further assumption on Frege's part and add to *Grundlagen*'s machinery the stipulation that all "cross-level" extension-identity claims are false. Or we could make on his behalf some other consistent assumption, one according to which some of these are true. Nothing in what happens later in *Grundlagen* will depend on how we decide such cases. The important point here is that Frege makes no attempt to provide such completeness, despite the fact that he takes himself clearly to have established that, on the analysis provided in §68, numerical terms are genuine singular terms.

The question in the air is whether Frege holds, at the time of writing *Grundlagen*, that a is a genuine singular term only if we have criteria that decide, for every genuine singular term b, the truth-value of the sentence $a = b$. The alternative is that, on his conception of singular terms (and hence of objects), there can be in the usual mathematical manner some "don't-care" cases that need not be assigned truth-conditions by an account of arithmetical discourse that is to be deemed adequate. If the latter is his view, then the point of the Caesar passages is not to establish the strong condition but to make vivid the failure of the recursive and Hume analyses to treat numerical terms as singular terms, that is, as occupying positions susceptible to first-level quantification.

Recall that the reason to ask this interpretive question is in order to answer the further question: why did Frege not adopt the Hume strategy in response to Russell's paradox? For this purpose, the most important interpretive question is not the question of Frege's attitude to the strong condition in 1884 but his attitude to it in 1902. And things become much clearer in the intervening years. In the next section, we look briefly at Frege's analysis of arithmetical discourse in *Grundgesetze*, with an eye toward his allegiance to the strong condition.

6.   *Grundgesetze*

In the years leading up to the *Grundgesetze*, Frege's view of functions and objects as the referents of predicative phrases and singular terms, respectively, becomes more precise as it is integrated into the mature semantic theory of sense and reference. His view of the semantic requirements on the kind of formal language he presents in *Grundgesetze* sharpens as well. One clear requirement on the formal language is that every well-formed string of symbols of that language must have a determinate reference, so that each well-formed sentence has a determinate truth-value. The rationale for this requirement in Frege's hands is, as Michael Dummett has argued, straightforward: without the satisfaction of this requirement, it is impossible to provide a syntactic criterion of axioms and of rules of inference guaranteed to sanction the inference from truths only of truths.[17] It is crucial, then, that each completion of a function-expression with argument-expressions of appropriate syntactic type have a determinate reference. Call this the "linguistic completeness requirement."

A distinct requirement, which Frege is often viewed as endorsing, is the requirement that every function referred to by a functional expression of the language be *total*, which is to say that it delivers a value for every object or function (or n-tuple thereof) of appropriate type as argument. Call this the *totality requirement*. The totality requirement implies that the function referred to by, for example, an addition-sign be defined over not just pairs of numbers but also over pairs of people, lamps, and shoe-strings. Because the expression left by removing a singular term from an identity-sentence is itself a functional expression, the totality requirement includes the requirement for every singular term t that there be a fact of the matter whether t co-refers with $t*$ for every singular term $t*$. Hence the totality requirement implies the strong identity condition noted above.

The totality requirement is considerably stronger than the linguistic completeness requirement: a formal language intended for the arithmetic of the integers will satisfy the linguistic completeness requirement with respect to its addition-sign if the addition-function it refers to is defined over all pairs of integers. In general, if a formal language is restricted to a particular subject-matter, then the linguistic completeness requirement as applied to that language is satisfiable without satisfaction of (SIC): the semantics of the language need have nothing to say about singular terms from outside the language.

The usual view of Frege's texts is that he intends the totality requirement, and there is considerable textual support for this view. There are also texts that don't fit well with this requirement, so that here again there

---

17 For a clear discussion of this point, see [Dummett (1991)] ch. 13, esp. pp. 155–159.

is an interpretive difficulty. I have argued elsewhere that Frege is most plausibly understood in the relevant passages as endorsing the weaker linguistic completeness requirement, and this only for formal rather than for natural languages.[18] For current purposes, the important question concerns Frege's view as confined to the formal language of *Grundgesetze*, since this is the language whose semantics bears on the question of integrating the neo-logicist proposal into Frege's mature logicist framework. The central question is this: what does Frege require of the function-terms and singular terms of the language of *Grundgesetze*?

The answer to this question is clear. *Grundgesetze* I §29 opens as follows:

We now answer the question: when does a name refer to something?

The answer is as follows:

- A name of a first-level function of one argument *has a reference* if every result of filling its argument-place with a *referring* proper name *has a reference*.
- A proper name *has a reference* if:
  - Whenever it fills the argument-place of a *referring* first-level function of one argument, the resulting proper name *has a reference*, and
  - Whenever it fills one or the other of the argument-places of a *referring* first-level function of two arguments, the resulting function-name *has a reference*.
- A name of a first-level function of two arguments *has a reference* if the result of filling both of its argument-places with *referring* proper names *has a reference*.
- A name of a second-level function of one argument of the second kind *has a reference* if every result of filling its argument place with a *referring* first-level function *has a reference*. (By "an argument of the second kind," Frege means a first-level function of one argument.)[19]
- A name of a third-level function *has a reference* if every result of filling its argument-place with a *referring* second-level function of one argument of the second kind *has a reference*.[20]

As Frege notes, these criteria cannot stand alone as an answer:

These propositions are not to be regarded as explanations of the expressions "to have a reference" or "to refer to something," since

---

18 See [Blanchette (2012b)].
19 See *Grundgesetze* I §23.
20 *Grundgesetze* I §29.

their application always presupposes that one has already recognised some names as referential; but they can serve to widen the circle of such names gradually. It follows from them that every name formed out of referential names refers to something.[21]

The names that Frege will "presuppose" to have reference are primitive names of truth-values; the §29 clauses then provide a characterization of a broader class of names as having reference on the basis of that presupposition.

The first question one might ask about these clauses concerns what might seem their fragmentary character. Why, for example, does the referential character of a singular term (proper name) turn just on its behavior when filling argument-places in names of one-place and two-place first-level functions? What about the cases in which it fills argument-places in names of, say, unequal-leveled functions or of three-place first-level functions? The answer is simply that there are no such primitive function-names in *Grundgesetze*. The standard for "having a reference" appropriate to a proper name is not that a value is determined when that proper name appears in the (or an) argument-place of *every* appropriately-leveled referring function-name. It is that a value is determined when that proper name appears in the (or an) argument place of every appropriately-leveled referring function-name *of Grundgesetze*. The extremely limited nature of this requirement is most vivid as applied to the important case of singular terms of the form "$\acute{\varepsilon}\,\Phi(\varepsilon)$." In accordance with the second of the clauses listed above the question of whether such a value-range name has a reference is the question of whether, given a referring name of a first-level function of one or of two arguments, the result of filling the/an argument-place with that value-range name is a referring name. One of those first-level functions of one argument is the result of filling in one of the argument-places of the identity function with a referring singular term n, and the important question in this sub-case is whether every sentence of the form

(ID) "$n = \acute{\varepsilon}\,\Phi(\varepsilon)$,"

for referring singular term n, itself refers. The answer to this question turns out to be very easy: the only singular terms in *Grundgesetze* are names of truth-values or names of value-ranges. If n is a name of a truth-value, then it refers either to the true or to the false and hence via the stipulations of Vol I §10 to a particular value-range. In the first case, the instance of (ID) refers to the true if $\Phi()$ is a concept under which only the true falls, and to the false otherwise, while in the second case, the instance of (ID) refers

<hr>

21 *Grundgesetze* I §30.

to the true if $\Phi()$ is a concept under which only the false falls, and to the false otherwise. And if n is the name of the value-range of a function $\Psi()$, then "$n = \acute{\varepsilon}\,\Phi(\varepsilon)$" refers to the true if the value of $\Phi(x)$ is the same as the value of $\Psi(x)$ for every argument, and refers to the false otherwise. And we are done with this case.

In short, then, the criterion to be met by the well-formed names of *Grundgesetze* in order that they collectively meet the condition that each "has a reference" is the criterion of linguistic completeness. It is critical, for each singular term n, that the sentence $n = m$, for every singular term m, have a determinate truth-value. But nothing about the semantics of *Grundgesetze* provides any answer to the question of whether any singular term of *Grundgesetze* refers to Julius Caesar or to England. And the condition that Frege requires his singular terms to meet does not include the condition that such questions have answers. There are, in short, an enormous number of "don't care" cases in Frege's *Grundgesetze* analysis of value-range terms and hence of numerical singular terms. The identity-sentences that matter to arithmetic are settled, as are those extras (like the identities between finite cardinals and real numbers) that are needed for logical hygiene. But nothing even approximating (SIC) is met.

The emphasis on linguistic completeness, and the significant distance between this requirement and the totality requirement introduced above, are clear as well in the §10 stipulations regarding value-range names, referenced as just mentioned in the §31 argument. In §10, Frege raises the question of the reference of value-range names. The stipulation that a name of the form "$\acute{\varepsilon}\,\Phi(\varepsilon)$" (for "$\Phi()$" a name of a first-level function of one argument) is to refer to the value-range of that function, together with Basic Law V, settles the reference of all sentences of the form "$\acute{\varepsilon}\,\Phi(\varepsilon) = \acute{\varepsilon}\Psi(\varepsilon)$" in the obvious way. But this does not account for identity-sentences of the form "$\acute{\varepsilon}\,\Phi(\varepsilon) = q$" for q not of the form "$\acute{\varepsilon}\,\Psi(\varepsilon)$." As Frege puts it,

> By presenting the combination of signs "$\acute{\varepsilon}\,\Phi(\varepsilon) = \acute{\alpha}\,\Psi(\alpha)$," as co-referential with "$\forall x(\Phi(x) = \Psi(x))$", we have admittedly by no means yet completely fixed the reference of a name such as "$\acute{\varepsilon}\,\Phi(\varepsilon)$". We have a way always to recognise a value-range as the same if it is designated by a name such as "$\acute{\varepsilon}\,\Phi(\varepsilon)$", whereby it is already recognisable as a value-range. However, we cannot decide yet whether an object that is not given to us as a value-range is a value-range or which function it may belong to.[22]

The reader who finds this discussion reminiscent of *Grundlagen* sections 56 and 66 might expect at this point that Frege will supplement his account of value-range terms by providing criteria by means of which one

---

22 *Grundgesetze* I §10. I have substituted modern universal-quantifier notation for Frege's.

might determine whether their referents are identical to or distinct from, for example, Julius Caesar or England. But he does nothing of the sort. Following a brief argument in support of the claim he has just made—that is, that the stipulations so far introduced do not "completely fix[ ] the reference" of value-range terms, Frege continues:

> Now, how is this indeterminacy resolved? By determining for every function, when introducing it, which value it receives for value-ranges as arguments, just as for other arguments. Let us do this for the functions hitherto considered. These are the following:
>
> $\xi = \zeta,\ -\xi,\ \neg\xi$
>
> [Demonstration that the last two are reducible to the first] . . . After having thus reduced everything to the consideration of the function $\xi = \zeta$, we ask which values it has when a value-range appears as argument. Since so far we have only introduced the truth-values and value-ranges as objects, the question can only be whether one of the truth-values might be a value-range.

Frege then demonstrates that a stipulation that an arbitrarily chosen value-range is the true, and a different one the false, will introduce no contradiction into the system. And, finally, he makes such a stipulation, the stipulation that the value-range of any function under which just the true falls is to be identical with the true, and the value-range of any function under which just the false falls is to be identical with the false. The section concludes:

> We have hereby determined the *value-ranges* as far as is possible here. Only when the further issue arises of introducing a function that is not completely reducible to the functions already known will we be able to stipulate what values it should have for value-ranges as arguments; and this can then be viewed as a determination of the value-ranges as well as of that function.

Two remarkable features of the §10 strategy are these. First of all, the stipulations of §10 do not solve what one might have thought to be the "Caesar problem," that is, the (pseudo-)problem that the semantic principles laid down to this point do not determine, of any value-range terms, whether they do or do not refer to Julius Caesar. The only unresolved question of reference with which Frege is concerned is the question of the truth-conditions of identity-statements linking value-range terms with terms for truth-values. Once this question is swiftly answered via stipulation, the formal language has achieved linguistic completeness (or so Frege thinks), as argued in §31. There is still no fact of the matter about whether any value-range term refers to Caesar. Should we want to define a new language by adding the term "Julius Caesar," and hence also the function-name "$\xi =$ Julius Caesar" to the *Grundgesetze* language, we would need to add stipulations sufficient to regain linguistic completeness, including stipulations regarding the truth-value of such sentences as "$\exists \Phi\,(\acute{\varepsilon}\,\Phi(\varepsilon) =$ Julius Caesar)." But without the presence of such a name in the language, there is nothing incomplete, in Frege's judgment, about the referential status of the terms in the language as it is.[23]

The second remarkable feature of the §10 strategy is the fundamental idea that the truth-values of the identity-sentences in question are matters for stipulation. Regarding the identity-sentences linking value-range names and truth-value names: one might have thought that, independently of any linguistic stipulation, there is a fact of the matter about whether truth-values are, or are not, value-ranges. Or, more modestly, one might have thought that *if* there are such things as value-ranges and truth-values, then either some value-ranges are truth-values or none are, independently of the semantic stipulations for any particular language. Cows and horses are this way; why not truth-values and value-ranges? But this is clearly not how Frege thinks of the matter. The question of whether "$\acute{\varepsilon}\,\Phi(\varepsilon) = \forall x(x = x)$" is true or false is clearly not, for Frege, a matter of the identity of the objects in question, where these can be in some sense "located" independently of their role as references of the singular terms in a variety of sentences. The order of explanation is the other way around: that value-range is identical with that truth-value if the sentence "$\acute{\varepsilon}\,\Phi(\varepsilon) = \forall x(x = x)$" is true; they are distinct if the sentence is false. Some identity-sentences are provided truth-values via the laws of logic and some via empirical fact. And some, like this one, are provided truth-values via semantic *fiat*. The really interesting identity-sentences for our purposes are those that are not part of the language of *Grundgesetze*, but are part of a more comprehensive language that we could engineer by, for example, adding the term "Julius Caesar," with its ordinary reference, to the *Grundgesetze* language. It would be a simple matter to stipulate that every identity-sentence linking this term with a value-range term

---

23 Dummett notes ([Dummett (1991)] ch. 13 that in *Grundlagen*, Frege does not complete the answer to the question whether Julius Caesar is a number, since he doesn't there answer the question whether Caesar is an extension. Dummett takes §10 of *Grundgesetze* Vol 1 to involve the attempt to finally answer the lingering question. But this cannot be right. §10 and its stipulations are brief and clear, and the only question answered there is very straightforward: it is the question whether any value-range terms co-refer with any terms for truth-values; and if so, which. This is the only question whose answer is required for linguistic completeness, and the only one in which Frege shows any interest. There is nothing like an attempt, anywhere in *Grundgesetze*, to answer questions regarding the identity of value-ranges, or numbers, with any of the kind of "outlying" objects that purportedly give rise to the Caesar question.

is false. It would also be a simple matter to choose a value-range term and stipulate that the identity-sentence linking it with "Julius Caesar" is true. The important point, for the purposes of coming to grips with Frege's view of object-reference, is that nothing about the semantics of the *Grundgesetze* language rules out either of these alternatives for later expansion. As we might put it, just as there was no fact of the matter, prior to Frege's stipulation, whether any of the value-ranges referred to in *Grundgesetze* was identical with a truth-value, so too there is no fact of the matter whether any of the value-range terms in *Grundgesetze* refers to Caesar.

Frege's most mature and most careful presentation of the logicist project involves nothing like the totality condition and nothing like the strong condition (SIC). The numerical singular terms used there provably satisfy the linguistic-completeness requirement,[24] and in particular each numerical singular term n satisfies the requirement that for every singular term t of the *Grundgesetze* language, the sentence $n = t$ has a determinate truth-value. Nothing about the semantics of the language determines, for objects o not referred to via terms of that language, whether any singular term of the language refers to o.

If at the time of writing *Grundlagen* Frege held a successful account of arithmetical discourse to the strong condition (SIC), then his view changed by the time of writing *Grundgesetze*. If (as I think more plausible) his view at the time of writing *Grundlagen* involved instead merely the weak condition (WIC), then his view is essentially uniform from *Grundlagen* to *Grundgesetze*.[25] The important point for our purposes is that by the time he received the letter from Russell, he did not take the logicist project to require for its success the provision of criteria by means of which one might determine whether its numerical singular terms refer, for example, to people or to countries. It therefore cannot be correct to understand his reason for failing to adopt the neo-logicist strategy to be that that strategy would fail to meet this condition.

The appeal to Hume's Principle in order to fix the reference of numerical singular terms would have posed two remediable problems for Frege. The first is the problem he presents in *Grundlagen* §66, which is that the account given thereby would not have demonstrated that the numerical terms are in fact singular terms. The remedy for this would have been to rely on the robust apparent singular-termhood of those terms in order to anchor the claim to genuine singular-termhood, in just the way that Frege actually does for extension-terms in *Grundlagen* and for value-range

---

24 Or would have done so if not for the difficulty revealed by Russell's Paradox.
25 The qualifier "essentially" is needed because the *Grundgesetze* view is given in terms of the mature theory of (sense and) reference, while the *Grundlagen* view is concerned with the earlier notion of content.

terms in *Grundgesetze*. That is, this part of the strategy would be to follow the main line of the neo-logicist proposal. The second remediable difficulty is the internal Caesar problem arising from the requirement of linguistic completeness. The language of *Grundgesetze* is intended not just for the development of a theory of cardinal numbers but also for a theory of the reals and perhaps also for a theory of "negative, fractional, irrational [and] complex" numbers.[26] Frege's actual procedure, as far as it goes, is to define all numbers as value-ranges of first-level functions, in terms of primitive vocabulary already available in *Grundgesetze*, with the result that the argument for linguistic completeness in §31 will apply to terms for all such numbers. A resort instead to Hume's Principle for the definitions of terms for cardinal numbers will be of no help in defining terms for the remaining numbers, so some entirely different procedure will be needed for them.[27] However these new terms are defined, we will have a failure of the linguistic completeness requirement: for example, identity-statements linking finite cardinals and reals will have no truth-conditions and no truth-values prior to the provision of additional stipulations. The remedy here will be more of the kind seen already in §10: these cases, which are of no mathematical significance, will need to be settled by arbitrary stipulations, in the choice of which Frege would have had wide latitude.

But the appeal to Hume's Principle in order to fix the reference of numerical singular terms was, as Frege notes in the letter to Russell, blocked entirely by a more fundamental difficulty. We turn now to that difficulty.

7.   The Paradox, Law V, and HP

The fundamental idea of Law V is that there is a very strong equivalence between the two sides of any instance of it. Frege sometimes says that the two sides "express the same sense"[28] and sometimes puts it in terms of a guaranteed identity of reference.[29] More generally, he says simply that "we can convert" one side into another[30] and that if the universally quantified claim is true, we can "also say that" the two functions have the same value-range[31] and that we engage in a "transformation of the generality of an equality into a value-range equality."[32]

---

26 *Grundlagen* §109.
27 For discussion of this issue as it applies to the neologicist project, see [Cook and Ebert (2005)].
28 "Function and Concept" p 11.
29 *Grundgesetze* I §3.
30 Ibid.
31 Ibid.
32 *Grundgesetze* I §9.

Frege makes similar remarks about abstraction principles in general. In *Grundlagen*: "The judgment 'line a is parallel to line b' . . . *can be taken as* an identity"; in so doing, we "carve up the content" differently.[33] Similarly for shape and orientation,[34] and for length and color.[35] In each case, the two sides of the biconditional have something very like the same content, and we can unproblematically "transform" one into the other.

The strong semantic equivalence between the two sides of abstraction principles does not fit entirely easily with Frege's mature theory of sense and reference. That it is a "fundamental law of logic" that we can "transform" one side into the other requires a much stronger relation between the two sides than mere identity of reference (i.e., of truth-value), but at least some of Frege's remarks on sense-identity conflict with the idea that the two sides have the same sense. These are the remarks on which most of the post-Fregean theory of Frege's notion of sense tends to be based, the remarks according to which the possibility of a competent speaker judging that a sentence S expresses a truth while (say) being unsure about whether S' does so suffices for a difference in sense between the two sentences. Things are less clear than they might at first seem, since (i) Frege also claims on occasion that sense-identity is not so fine grained, and (ii) even the initial test for sense-distinction provides no clear result, since a gray area surrounds the question of whether a speaker who affirms one side of an abstraction-principle while failing to do so for the other side can in fact be said to understand the sentences.[36] The difficulty here is that Frege simply has no clear criterion of sense-identity. The crucial feature of Basic Law V is that the two sides bear to one another the strong semantic similarity relation that Frege in *Grundlagen* would have called the result of a "re-carving" of their shared content, and by the time of *Grundgesetze* seems sometimes to have viewed as sense-identity and sometimes as a somewhat weaker relation of similarity, but in any case as strong enough to justify the inference from one side to the other on purely logical grounds.

Michael Dummett has argued that Frege's claim in "Function and Concept" that the two sides of V share sense is inconsistent with other central aspects of Frege's view of sense.[37] Frege claims that the senses of sub-sentential pieces of language are in some sense "parts" of the sense of the whole sentences in which they appear. How exactly we are to understand the part-whole metaphor is never made entirely clear by Frege, but Dummett argues that one fixed point is this: that if the sense of a phrase is *part* of the sense of a sentence, then grasp of the sense of that sentence must

---

33 *Grundlagen* §64.
34 Ibid.
35 *Grundlagen* §65.
36 For a fuller discussion of these issues, [Blanchette (2012a)] ch. 2.
37 [Dummett (1991)] ch. 14.

involve a prior grasp of the sense of that phrase. If this is the case, then indeed, there is a problem: for it is presumably clear that a grasp of the sense of an instance of "Every F is a G and vice-versa" does not require a grasp of the sense of "the extension of F," with the result that the two halves of Law V cannot, on Dummett's understanding of the requirements on sense-parthood, express the same sense. But it is difficult to see why we should take Frege to have intended the connection between "parthood" and grasping in this way. In keeping with the *Grundlagen* idea that on recognizing a "re-carving" of a content already grasped, one can come to discover new "concepts" (as he then calls the relevant content-parts), one might (perhaps more plausibly) take Frege to have intended by his talk of "parthood" merely that when a complete sense can be broken down into parts in a particular way, then *one* way of coming to grasp that complete sense is via a grasp of those parts and of their mode of composition. That the sense of "line *a* is parallel to line *b*" can be "re-carved" to yield the concept of *direction* as "the direction of *a* = the direction of *b*" does not, on this account, entail that only a speaker already in possession of the concept *direction* can understand the original sentence. On this way of understanding Frege's talk of senses and their parts, one might plausibly claim that while a grasp of the senses of *both* sides of an instance of Law V requires a fairly immediate recognition that they express the same (or very closely similar) senses, a grasp of the universally quantified half is possible for someone who has not yet mastered the notion of *value-range*. In any case, the crucial feature of Frege's view of the relationship between the senses expressed by the two sides is not whether he took them to be strictly identical but that he took them to be sufficiently similar to underwrite an immediate inference from one to the other. This, prior to reading Russell's letter, would seem to have been Frege's view of the relationship between the two sides of abstraction principles in general, and it was certainly his view of the two sides of Basic Law V.

Frege's immediate reaction to Russell's letter is that the paradox shows that the "transformation of the generality of an identity into an identity of ranges of values . . . is not always permissible. . . ."[38] A month later, he discusses the strategy of turning to a numerical abstraction principle, that is, to Hume's Principle, for the "possibility of placing arithmetic on a logical foundation."

> We can also try the following expedient, and I hinted at this in my *Foundations of Arithmetic*. If we have a relation $\Phi(\xi, \zeta)$ for which the following propositions hold: . . . [symmetry and transitivity] . . ., then this relation can be transformed into an equality (identity), and $\Phi(a, b)$ can be replaced by writing, e.g., $\S a = \S b$. If the relation is e.g. that

of geometrical similarity, then "a is similar to b" can be replaced by saying "the shape of *a* is the same as the shape of *b*." This is perhaps what you call "definition by abstraction." But the difficulties here are the same as in transforming the generality of an identity into an identity of ranges of values.[39]

If the account sketched above is correct, then it is clear why the difficulties are "the same." The paradox undermines the core principle that Frege thought grounded all such biconditionals: the principle that the two halves are merely alternative expressions of essentially the same content. It is a natural thought, and it was certainly Frege's thought, that where $\Phi(\xi, \zeta)$ is an equivalence relation, then

- $\Phi(a, b)$

makes essentially the same claim as does

- The ϕ of a = the ϕ of b,

for the obvious choice of a function ϕ. For equivalence in height, we have *the height of . . .*; for equivalence in shape, *the shape of . . .*; and for equivalence in extension, we have *the extension of. . . .* The lesson of the paradox, as Frege sees it, is that this simply isn't true.[40] We cannot assume that each equivalence-relation comes along with a function delivering objects whose identity-conditions are given by the holding of that equivalence-relation. And if this is not a general and obvious principle, then it's not a law of logic that the pairs of claims linked in this way are equivalent in truth-value. If it's true that there are, say, shapes and numbers meeting these criteria, then their existence is grounded in principles peculiar to their cases and not in general laws of logic. Logic, as Frege now sees it, is insufficient to give us objects.

8. Conclusion

Frege's view in 1884 of the necessary conditions on an acceptable analysis of arithmetical discourse, especially those parts of his view expressed in the "Caesar passages" of *Grundlagen*, are difficult to pin down with any precision. It is possible, though not mandated by the texts and inconsistent with some of them, that Frege held, at that time, the condition (SIC) as a requirement on the successful analysis of arithmetical discourse. But it is entirely clear that by the time of writing *Grundgesetze*, he had no such view. Frege's account of the semantics of the *Grundgesetze* language,

---

39 Letter to Russell 24 July 1902.
40 For a similar diagnosis of the difficulty posed by the paradox, see [Dummett (1991)], ch. 18.

and the content of the claims he takes himself to have proven about that language, make it clear that by 1893, he did not view it as a necessary condition on an adequate analysis of arithmetical singular terms that the analysis provide truth-conditions for such "outlying" sentences as "Julius Caesar is the number of planets." And this is all to the good, from the point of view of the plausibility of Frege's view of the semantics of mathematical language. The idea that an understanding of mathematical discourse requires an assignment of truth-conditions to non-mathematical sentences, for example, to sentences that identify real numbers with particular rationals, cardinals with people, or line-directions with countries, is an idea contrary to the ordinary successful functioning of mathematical discourse. I have suggested here that we should understand Frege's treatment of the few "don't-care" cases that arise for him in *Grundgesetze*— for example, the identification of truth-values with value-ranges—in just the way we understand the decision to identify, or not, rationals with pairs of integers: of purely housekeeping significance.

Frege's own mature analysis of arithmetical language does not provide truth-conditions to sentences like "Julius Caesar is the number of finite cardinals." So his rejection of the HP strategy as a way to resurrect the logicist project in the light of Russell's paradox cannot be understood as motivated by the failure of Hume's Principle to determine truth-conditions for such sentences. I have suggested here that the inadequacy of the Hume's Principle strategy, from Frege's 1902 perspective, is rather that the treatment of Hume's Principle as fundamental within a logicist reduction requires what Russell's paradox shows to be false: that abstraction principles can be trusted, in virtue of their form, to be essentially truths of logic.[41]

References

Blanchette P. (2012a), *Frege's Conception of Logic*. Oxford University Press.

Blanchette P. (2012b), Frege on shared belief and total functions. *The Journal of Philosophy*, 15(3): 9–39.

Blanchette P. (2015), The breadth of the paradox. *Philosophia Mathematica*, 24(1): 30–49.

Boolos G. (1990), The standard of equality of numbers. In George Boolos, editor, *Meaning and Method: Essays in Honor of Hilary Putnam*, pages 261–277. Harvard University Press. Reprinted in [Boolos (1998)] 301–314. Page citations are to this reprint.

---

41 This raises the question of whether there is a way to defend the logical or analytic status of Hume's Principle from this skeptical response. I argue in [Blanchette (2015)] that the answer to this question, from Frege's point of view, is "no".

Parts of this material were presented at the University of Connecticut Analytic Philosophy Workshop. Many thanks to the organizers of that workshop and to its participants for helpful feedback, especially to Marcus Rossberg, Philip Ebert, Junyeol Kim, Richard Kimberly Heck, Robert May, Kai Wehmeier, Roy Cook, Jamie Tappenden, Michael Hallett, and Sanford Shieh. Thanks especially to Philip Ebert for comments on an earlier draft.

Boolos G. (1998), *Logic, Logic and Logic*. Harvard University Press, Cambridge. Edited by Richard Jeffrey.

Cook R. and Ebert P. (2005), Abstraction and identity. *Dialectica*, 59(2): 121–139.

Demopoulos W. (Ed.) (1994), *Frege's Philosophy of Mathematics*. Harvard University Press.

Dummett M. (1991), *Frege: Philosophy of Mathematics*. Harvard University Press.

Ebert Philip A. and Rossberg M. (2016), Introduction to abstractionism. In Philip A. Ebert and Marcus Rossberg, editors, *Abstractionism*. Oxford University Press.

Frege G. (1979), *Posthumous Writings*. Chicago University Press. This is an English translation of most of [Frege (1983)]; edited by H. Hermes, F. Kambartel, and F. Kaulbach; translated by P. Long and R. White.

Frege G. (1983), *Nachgelassene Schriften* (2nd rev ed). Felix Meiner Verlag. Edited by H. Hermes, F. Kambartel, and F. Kaulbach.

Hale B. and Wright C. (2001a), *Reason's Proper Study*. Oxford: Clarendon Press.

Hale B. and Wright C. (2001b), To bury Caesar . . . In [Hale and Wright (2001a)]. Oxford: Clarendon Press, 2001b.

Heck R.K.(1997), The Julius Caesar objection. In *Language, Thought and Logic: Essays in Honour of Michael Dummett*, pages 273–308. Oxford: Clarendon Press. Revised version in [Heck (2011)] 127–155. (Originally published under the name "Richard G. Heck, Jr.").

Heck R.K. (2005), Julius Caesar and Basic Law V. *Dialectica*, 59: 161–178. Reprinted in [Heck (2011)] 111–126. (Originally published under the name "Richard G. Heck, Jr.").

Heck R.K. (2011), *Frege's Theorem*. Oxford University Press.

Heck R.K. (2012), *Reading Frege's Grundgesetze*. Oxford University Press.

Parsons C. (1965), Frege's theory of number. In Max Black, editor, *Philosophy in America*, pages 180–203. Cornell University Press. Reprinted with postscript in [Parsons (1983)]. Reprinted (with same postscript) in [Demopoulos (1994)].

Parsons C. (1983), *Mathematics in Philosophy: Selected Essays*. Cornell University Press.

Schirn M. (Ed.) (1996), *Frege: Importance and Legacy*. Walter de Gruyter.

Tait W. (1996), Frege versus Cantor and Dedekind: On the concept of number. In *Frege, Russell, Wittgenstein: Essays in Early Analytic Philosophy (in honor of Leonard Linsky)*, pages 213–248. Open Court. Reprinted in [Schirn (1996)] pp 70–113.

Wright C. (1983), *Frege's Conception of Numbers as Objects*. Aberdeen University Press.

3 The Role of Truth

Robert May

1. Introduction

Truth is an object, not a concept. This well-known postulate sits squarely at the core of Frege's conception of logic and meaning. It is to claim, in Fregean terminology, that truth is saturated, not unsaturated; in the context of Frege's semantic model, it is to say that the truth-values are 0-level entities of the conceptual hierarchy. To be sure, this claim has not been without its notoriety, but it is important to understand that with respect to Frege's core concerns, there is nothing more that needs to be said about truth or, for that matter, that can be.

In considering Frege's thesis, it is natural to assume that his intention was to posit a *substantive* claim about truth: a theory of what truth *is,* which would thereby give an analysis that justifies that "the True" and "the False" are referential terms. If so, questions arise directly: If truth-values are objects, then what objects are they? What *sort* of objects are they, and of that sort, which are they in particular? But asking these questions comes with an attached presupposition: That Frege is proposing a metaphysical thesis about truth. That, however, would be mistaken. Frege *denies* that there is any metaphysical basis for deciding what truth-values are; whether some object is a truth-value is *not* a metaphysical judgment.

Rather, Frege's position is that there are no inherent properties of objects by virtue of which they qualify *as* truth-values. This is to say that there are no properties in terms of which the truth-values can be defined, and that is to say that the truth-values are indefinable. But what this is *not* to say is that the truth-values cannot be *identified* as objects. The inference is only if an object is to be a truth-value, it is because it is *stipulated* to be so, independently of its native characteristics; it is an arbitrary matter which objects are the truth-values. This is as Frege proceeds in the context of the logical language, stipulating that certain value-ranges are the truth-values. This is set forth in §10 of

Appreciation is due to Rachel Boddy, Riki Heck, Marco Panza, Marcus Rossberg, Sanford Shieh, Jamie Tappenden, and Ed Zalta for discussion of the topics addressed in this chapter.

DOI: 10.4324/9780429277894-4

Grundgesetze, in the midst of Frege's initial presentation of the notions on which logic is built.[1]

The reason Frege is satisfied with this weaker identifying condition is that the significance of truth-values is located not with anything inherent to them by which they are identified as objects but rather in the *role* that they play in logic and language. By taking this position, Frege is telling us that the question is not "What is truth?" but rather "What role does truth play?" Explicating its role, Frege maintains, does not require a presupposed metaphysical thesis about truth. But it does require that truth-values be objects for this role to be successfully carried out. In turn, that truth-values are objects is justified by how successfully they carry out this role.[2]

What is this essential role? It is this: To found the fundamental *concepts* on which logic is built. If truth-values are 0-level entities, then the logical concepts can be characterized as truth-functions. The conditional, negation, and identity—the predicates of elementary propositions—will be concepts of the first level, mapping truth-values *qua* objects to truth-values. The importance of this characterization is that it sets in place the base for the "proof of referentiality" found in §§29–32 of *Grundgesetze*, by which Frege intends to show that the language of logic is a referential language, one in which every term has a determinate referential value. Referential languages, by virtue of their consistency, are by Frege's lights properly suited for the development of scientific applications. Given that in such a language, the logicist thesis—as conceived by Frege, that arithmetic is a conservative definitional extension of logic—can be carried through; the thesis thereby gains its scientific credence.

In contrast to this meta-logical role, that truth-values are objects has no particular role to play logically. The domain of logic is proof, formalized as logical derivations, and for this, it is inconsequential that truth-values are objects. All that logic requires, as Frege sees it, is that the propositions that occur in proofs be judgments, that is, that they are established truths. But that they are truths is independent of any assumptions about what truth-values might be, in particular of whether truth-values are objects. The "logical" role of truth-values *qua* objects lies not in logic but rather in its conceptual foundation.

But if truth-values fail to have a logical role, they do have a semantic role to play, as the referents of sentences. Frege states this succinctly in "Function and Concept":[3]

---

1 Frege (1893/1903); English translation, Frege (2013). Throughout, references are to the English translation, with text citations as *Gg*.
2 To look ahead, what Frege is not saying is that truth-values are whatever play the specified role; he is not offering some form of implicit definition. Rather, there must be objects that can serve as truth-values, and it is incumbent that these objects be identified.
3 Frege (1970), p. 32.

A statement contains no empty place, and therefore we must regard what it stands for as an object. But what a statement stands for is a truth-value. Thus the two truth-values are objects.

A sentence refers to either the True or the False by virtue of the thought it expresses. This semantic thesis is parasitic on the core role, for it is to identify the references of sentences with the values of concepts: Terms that refer to objects can compose with terms that refer to concepts to form a complex, saturated term that refers to the value of that concept for that object as argument.[4] The importance here is for judgment: The judgment of thoughts, Frege tells us, resides in the recognition that this composition *in fact* refers to the True. By this thesis, judging—recognizing that a thought is true—is a form of object recognition—recognition of a referent.[5] This carries the core epistemic payload, as to make a judgment is precisely, in Frege's view, to gain knowledge.

Note that this semantic role of truth-values is no more metaphysically loaded than their logical role: Whatever objects are the True and the False, they will be adequate so long as they are recognizable as referents and so can afford the separation of truths from falsities. In principle, any two distinct objects that can be recognized as referents of sentences could serve this purpose and so by stipulation play the role of truth-values for the purpose of judgment.[6]

There is much to fill in with this sketch of Frege's view of the role of truth in founding the conceptual structure of logic and language. Much of this is carried out in Heck and May (2018) and Heck and May (2020). Here, my goal is limited to addressing two matters that Frege believed deserved his explicit attention in *Grundgesetze*. The first is the identification of truth-values as 0-level entities, as objects. This is taken up by Frege in §10 of *Grundgesetze*. Second is how truth-values so identified play their role in characterizing a fully referential language. *Grundgesetze*, §§29–32 is devoted to this topic. My discussion will reverse the order of address; I discuss first the proof of referentiality and second Frege's choice of objects to be the truth-values. The former will lead to remarks on the relation of full referentiality to Frege's dicta on proper definition, the latter to remarks on what shall be dubbed Frege's quasi-paradox of truth. Before turning to these matters, however, some stage setting is needed in the form of discussion of Frege's hierarchy of concepts, which serves as the backbone of his conception of logic and hence of his logicist project.[7]

---

4 The thesis generalizes to higher levels; see discussion in Section 2.
5 This is what Textor (2010) labels ontic recognition.
6 For discussion of Frege on judgment, see May (2018).
7 I will not address here the other half of the opening claim, Frege's rejection that truth is a concept. But, briefly, Frege's argument is that while thoughts do have the property of being true, this is not something that can be asserted by predicating truth of a thought.

2.   The Conceptual Hierarchy

For Frege, the logical laws are the laws of a conceptual universe. A conceptual universe is one that conforms to the conceptual hierarchy; any universe so conceived is conceptual. The Basic Laws of logic are the laws that hold of *any* conceptual universe, and the judgments derived from the Basic Laws hold of any such universe. For a purely conceptual universe, there need to be at least two objects to allow for the specification of the foundational logical concepts as truth-functions; identify these as the True and the False. Accordingly, these concepts—the horizontal, the conditional, negation, and identity—are concepts of the first level of the conceptual hierarchy. The True and the False need not be the only entities of the first level; there may be others. But if there are others, they are justified only insofar as they allow logic to be applied.[8]

A *conceptual hierarchy*, as Frege constructs the notion, is based on distinguishing functions from objects and then organizing the concepts—functions whose values are always truth-values—into successively higher levels. In this conception, a conceptual universe $\mathcal{U}$ consists of a sequence of levels $D^0, \ldots, D^n$, where for every $D^i (0 \leq i \leq n)$, $D^i \subseteq \mathcal{U}$. Objects are the entities that make up D^0, the base level; each successive level D^n is a class of functions $C^n : D^{n-1} \rightarrow T$, where $T = \{the\ True,\ the\ False\}$ and $T \subset D^0$. So, D^1 will be the class of concepts $C^1 : D^0 \rightarrow T$. These are the first-level concepts. Second-level concepts map $D^1 \rightarrow T$, and so on for the successive levels of concepts.[9]

Embedded within this conception is a compositionality principle. A common illustration is with first-level concepts that take 0-level entities as arguments:[10]

$$\varphi(\ ) + a \Rightarrow \varphi(a)$$

That the *sentences* of the logical language (in fact, of languages in general) represent the application of a function to an argument as a structural

---

What can be asserted is rather a thought that refers to the True; there is no other way, according to Frege, to express that a thought is true. This argument, the so-called regress argument, is found most notably in "The Thought" (Frege 1977). For an in-depth discussion, see Heck and May (2018), section 7.5.

8   For discussion of Frege's conception of science as applied logic, see May (2018). As discussed there, from the perspective of applied logic, logicism can be understood as the thesis that arithmetic is the scientific application of logic that requires no axioms in addition to the Basic Laws of logic. In the context of this application, value-ranges are justified as logical objects precisely because they allow logic to be applied as the science of number, as arithmetic.

9   While this construction allows for infinite levels of concepts, this is significant in principle only. Logic, as it is equipped to apply to scientific inquiry, need avail itself just of the first three. The primary concepts at the two higher levels are generalizations: First-order generalization is a second-level concept, second-order generalization a third-level concept.

10   For illustrative purposes, I have used a blank space to indicate unsaturated argument positions. I will keep to Frege's convention of using the Greek letters ξ and ζ in the following.

composition is perhaps Frege's key insight about logical form, sitting at the heart of his conception of logic. These sentences, as expressions of a language, have a sense—a thought—and a reference. The reference Frege identifies with the value of the constituent concept for the constituent argument. Accordingly, sentences refer to truth-values, 0-level entities, by virtue of their composition.

The compositionality principle naturally generalizes through the conceptual hierarchy: For a function of level n, its arguments will be entities of level $n - 1$.[11] Thus, a second-level function has a first-level function as argument; a third-level function, a second-level function as argument; and so on. Frege's notation suggests this sort of composition. If:

$$\Phi(\ )$$

is a second-level function and if:

$$\varphi(\ )$$

is a first-level function, then their composition is:

$$\Phi(\ ) + \varphi(\ ) \Rightarrow \Phi(\varphi(\ )).$$

But if this is how composition is to be understood, then there is a curious result: *All* concepts are of the first level. There are no concepts at the higher levels, only functions. The second-level *function* maps a first-level concept to a first-level *concept*; it does not map onto truth-values as its value.[12] There would be a hierarchy of functions, but it would not be a hierarchy of *concepts,* as Frege desires.

The practical consequence of this observation is that if a higher-level concept is to compose with a lower-level concept as argument, doing so must close any open positions. In Fregean terminology, even though the argument is unsaturated, what results must be saturated. This is accomplished by requiring that binding be part of concepts of the second level or greater.[13] The central illustration is the generalization concept:

$$\text{\textbardbl}^{\mathfrak{a}}\ \Phi(\mathfrak{a})$$

11 More precisely, at least one of its arguments must be of level $n - 1$. If there are others, they may be of levels lower than n. This accommodates relations.
12 Frege makes this observation in *Gg* I, §21. He further remarks in Frege (1979) that "$\phi(\ ) = \varphi(\ )$" does not assert the identity of concepts but rather is itself a first-level concept.
13 See *Gg* I, §§8, 17, and 19–25. In "Function and Concept", Frege says that "just as functions are fundamentally different from objects so also functions whose arguments are and must be functions are fundamentally different from functions whose arguments are objects and can not be anything else" (Frege 1970, p. 38). Note that concepts having a binder as a constitutive part is inherited from their being functions that have 0-level values. Thus, the value-range function $\grave{\varepsilon}\Phi(\varepsilon)$ has the same characteristic, but it is not a concept.

This composes with a concept of the first level:

$$\text{-}^{\mathfrak{a}}\text{-}\ \Phi(\mathfrak{a}) + \varphi(\ ) \Rightarrow \text{-}^{\mathfrak{a}}\text{-}\ \varphi(\mathfrak{a})$$

The result is now as desired: In general, if a concept C^{n-1} is to be an argument of a concept C^n, any open position in C^{n-1} will become closed by their composition. Accordingly, there is a conceptual hierarchy, a hierarchy of functions that, regardless of level, map onto the truth-values, entities of the 0 level.[14]

The effect of the conceptual hierarchy is to impose a compositional structure on the complex expressions of the logical language; any analysis of its logical form must be in conformance with the hierarchy. Compositionality is thus the syntactic reflection in the logical language of the semantic structure characterized by the hierarchy. Accordingly, the Begriffsschrift is a *conceptual* language, per its name, representing logical form as a functional structure of concepts.[15]

This notion of compositionality is novel in *Grundgesetze,* an innovation over Frege's initial presentation of logic in *Begriffsschrift.*[16] There, Frege lays out a different view, starting with the idea that content is constituted as a class of judgments or, as he has it there, of facts. As the goal is to characterize the logical relations that hold among these judgments—how they are inferentially related—such contents, Frege argues, must be subject to a conceptual analysis: They must be recognized as being made up of function and argument as their *conceptual content.* Thus, in the language of *Begriffsschrift*, $f(a)$ is a proposition made up of a function and an argument, but adjudging which is function and which is argument depends on whether $f(a)$ is taken in contrast to $f(b)$, $f(c)$, $f(d)$, and so on, in which case f is function and a is argument, or in contrast to $g(a)$, $h(a)$, $k(a)$, and

---

14 While concepts, regardless of level, map onto the 0 level, concepts of the first level nevertheless have a special place, as their composition with their arguments forms *elementary* propositions. For the importance of this, especially as Frege contrasts his logical innovations to Boolean logic, see Heck and May (2013), section 3.

15 Compositionality is not to be confused with the methods of specifying a taxonomic classification of expressions of the language that Frege postulates in §§26 and 30 of *Grundgesetze.* Frege's primary concern in these sections is the class of first-level function names, which is characterized by two procedures: Removal of a proper-name from a proper name or by addition of a proper name to a first-level function/relation name. But regardless of how category membership is established, propositional occurrence will be determined by conformance to the conceptual hierarchy. Looking ahead to the section to come, the proof of referentiality proceeds by showing that each of the classes of expressions is made up of referential terms, from which it follows that the language is referential, given that composition comports with the conceptual hierarchy.

16 Frege (1879); English translation, Frege (1972). Throughout, references are to the English translation, with text citations as *Bg.* Note that the name of the book will be italicized but not the name of the logical language.

so on, in which case *a* is function and *f* is argument.[17] Whichever analysis it is, however, in itself, *f*(*a*) is not unified in a manner that would allow it to be inferentially related to otherwise unified conceptual contents. For this, the parts have to be tied together. Accomplishing this is the role of the horizontal content stroke: Specifically, Frege tells us that *"The horizontal stroke . . . ties the symbols which follow it into a whole"* (*Bg*, 112; emphasis in original) and that "the content which follows it is unified, so that other symbols can be related to it [as a whole]" (*Bg*, 94)—*f*(*a*) is a conceptual content, which may be affirmed or denied; if affirmed, and so judged to be a fact, it can be inferentially related to other judgments.[18]

The important take-away here is this: The logical system of *Begriffsschrift* is deeply *non*-compositional. In *Grundgesetze*, this changes; the key advance is that the logical language is fully compositional.[19] *f*(*a*) represents functional composition (as described previously), and, importantly, the horizontal stroke is reconceived as a concept, mapping only truths to truths. If in *Begriffsschrift*, *f*(*a*) required external unification (by the content stroke), in *Grundgesetze*, unification is internal, a matter of functional application. This is what Frege is parsing out when he distinguishes functions, inclusive of concepts, as unsaturated and their arguments as saturated.[20] Conceptual content as conceived in *Begriffsschrift* is not compositional structure as conceived in *Grundgesetze*. In *Begriffsschrift*, the concern is only that a proposition *have* conceptual content, that it can be analyzed in terms of function and argument, and so has the requisite logical form to afford the construction of rigorous, gap-free proofs via logical derivation. The concern is not with how it *comes* to have that content.[21] That, however, is just the concern in *Grundgesetze*. The introduction of compositionality, and the alignment with the hierarchy of concepts that

17 See *Bg*, §9. For a discussion of Frege's view of function and argument in *Begriffsschrift*, and the transition to the view of *Grundgesetze*, see Heck and May (2013), section 2.

18 In *Begriffsschrift*, not only is the horizontal not a function, neither are the other logical connectives. A conditional is affirmed, and hence a judgment, just in case the possibility that the antecedent is affirmed and the consequent is denied "does not occur". See Frege (1972), §§5–7, and Heck and May (2018), section 7.3, for discussion.

19 The only symbol in the logic of *Grundgesetze* that does not refer to a concept is the judgment stroke: "The judgement stroke cannot be used to construct a functional expression; for it does not serve, in conjunction with other signs, to designate an object" (Frege (1970), p. 34). In contrast, in *Begriffsschrift*, the judgment stroke is a predicate of sentences, with the meaning "is a fact". (See *Bg*, §3.) It is the sole compositional term in the language.

20 This change is an aspect of what Tappenden (2019) has documented as Frege's development from mechanical to organic accounts. For discussion of the articulation of Frege's understanding of concepts, see Heck and May (2013).

21 Correlatively, Frege's concern in *Begriffsschrift* is only that a proposition *be* a judgment, and not how it comes to be recognized as such. The latter only becomes prominent in Frege's later work, where he is careful to distinguish making a judgment from logical derivation. Failure to do so is to commit the sin of psychologism. See May (2018).

it affords, is crucial; it is the central assumption required for Frege's demonstration that the logical language of *Grundgesetze* is a fully referential language. This is the proof of referentiality, to which we now direct our attention.[22]

3. The Proof of Referentiality

In *Grundgesetze* §28, Frege sets forth the following principle:

Correctly formed names must always refer to something

In the three sections to follow, Frege sets out to answer the question "When does a name refer to something?" by presenting an informal demonstration—the *proof of referentiality*—that the logical language of *Grundgesetze* is a language that consists exclusively of correctly formed names. That is, every expression of the Begriffsschrift is a referential name, either by being a primitive referential name or by being composed of referential names. No name that is not correctly formed in this way is an expression of the language.

For the proof of referentiality, the conceptual hierarchy is a sufficient model to establish that the logical language is a referential language.[23] The proof is inductive, based on the assumption that there are at least two objects. These objects are the True and the False. Given that there are correctly formed names for these objects,[24] it can be shown that every term of the Begriffsschrift is a properly formed name. This role, the founding of referentiality, is the core role of truth-values in Frege's conception of logic.

Before turning to the proof itself, it is important to be oriented to its position in the presentation of *Grundgesetze,* and for this we consult the book's table of contents. Volume I from 1893 contains two parts, "Exposition of the concept-script" and "Proofs of the basic laws of cardinal number". The former is broken down into three sections, and it is in the second of these, entitled "Definitions", that the proof of referentiality is located. Frege brackets the sections containing the proof under the title "General remarks", encompassing §§26–32 of *Grundgesetze.* In

---

22 In the *Tractatus*, Wittgenstein takes exception to Frege's departure from the logic of *Begriffsschrift* to that of *Grundgesetze*, for what he takes as the latter's introduction of a realist perspective on logic. Wittgenstein explicitly rejects that the logical connectives refer to concepts; see paragraphs 4.0312 and 5.4 for forthright statements to this effect. See May (2018), where this point is elaborated.

23 Note that this is inclusive of *all* terms of the language and is not limited only to terms that find their reference in the conceptual hierarchy. In particular, non-conceptual functions are included, notably, the value-range function. See discussion in the following.

24 Which, it turns out, are not "the True" and "the False", which are not names in the Begriffsschrift. More on this in Section 5.

introducing the canons of definition in these sections, Frege specifies that they are explicit and conservative, tying these properties to the correct formation of names as a prerequisite for the "principles of definition" that he specifies in the final section of the group. What is to be noted is that Frege places the proof of referentiality squarely as a centerpiece in introducing definitions: Clearly, he intended the proof to be understood as essential to legitimizing the standards of definition that he is laying out.[25] With this in mind, we turn to the proof itself.

3.1. *The Proof of Referentiality*

The section of Part I of *Grundgesetze* that precedes the section on definitions is entitled "The Primitive Signs". Under this heading, in §9, Frege introduces the notion of a value-range of a function, setting these as the logical objects.[26] In §§10 and 11, Frege's purpose is to establish that there are two truth-values and that they are 0-level entities, that is, objects, and as such are to be identified with value-ranges. He sets the True to be $\acute{\varepsilon}(\varepsilon = (\varepsilon = \varepsilon))$ and the False as $\acute{\varepsilon}(\varepsilon = \neg\mathfrak{a}\, \mathfrak{a} = \mathfrak{a})$. What is remarkable about these value-ranges is that they are the value-ranges of concepts under which one and only one entity falls, and they are the True and the False, respectively. From this, Frege derives that there is just one entity that is the True, and likewise the False. There will be more later on Frege's reasons for the choice of these value-ranges to be the truth-values. For now, it suffices to say that by making these stipulations, Frege is not *defining* truth and falsity. Rather, he is just picking out objects to *act* as truth-values. Other value-ranges could have served the purpose; the decision, at heart, is arbitrary.

With this much in his quiver, Frege turns in §29 to setting the conditions that are to be met for being a referential name. There are two main cases, function names and proper names:

I.   A *function name* is a referential name only if, when its argument-places are filled by referential names, the resulting proper name is also a referential name.

II.   A *proper name* is a referential name only if, when it fills the argument place of a referential function name, the resulting proper name is also a referential name.

Simply put, for $f(a)$, f is referential only if a is referential and $f(a)$ is referential; a is referential only if f is referential and $f(a)$ is referential.

---

25 This positioning of the proof is observed in Boddy and May (2021).

26 This includes not only single-value ranges but also double-value ranges; accordingly, there are value-ranges of concepts and relations. See Heck and May (2018), section 7.4, for a discussion of how double value-ranges depend on the assumption that truth-values are objects.

(The conditions naturally extend to cover names of functions of higher levels.) Note that Frege in effect builds the Context Principle into these conditions: The referentiality of f and a is established in the context of $f(a)$.[27]

Bear in mind that by the proof of referentiality, Frege's intention was to show that the logical language is a referential language. This language consists of eight primitive functions, and in §31, Frege raises the question of whether the names of these primitive functions are referential. For the truth-functions, this follows directly. Since they map truth-values to truth-values—they are concepts—it follows that their argument places are filled by referential proper names of truth-values and that the resulting complex proper names are also referential, referring to truth-values as well. Thus, if the 0-level entities are truth-values, then names of truth-functions are referential names. This establishes that four primitive functions of *Grundgesetze* are referential—the horizontal, negation, conditional, and identity. All of these are first-level functions.

We make the following observation: If the truth-values are 0-level entities, and there are referential names of 0-level entities, then concept-names of *any* level are referential. If 0-level names are referential, then so, too, are first-level concept-names, since the corresponding concepts map from the 0-level to the 0-level. But then second-level concept-names are also referential, as they map first-level concepts, which have referential names, to 0–level entities (so far, just the True and the False), which also have referential names. And so on for names of increasingly higher-level concepts. This increases the inventory of referential names of primitive functions by two, to include first- and second-order generalization, as these are second- and third-level functions, respectively.

The next primitive function name is *value-range*: $\acute{\varepsilon}\varphi(\varepsilon)$. This function maps functions (of n-level) to 0-level objects, so to establish that $\acute{\varepsilon}\varphi(\varepsilon)$ is a referential name, it must be shown by I that the names of its arguments and values are referential. By the observation just made, we need only consider *regular* value-range names, that is, those whose argument places are filled by names of first-level functions, which have been established to be referential names. The question is, then, whether for $\Phi(\xi)$, the referential name of a first-level function as argument, $\acute{\varepsilon}\Phi(\varepsilon)$ is a referential name. This is decided by II. Accordingly, if the result of this value-range name filling the argument-places of the referential names of the first-level functions is itself a referential name, then $\acute{\varepsilon}\Phi(\varepsilon)$ is also a referential name. This is to say if this name fills the argument-places of the primitive functions that are already established as having referential names (the horizontal,

---

27 For discussion of the relation of the Context Principle and the proof of referentiality, but with a rather different take, see Linnebo (2004) and Linnebo (2019). For a rejoinder, see Heck (2012), section 5.2.

negation, conditional, and identity), and the result is a referential name, then $\acute{\varepsilon}\Phi(\varepsilon)$ is also a referential name. And if this is a referential name, then so too is the function $\acute{\varepsilon}\,\varphi(\varepsilon)$.

To show this, Frege first considers identity. If value-range names fill the argument positions, then the result will be guaranteed to be the referential name of a truth-value, as "$\acute{\varepsilon}\Psi(\varepsilon) = \acute{\varepsilon}\Phi(\varepsilon)$" is coreferential with $\neg\mathfrak{a}\,\Psi(\mathfrak{a}) = \Phi(\mathfrak{a})$ (which has already been established to be a referential name if $\Psi(\xi)$ and $\Phi(\xi)$ are referential names). The horizontal follows, given that it always has the same value for the same argument as $\xi = (\xi = \xi)$, as do negation and the conditional, since they are referential names just in case the horizontal is a referential name. Hence, $\acute{\varepsilon}\Phi(\varepsilon)$ is a referential name, and so too is $\acute{\varepsilon}\varphi(\varepsilon)$, since the former will be a referential name for any substitution of a referential function name for $\Phi(\xi)$. So now another primitive function name is added to the list of referential names, and so too are the names of all the 0-level entities, that is, all the value-range names. Note that if the truth-values are value-ranges, then the names of those value-ranges are the names of the truth-values, and they too are referential names. (More on this anon.)

The last primitive function name is *description*: $\backslash\xi$. It will be a referential name, since it will form a referential name for any referential name as its argument. For $\backslash\Delta$, if Δ is the value-range name $\acute{\varepsilon}(\varepsilon = \Gamma)$, then $\backslash\Delta$ will be a referential name of Γ; if not, it will be a referential name of Δ. Hence, $\backslash\xi$ is a referential name, and accordingly all of the primitive names are referential names.

The last step of the proof is a consequence of the closure principle that every name formed from referential names is also a referential name. Then it immediately follows by compositionality that every well-formed name of the Begriffsschrift is a referential name and hence that the Begriffsschrift is a fully referential language.[28]

So, we have the following result: So long as the True and the False are 0-level entities, then it can be proven that the Begriffsschrift is a fully referential language. The structure of the proof depends on the compositionality of the conceptual hierarchy, with the truth-values being objects sitting at its base. Note that what is proven is that every expression of the logical language is referential, *not* that every expression has a reference in the conceptual hierarchy: The language includes correctly formed names that refer to value-range functions, which are not in the conceptual hierarchy. Nevertheless, the conceptual hierarchy is sufficient to show that value-range terms are referential and so that every term of the Begriffsschrift is referential.

---

28 For a more detailed critical account of the proof of referentiality, see Heck (2012), chapters 3 and 5.

It was assuredly not lost on Frege that if the Begriffsschrift were a referential language, then it would be consistent. Frege saw immediately that Russell's paradox showed that the proof failed: That not every term is referential. The locus of the problem is the reification of concepts as value-ranges: The effect of this is the collapse of the conceptual hierarchy into the 0-level. The consequence is that the 0-level is specified as entities whose existence depends on higher levels of the hierarchy: Without concepts, there are no value-ranges. But there can only be concepts if there are 0-level entities, that is, value-ranges. The circularity is obvious and presents an open invitation to inconsistency. Frege's reason for giving the proof, however, was not to give a consistency proof as such. It was rather to legitimize definitions. This judgment has already been prejudiced in noting the location of the proof of referentiality in Frege's exposition. Indeed, Frege labels the principle in §28 of *Grundgesetze* cited previously, that correctly formed names must always refer to something, the "governing principle for definitions" and immediately follows the proof, in §33, by adding six additional "principles [that] govern the use of definitions". Among these are that a defined term must have a unique definition, that defined terms are simple, and that a defined term inherits the meaning of the *definiens*. So our conclusion must be that Frege saw the primary purpose of the proof to pertain to the legitimization of definitions.

The point here is due to Boddy and May (2021). Already in *Begriffsschrift*, and carried through to *Grundgesetze*, Frege is firm that logically, definitions are explicit and accordingly are semantically conservative over the language. Definitions are not innovative of meaning, but they are in their coinage of new terms as abbreviations, replacing (complex) terms whose meanings have had prior establishment. Thus, under Frege's canons of definition, the *definiendum* inherits its reference, and its status as a referential term, from the *definiens*. That this inheritance obtains can be assured if definitions introduce novel terms into contexts that are established as referential. Frege neatly sums this up:[29]

> By means of a *definition* we introduce a new name by determining that it is to have the same sense and the same reference as a name composed of already known signs. The new sign thereby becomes co-referential with the explaining sign

If the Context Principle of *Grundlagen* sets the context for the meaning of a term as its propositional context, then what Frege is saying with the proof of referentiality in *Grundgesetze* is that the context is wider than that, constituted by the language as a whole. Definitions explicitly rendered are properly admissible only in the context of a referential language.

29 *Gg*, §27.

The proof of referentiality is meant to establish that the *Begriffsschrift* is an appropriate milieu for the introduction of definitions, none more notably than definition (Z), the definition of number.[30]

4. The Truth-Values

For Frege, the language of logic is a representational language; its terms are referential terms. Among these terms are names of concepts and objects, which are anchored to referents in the conceptual hierarchy. This relation raises an immediate question: What, for logic, are the 0-level entities? In *Grundgesetze*, Frege believed he had hit on a stratagem to answer this question. It was to populate the 0-level with the value-ranges of functions: "Value-ranges of functions are objects, whereas functions themselves are not. . . . Extensions of concepts likewise are objects, although concepts themselves are not".[31] As such, the value-ranges are logical objects; the reification of functions, including concepts, is the base level of the conceptual hierarchy.

Having set out this thesis in §9 of *Grundgesetze*, Frege poses a question in §10: Are the truth-values among these entities? Are these, too, logical objects? His concern with this question is married to a prior concern that he poses at the opening of this section, that the "criterion of recognition" for value-ranges—the condition that:[32]

$$\grave{\varepsilon}\Phi(\varepsilon) = \acute{\alpha}\,\Psi(\alpha)$$

is coreferential with (has the same truth-value as):

$$\text{─}^{\mathfrak{a}}\text{─}\,\Phi(\mathfrak{a}) = \Psi(\mathfrak{a})$$

and underdetermines the referents of value-range names. This is shown, Frege observes, by there being functions whose values meet the criterion

---

30 As argued in Boddy (2019) and Boddy (forthcoming), Frege distinguishes two roles of definitions, between explicit definitions (those under discussion) and analytic definitions, whose role is to introduce concepts. These may look the same, but they are distinguished by Frege. From the perspective of logic, definitions are explicit and abbreviatory, and from that perspective, nothing more need be said. But their justification in the theory ultimately resides in their analytic role. Thus, the definition of number from the analytic perspective explicates the notion of number, but from the logical perspective, it merely replaces a complex term of the language. In *Begriffsschrift* and *Grundgesetze*, Frege's concern is with logical definition; in *Grundlagen*, his attention is on analytic definition.

31 Frege (1970), p. 32. Bear in mind that in the logicist thesis, the names of logical objects includes the names of numbers, given that numbers are defined as being value-ranges.

32 This is *not* Frege's Basic Law V (as Heck 2012 emphasizes). Rather, Basic Law V is the *assertion* of the recognition condition; it asserts their identity of reference as a logical law.

when taking value-ranges as arguments, but yet this does not in itself determine whether these values are value-ranges. More specifically, for any one-one function $X(\xi)$ from objects to objects, whether they have the same value for value-ranges as arguments is determined by the criterion. This is because if their arguments are so identified, so too are their values: "For then '$X(\acute{\varepsilon}\Phi(\varepsilon))' = X(\acute{\alpha}\Psi(\alpha))$' too is coreferential with '$\text{---}^{\alpha}\Phi(\mathfrak{a}) = \Psi(\mathfrak{a})$'." But this in itself gives us no clue to what that value *is*, other than whatever is named by '$X(\acute{\varepsilon}\Phi(\varepsilon))$' is the same as what is named by '$X(\acute{\alpha}\Psi(\alpha))$'. It may be a value-range, but this is not determined; no guidance is provided by the recognition condition.

This underdetermination can be ameliorated, however, by definitively stipulating the values of functions upon their introduction. By the stipulation, there ought not be any question whether the values are value-ranges. At this juncture, Frege turns to the functions that have been thus far introduced in the narrative—the horizontal, negation, and identity[33]—whose values have been stipulated to be truth-values. Are these value-ranges?[34]

> Since so far we have only introduced the truth-values and value-ranges as objects, the question can only be whether one of the truth-values might be a value-range.

The importance of Frege's question resides in that these functions are to be the basic *logical* functions by virtue of their being mappings from logical objects to logical objects. What is to be secured is their logical status, and it will be if their values—the truth-values—are identified with logical objects that have otherwise been introduced. But the underdetermination is persistent. The stipulation of their values is *not* definitive one way or the other whether they are value-ranges if all we have to rely on is the condition by which value-ranges are identified:

> Now, the question whether one of the truth-values is a value-range cannot possibly be decided on the basis of '$\acute{\varepsilon}\Phi(\varepsilon) = \acute{\alpha}\Psi(\alpha)$' having the same reference as '$\text{---} \Phi(\mathfrak{a}) = \Psi(\mathfrak{a})$'.

Frege's question thus amounts to asking whether (*i*) is true (for some concept φ):

$$\text{the True} = \acute{\varepsilon}\varphi(\varepsilon). \qquad\qquad (i)$$

---

33 In §10, Frege reduces the first two to the third and so sets the question as the value of the identity-function. We follow suit.
34 *Gg*, §10. *Ibid.* for the quotes to follow in this section unless noted.

The answer is that even though the True must be a value-range in order for (*i*) to be true, the truth of (*i*) cannot be established by a value-range identity being coreferential with (*ii*):

$$\unicode{x2015}\!\!\text{--}\,\, \Phi(\mathfrak{a}) = \Psi(\mathfrak{a}) \qquad\qquad (ii)$$

For this to be so, the answer to the question being posed would have to be already presupposed; that is, there would need to be prior establishment one way or the other whether truth-values *are* value-ranges. Then, (*i*) would be true if the True is a value-range and false if it is not.[35]

Frege does show, however, that a somewhat different result can be obtained that is nevertheless sufficient for his purposes:

> Thus, without contradicting our equating '$\acute{\varepsilon}\,\Phi(\varepsilon) = \acute{\varepsilon}\,\Psi(\varepsilon)$' with '$\unicode{x2015}\!\!\text{--}\,\,\Phi(\mathfrak{a}) = \Psi(\mathfrak{a})$', it is always possible to determine that an arbitrary value-range be the True and another arbitrary value-range be the False.

Frege obtains this result by specifying a function whose values are identified in the same way as value-ranges and which has as values the truth-values. It will then follow that stipulating that the truth-values are value-ranges will be consistent with the recognition condition. Frege arrives at this by observing that the criterion of recognition holds generally for "objects of conception", that is, values of functions from concepts to objects. This allows for objects "whose differentiation and recognition the same criterion would hold as for the value-ranges" (leaving open whether these objects are to be identified with value-ranges). Frege then introduces a function $X(\xi)$, as above one-one, that has objects of conception as arguments, and for one such argument the True as its value and for another the False.[36] Again, as previously, since the arguments of this

---

35 The argument can be put meta-linguistically, namely that references must be presupposed for the terms "the True" and "the False." Frege puts the issue very much this way in commencing §10, when he says that "We have a way always to recognize a value-range as the same if it is designated by a name such as '$\acute{\varepsilon}\,\Phi(\varepsilon)$', whereby it is always recognizable as a value-range." But no such guarantee is forthcoming if value-ranges are given in any other way; the recognition condition only allows for recognition of value-ranges given as such by value-range names. The reminiscence of this form of the argument to the Julius Caesar problem in §§66–67 of *Grundlagen* has been well recognized; see Dummett (1991), chapter 17, and Heck (2012), chapter 4, on the parallels.

36 This aspect of Frege's discussion in §10 has become known as the permutation argument, given that $X(\xi)$ is specified as mapping an object of conception to the True, the True to that object, and an identity map for all other arguments. So specified, this is the simplest function over the domain that suffices for Frege to make his point about the role of the recognition condition in justifying whether truth-values are value-ranges. Although not addressed here, the logical structure of the argument has been of additional interest in

function are identified by the recognition condition, so too are its values, and since the truth-values are values of this function, it follows that they are identified by the recognition condition. Stipulating that the truth-values are value-ranges obviously could not upset this result. Thus, even if it cannot be determined by the recognition condition that truth-values are value-ranges, it can nevertheless be shown that it is not incompatible with the condition if they are.[37]

Frege's argument shows no more than that truth-values *can* be value-ranges. It yields no grounds for any principled decision as to which value-ranges *are* the truth-values, although any such stipulation will be compatible with the recognition condition, as Frege states in the quotation previously. Frege takes this as providing license to make a specific choice. The True is set to be the value-range of a concept under which only the True itself falls, namely $\acute{\varepsilon}(\varepsilon = (\varepsilon = \varepsilon))$; comparably, the False is to be the value-range of a concept under which only falls the False, that is, $\acute{\varepsilon}(\varepsilon = \neg^{a} a = a)$. In effect, this is to stipulate that the True is the value-range of the concept of being identical to the True and the False the value-range of the concept of being identical to the False.[38] Importantly, by these stipulations, the truth of (*iii*), an instance of (*i*), is now settled:

$$\text{the True} = \acute{\varepsilon}\big(\varepsilon = (\varepsilon = \varepsilon)\big), \tag{*iii*}$$

and moreover, it is ensured that the True and the False are distinct. Thus, (*iv*) is false:

$$\acute{\varepsilon}(\varepsilon = (\varepsilon = \varepsilon)) = \acute{\varepsilon}(\varepsilon = \neg^{a} a = a) \tag{*iv*}$$

determined by its being coreferential with the appropriate instance of (*ii*), since this is an identity of value-ranges. And with this, the task of §10, to eliminate the underdetermination of value-ranges relative to the logical concepts, has been accomplished, as Frege explicitly states in closing the section: "We have hereby determined the *value-ranges* as far as possible here".

Having concluded the core dialectic of the text of *Grundgesetze* §10, Frege appends a footnote at the end of the section. In it, he explores

the literature; see Dummett (1991); Ricketts (1997); Wehmeier and Schroeder-Heister (2005); Heck (2012).

37 That is, the recognition condition does not entail that truth-values are value-ranges; the question whether they are is independent of the condition.

38 Strictly, Frege stipulates that the True is the value-range of the horizontal-concept that maps the True to the True and everything else to the False. Frege points out, however, that this concept and $(\xi = (\xi = \xi))$ map the same arguments to the same values; hence its value-range is also the True, with the advantage that the parallelism with the False is clearly displayed.

whether the truth-value stipulations he has made are arbitrary after all: If all objects are to be identified with the value-ranges of concepts under which only they fall as a matter of course, then there is nothing unusual about this being so for the True and the False.[39] Frege rejects this as a general claim, however; it fails, he observes, in those cases in which an object is given as a value-range. In that case, a value-range itself will be identified with the value-range of a concept under which only it falls. (The concept of being identical to that value-range.) This is problematic because the resulting identities will have their truth determined by being coreferential with an instance of (*ii*), and this will require that $\grave{\alpha}\Phi(\alpha)$ be the only object that falls under $\Phi(\xi)$. But "this is not necessary", Frege observes: Falling under the concept of being identical to the value-range of a concept does not depend in any way on how many things fall under that concept. It only depends on concepts having one and only one value-range. Accordingly, "our stipulation cannot be upheld in its generality".

This does not mean, however, that a function cannot be introduced mapping to an object just in those cases in which that object *is* the unique object that falls under a concept and which then could be asserted. Frege does just this in *Grundgesetze* §11, introducing the function $\backslash\xi$ with the following stipulation: It returns Δ as a value for a value-range as argument if Δ is the unique object falling under the corresponding concept (i.e., the number of the concept is 1), and it returns the value-range argument itself as a value if Δ is not the unique object falling under the corresponding concept (i.e., the number of the concept is other than 1). Given this specification, it now follows that:

$$\backslash\grave{\varepsilon}\left(\varepsilon=(\varepsilon=\varepsilon)\right) \tag{v}$$

has the True as its value, for only the True falls under the corresponding concept.[40] Accordingly, we have the following judgment:

$$\vdash \text{the True} = \backslash\grave{\varepsilon}\left(\varepsilon=(\varepsilon=\varepsilon)\right) \tag{vi}$$

39 Part of the intuitive interest to Frege of this claim is that if all objects could be identified with their unit-class, then the 0-level entities of any scientific application of logic would be value-ranges. This is to be distinguished from the subject-matter of the science, which are the corresponding objects. In this passage, Frege, speaking in a loosely metalogical fashion, means to encompass the subject-matter of any logical application. This usage, however, is independent of the question of what the 0-level entities are for any particular application. The point here is subtle, but it is important not to mistake Frege's discussion of what holds of all objects with the claim that the logical 0-level has a universal domain. See May (2018) for discussion of Frege's view of science as logical application.

40 To look ahead to the next section, note that (*v*) is rather curious. If $\grave{\varepsilon}(\varepsilon=(\varepsilon=\varepsilon))$ is stipulated to be the True, then $\grave{\varepsilon}(\varepsilon=(\varepsilon=\varepsilon))$ is the value of (*v*). But this is also the value if $\grave{\varepsilon}(\varepsilon=(\varepsilon=\varepsilon))$ is not the True, since in that case nothing falls under the concept $\xi = (\xi = \xi)$.

Parallel remarks hold for the False.

At this point, we can comprehend the answer to the question that has been lurking: Why does Frege opt for the particular choice of value-ranges to be the truth-values? From the stipulations, it follows that there can be one and only one object that is the True and, distinctly, one and only one object that is the False; the True and the False are unique. Other choices of value-ranges would not lead to this outcome. Specifically, the choice of $\acute{\varepsilon}(\varepsilon = \varepsilon)$, the only alternative at play at this stage of *Grundgesetze*, would not. In this case, the value of:

$$\backslash\acute{\varepsilon}(\varepsilon = \varepsilon) \tag{vii}$$

will not be the True, as everything falls under the universal concept (including the False), and accordingly, (*viii*) is false:[41]

$$\text{the True} = \backslash\acute{\varepsilon}(\varepsilon = \varepsilon) \tag{viii}$$

What Frege intends to accomplish by the stipulations is to characterize that it is truths, and only truths, that fall under the concept of being true. This is modeled by a function that maps the True to the True and everything else to the False, and the True is the value-range of this concept. (Comparably, the False is the value-range of the concept that maps the False to the True and everything else to the False.) Because the truth-values are unique in this way, it follows that all concepts map to the same values and that all true (false) sentences have the same reference. Just as there is only one concept of truth under which all truths fall, correspondingly, there is only one object to which all truths refer, and this is the True. In this regard, the True is the abstract of truth; it is what all truths have in common.

What, then, underlies Frege's concerns in these sections of *Grundgesetze*? It is this: To justify the functions on which logic is built, by establishing that they are first-level concepts, mapping logical objects to logical objects.[42] Given this, the truth-values must be entities of the 0-level if they are to be arguments and values of the logical functions. With this placement, Frege has the core building block for showing that every expression of the Begriffsschrift has a reference (the proof of referentiality), the core property that justifies the Begriffsschrift as a language suitable for

---

41 Frege is thus distinguishing being self-identical holding of the True from being identical to the True holding of the True. The True is the unique bearer of this latter property; the former it shares with all other objects. The True, Frege stipulates, is the value-range of the latter concept, not the former. Also, note here the same curiosity as that observed in fn. 40: $\acute{\varepsilon}(\varepsilon = \varepsilon)$ will be returned as the value, although *not* as the True.

42 Bear in mind that at this juncture, Frege has not yet introduced generalization as a second-level logical function.

scientific applications. This demonstration, as we have seen, is predicated on the compositionality of the language. In turn, this depends on the characterization of the horizontal-concept as mapping the True to the True and everything else to the False, so that the classical logical functions are truth-functions.[43] All this, however, raises a foundational question: Are the truth-values recognizable as logical objects, as befitting the values of logical functions? For certain, they are, if they are identified with value-ranges. And this is what Frege does by the stipulations, given that there is no logical prohibition against doing so. The criterion of recognition cannot determine whether truth-values are value-ranges; the condition is independent of this matter. But if they are value-ranges—stipulated to be so—then they will be recognized by the condition by which value-ranges are recognized.

In taking this stance, Frege is showing a metaphysical conservatism. There is no reason to take truth-values to be anything other than the sort of things we already have. The truth-values are among the value-ranges, governed by the identity condition for value-ranges, and there is nothing that pressures us into thinking that they might be anything else. In this regard, they are among the "ordinary" objects, not special objects distinct from what there otherwise is: The truth-values can be encompassed among the logical objects without saying anything more than is already been established about logical objects. That value-ranges can play the role required for truth-values is thus the outcome of §§10 and 11 of *Grundgesetze*. It is unclear, however, that Frege's metaphysical parsimony can be maintained.

5. Frege's Quasi-Paradox of Truth

In *Grundgesetze*, §10, Frege places the following adequacy condition on truth-values being value-ranges: "Thus — $\acute{\varepsilon}\Phi(\varepsilon)$ is the True only if the function $\Phi(\varepsilon)$ is a concept under which only the True falls; in all other cases — $\acute{\varepsilon}\Phi(\varepsilon)$ is the False." If this condition is satisfied, then $\acute{\varepsilon}\Phi(\varepsilon)$ is the True. Since the horizontal function is specified as mapping the True and only the True to the True, it follows directly that Frege's dictum is satisfied by the value-range of the horizontal itself, and it will be stipulated to be the True. Thus, by this stipulation, Frege *has* "decide[d] . . . whether an object that is not given to us as a value-range is a value-range" and "which function it may belong to".[44]

---

43 That is, as functions from truth-values to truth-values. Frege notes this for negation in *Gg*, §10, and in "Function and Concept" explicitly notes for the conditional that "we can always regard as the arguments of our function -x and -y, i.e. truth-values" (Frege 1970, p. 39).

44 *Gg*, I, §10. On the importance of the latter concern, see Heck (2012), ch. 4.4.

The value-range of the horizontal, as observed, is equivalent to the value-range of $\xi = (\xi = \xi)$, as they each return the same values for the same arguments. So it too is the True, and this is how we have been considering it. But now we have a question: Does this value-range contain itself? Does $\grave{\varepsilon}\big(\varepsilon = (\varepsilon = \varepsilon)\big)$ fall under its corresponding concept? It does and it does not. It does if it is stipulated to be the True; it does not if it is not so stipulated.[45] This is the *quasi-paradox of truth.*

The matter can be put this way: Is the following true?

$$\grave{\varepsilon}\big(\varepsilon = (\varepsilon = \varepsilon)\big) \cap \grave{\varepsilon}\big(\varepsilon = (\varepsilon = \varepsilon)\big) \tag{ix}$$

Frege defines the relation $\xi \cap \zeta$, in the case of concepts, as membership: *a* is a *member* of a value-range if and only if it falls under the concept corresponding to the value-range. Thus, (*ix*) is the proposition that $\grave{\varepsilon}\big(\varepsilon = (\varepsilon = \varepsilon)\big)$ falls under the concept $\xi = (\xi = \xi)$, of which it is the value-range. Clearly, this is false: Compositionally, the identity of the value-range to itself refers to the True, but this value-range is *not* identical to the True, only to itself, so (*ix*) is false. Frege notes this: "the value of the function $\xi = \zeta$ is always the False when a truth-value is taken as one of the arguments and a value-range as the other" (*Gg*, §10). On the other hand, if this value-range is stipulated to be the True, then (*ix*) is true. Again, compositionally, the identity of the True with itself refers to the True, and this in turn is identical to the True. So which is it? Is (*ix*) true or false? The answer, apparently, is that it depends: Whether a value-range falls under the concept $\xi \cap \grave{\varepsilon}\big(\varepsilon = (\varepsilon = \varepsilon)\big)$ depends *completely* on whether it is stipulated to be the True. That very value-range not so stipulated does not fall under this concept.

This is all very peculiar. The stipulation, as Frege is deploying it here, is creative. By stipulating that a value-range is the True, that value-range is taking on properties that it does *not* have, nor could it have, by virtue of being the value-range of a concept. It is turning value-ranges into something that natively they are not, with novel characteristics by virtue of which they fall under concepts that they otherwise would not; accordingly, false propositions turn into true. In effect, as truth-values, they are new objects. But how is it to be discerned which it is in (*ix*) if value-ranges are to be stipulated to be the truth-values without regard for their constitutive properties as value-ranges? Where is that encoded in the logical form? How do we know that (*ix*) is to be (*x*) outside of its being stipulated to be so?

$$\text{the True} \cap \grave{\varepsilon}\big(\varepsilon = (\varepsilon = \varepsilon)\big) \tag{x}$$

---

45 Or does the value-range of the horizontal fall under the horizontal? It does, but only if it is stipulated to be the True.

But recognize that there is a cost to the stipulation: The relation $\xi \frown \zeta$ is *not* membership if (x) is to be accommodated. (x) is not true because $\grave{\varepsilon}\big(\varepsilon = (\varepsilon = \varepsilon)\big)$ falls under the concept $\xi = (\xi = \xi)$; to the contrary, it would be false. It is true only because $\grave{\varepsilon}\big(\varepsilon = (\varepsilon = \varepsilon)\big)$ is stipulated to be the True.

It might be thought that there is a way out of the dilemma. Rather than stipulating that $\grave{\varepsilon}\big(\varepsilon = (\varepsilon = \varepsilon)\big)$ is the True, *define* it to be the True:

$$\Vert{-}\,\grave{\varepsilon}(\varepsilon = (\varepsilon = \varepsilon)) = \text{the True} \qquad\qquad (xi)$$

But that hardly helps. As already noted, by Frege's canons, definitions are explicit and conservative and hence non-creative: All that is accomplished by a definition is to introduce an abbreviation into the language; they create only names, not objects.[46] Accordingly, the True so defined has exactly the properties of the *definiens*. The definition cannot not *make* the value-range be the True by imbuing it with properties that it does not have; all that is being done is assigning to it a label, "the True". The effect of the definition would only be to legitimize (x) as a sentence of the language, but it would not resolve its truth-conditions. Conspicuously, Frege does *not* offer a definition of the True. To the contrary, in many places, he asserts that truth is indefinable, most notably in *The Thought*.[47] But without a definition, it is not possible within the Begriffsschrift to assert anything about the True: How would we know *in that context* whether (ix) is an assertion about the True if it has no name?[48] The quasi-paradox remains.[49]

So, where are we? In §10 of *Grundgesetze*, Frege mounts an argument that the choice of value-ranges to be the truth-values is arbitrary relative to the condition by which value-ranges are recognized as the same again. He further maintains that if the choice is going to be able to play the role demanded of truth-values in the overall theory, then there can be one and only one value-range that is the True (and one and only one that is the False). This is accomplished by stipulating that the True is the value-range of a concept under which only the True itself falls; this value-range is marked out to play the role of a truth-value. But *no* value-range has this

---

46 If (xi) were a definition, then (x) [and hence (ix)], would be judgments. This is the creativity of the putative definition on display. Like the stipulation, it would turn falsehoods into truths.

47 For discussion of the indefinability of truth, see Heck and May (2018).

48 So, strictly speaking, none of the previous that use the term "the True" are well formed; they are not Begriffsschriftsätze.

49 An implicit definition hardly fares any better. Let the True be whatever meets the condition of being the only thing that is identical to being identical to itself. But there is no such thing, at least not among the value-ranges. So if the definition is to be satisfied, there must be logical objects that are not value-ranges. But this implies a metaphysical thesis that Frege did not entertain.

property; no value-range inherently has the property of being the True. There is a value-range, however, that if it is *stipulated* to be the True—the value-range of the horizontal—then it will have this property, but *only by virtue of the stipulation*. So the stipulation, unlike a definition, imbues objects with novel properties; it is *creative*. But, again unlike definitions, they are not linguistically innovative: The language does not gain the term "the True" by warrant of the stipulation. But now we are left with the question: Is (*ix*) true or false? How is it to be determined whether the value-range is functioning in terms of its native properties or because of the role it is playing as the True? Does it express that the True falls under the True? Are we to conclude that it does and does not?

6.   Conclusion

At the end, we face a culminating question: What justifies that truth-values are objects? The answer is that being objects is what is demanded by the role they play in the overall theory. Frege's mode of explanation is holistic; to understand truth is to comprehend how it is interwoven into the fabric of the theory. As Frege remarks in the Introduction to *Grundgesetze*:[50]

> Only a thorough engagement with the present work can teach how much simpler and more precise everything is made by the introduction of truth-values. These advantages alone already weigh heavily in favor of my conception.

The core role of truth in this context is witnessed by its integration into the characterization of the language of logic as a conceptual language, that is, a language that is referentially anchored to the conceptual hierarchy. The proof of referentiality provides this anchor. To Frege, there is no sense to the question of justification pulled out of this context; there is no establishing necessary and sufficient conditions for being a truth-value independently of the theoretical context. But if truth is indefinable in this sense, nevertheless truth-values have to be identified to play the appointed roles. But if the choice is in principle arbitrary, the field is narrowed if it is to be that one and only one object is the True and one and distinctly only one object the False. For a universe of value-ranges—the logical universe—this criterion is met by the value-ranges of concepts under which the True and the False each uniquely fall, so long as those value-ranges are stipulated to be the True and the False.

But this leaves us in a muddle. By stipulating the truth-values to be value-ranges independently of their native characteristic marks, Frege has

---

50 *Gg*, p. x.

to buy into a notion—the theoretical creation of objects—that he categorically rejects for definitions. Truth-values, it turns out, can be created, but numbers pointedly cannot be. And this leaves us, in the end, on the horns of a dilemma: How are we to make the judgment of whether the True, and only the True, falls under the True?

References

Boddy, R. (2019). *Fregean Definition: Content Without Creativity*. PhD thesis, University of California, Davis.

Boddy, R. (forthcoming). "Frege on the Fruitfulness of Definitions". *Journal of the History of Analytic Philosophy*.

Boddy, R. and Robert M. (forthcoming). "Frege on Reference". In Heimer Geirsson and Stephen Briggs, editors, *Routledge Handbook on Linguistic Reference*. Routledge, London, 2021.

Dummett, M. (1991). *Frege: Philosophy of Mathematics*. Harvard University Press, Cambridge, MA.

Frege, G. (1879). *Begriffsschrift. Eine der Arithmetischen Nachgebildete Formelsprache des Reinen Denkens*. L. Nebert, Halle.

Frege, G. (1893/1903). *Grundgesetze der Arithmetik. Begriffsschriftlich abgeleitet*. Hermann Pohle, Jena.

Frege, G. (1970). "Function and Concept". In Peter Geach and Max Black, editors, *Translations from the Philosophical Writings of Gottlob Frege*. Basil Blackwell, Oxford.

Frege, G. (1972). *Conceptual Notation and Related Articles*. Translated by Terrell Ward Bynum, Oxford University Press, Oxford.

Frege, G. (1977). "The Thought". In P. T. Geach, editor, *Logical Investigations*. Basil Blackwell, Oxford.

Frege, G. (1979). "Comments on Sense and Meaning". In Hans Hermes, Friedrich Kambartel, and Friedrich Kaulbach, editors, *Posthumous Writings*. Basil Blackwell, Oxford.

Frege, G. (2013). *Basic Laws of Arithmetic. Derived Using Concept-Script*. Translated and edited by Philip A. Ebert and Marcus Rossberg. Oxford University Press, Oxford.

Heck, R.K. (2012). *Reading Frege's Grundgezetze*. Oxford University Press, Oxford. (Originally published under the name Richard G. Heck, Jr).

Heck, R.K. and May R. (2013). "The Function Is Unsaturated". In Michael Beaney, editor, *The Oxford Handbook of the History of Analytic Philosophy*. Oxford University Press, Oxford. (Originally published under the name Richard G. Heck, Jr).

Heck, R.K. and May R. (2018). "Truth in Frege". In Michael Glanzberg, editor, *The Oxford Handbook of Truth*. Oxford University Press, Oxford.

Heck, R.K. and May R. (2020). "The Birth of Semantics". *The Journal for the History of Analytic Philosophy*, 8: 1–31.

Linnebo, Ø. (2004). "Frege's Proof of Referentiality". *Notre Dame Journal of Formal Logic*, 45: 73–98.

Linnebo, Ø. (2019). "The Context Principle in Frege's *Grundgesetze*". In Philip Ebert and Marcus Rossberg, editors, *Essays on Frege's Basic Laws of Arithmetic*. Oxford University Press, Oxford.

May, R. (2018). "Logic as Science". In Paolo Leonardi Annalisa Coliva and Sebastiano Moruzzi, editors, *Eva Picardi on Language, Analysis and History*. Palgrave Macmillan, London.

Ricketts, T. (1997). "Truth-values and courses-of-value in Frege's *Grundgesetze*". In W. Tait, editor, *Early Analytic Philosophy: Frege, Russell, Wittgenstein*, pages 187–211. Chicago, IL, Open Court.

Tappenden, J. (2019). "Frege, Carl Snell and Romanticism: Fruitful Concepts and the Organic/Mechanical Distinction". Ms, University of Michigan.

Textor, M. (2010). "Frege on Judging as Acknowledging the Truth". *Mind*, 119: 615–655.

Wehmeier, K. and Schroeder-Heister P. (2005). "Frege's Permutation Argument Revisited". *Synthese*, 147: 43–61.

4 *Anzahlen* and *Werthverläufe.* Are There Objects Other Than Value-Ranges in Frege's Universe?

An Argument From *Grundgesetze*, §§I.34–40[*]

Marco Panza

It has been observed several times that the formal system in Frege's *Grundgesetze* (1893–1903) is, in a way, an interpreted system, in the sense that it assigns invariant meanings to its primitive functional constants and assumes that the quantifiers, both first- and second-order, have a fixed and universal domain. This implies that its definitions and theorems are also supposed to have invariant meanings. It remains, however, that these meanings depend on how one conceives of truth-values—which are, for Frege, two primitively distinct objects, the True and the False, to which I will refer by 'T' and 'F', respectively—and of *Werthverläufe*, or value-ranges. Frege never tells us explicitly how he conceives of these objects. He just stipulates (in §I.10) that the former are among the latter without clarifying whether it should be admitted that there exist objects that are not value-ranges. It follows that the invariant meanings that Frege seems to assign to his definitions and theorems are not, in fact, as precise as would be expected.

The aim of this chapter is to illustrate this ambiguity by focusing on the system of definitions that leads to the definition of *Anzahlen*[1] that

[*] This chapter is a translated and partially revised version of *"Anzahlen et Werthverläufe. Quelques remarques sur les §§I.34-40 des Grundgesetze de Frege"*, which originally appeared in *L'Épistémologie du dedans. Mélanges en l'honneur de Hourya Benis-Sinaceur*, edited by Emmylou Haffner and David Rabouin, Classiques Garnier, 2021, pp. 141–189. Many thanks to Pierre Chardot and Librairie Classiques Garnier for their permission to translate and reprint it. Thanks also to Marina Immocrante (who translated the French paper, before revisions) and to Hourya Benis-Sinaceur, Joan Bertran-San Millán, Francesca Boccuni, Méven Cadet, Emmylou Haffner, Gregory Landini, Paolo Mancosu, Robert May, David Rabouin, Erich Reck, Andrea Sereni, and Timothy Williamson, for comments and suggestions.

1 I will not translate Frege's term Anzhal, used in both the *Grundgesetze* and the *Grundlagen* (1984), as any English translation seems unsuitable to me. However, it is worth remembering that, following Frege, *Anzahlen* are a kind of numbers which include natural numbers but are not exhausted by them. More precisely, Frege defines the latter as finite *Anzahlen*: those that are ancestral to the successor relation with zero.

DOI: 10.4324/9780429277894-5

Frege presents in the *Grundgesetze* and by showing how it is particularly problematic in the case of this definition. In conclusion, I will consider two possible ways of limiting this ambiguity.

1.   Preliminaries

Let Φ be a one-argument function, Ψ a two-argument function,[2] Γ and Δ whatever two objects, and **T** and **F** the True and the False, respectively, as stated previously. Let's use '$\mathcal{E}\Phi$' and '$\mathcal{E}\Psi$' to refer to the value-ranges of Φ and Ψ respectively—also called 'extensions', if Φ is a concept and Ψ is a relation in Frege's sense.[3]

For brevity, let's write

$$\text{'}\mathfrak{C}_\Phi\Gamma\text{'} \quad \text{for} \quad \text{'}\Phi(\Gamma) = \mathbf{T}\text{'}$$

and

$$\text{'}\Gamma\mathfrak{R}_\Psi\Delta\text{'} \quad \text{for} \quad \text{'}\Psi(\Gamma, \Delta) = \mathbf{T}\text{'}.$$

It follows that '$\mathfrak{C}_\Phi$' and '$\mathfrak{R}_\Psi$' denote the concept ⌐argument for which $\Phi(\xi)$ takes the value **T**⌐ and the relation ⌐pair of arguments for which $\Psi(\xi, \zeta)$ takes the value **T**⌐, respectively. If Φ is a concept and Ψ is a relation, '$\mathfrak{C}_\Phi\Gamma$' and '$\Gamma\mathfrak{R}_\Psi\Delta$' are mere rewritings of '$\Phi(\Gamma)$' and '$\Psi(\Gamma, \Delta)$' respectively. On the other hand, if Φ is not a concept and Ψ is not a relation, then '$\mathfrak{C}_\Phi\Gamma$' and '$\Gamma\mathfrak{R}_\Psi\Delta$' have different meanings from '$\Phi(\Gamma)$' and '$\Psi(\Gamma, \Delta)$'. While they

---

2 Frege distinguishes functions according to their levels: the arguments of a first-level function are objects, those of a second-level function are first-level functions, and so on. In this chapter, I will consider only first-level functions: for the sake of simplicity, therefore, I will speak of functions to refer to first-level functions. When this does not cause any ambiguity, I will also take the liberty to designate functions without explicitly indicating the (empty) places to be filled by their arguments; for example, I will use 'Φ' and 'Ψ' to refer to functions rather than using '$\Phi(\xi)$' and '$\Psi(\xi, \zeta)$' as Frege would do. Recall that the role of 'ξ' and 'ζ' in an expression of Frege's system is to keep two places free for the names of the objects that can be the arguments of the first-level function of which the given expression is in turn the name. If 'ξ' and 'ζ' are replaced by such names, the expression in question turns into the name of the value of the function that takes those objects as arguments. In Frege's German, 'ξ' and 'ζ' are thus *Offenhalten*, or place-holders (§I.1). This means that 'ξ' and 'ζ' cannot be conceived of as variables in the current sense. For example, following Frege, it makes no sense to write an identity such as '$\zeta = \Phi(\xi)$' or, more generally, '$\Phi(\xi) = \tilde{\Phi}(\zeta)$' (where Φ and $\tilde{\Phi}$ are two one-argument functions), because for him the identity relation only holds between objects, and $\Phi(\xi)$ and $\tilde{\Phi}(\zeta)$ are not objects but functions. The current notion of variable is absent from Frege's conceptual horizon. Although Frege's Latin letters have often been matched to variables in the current sense, I do not think this is correct; however, this question is not relevant for the arguments advanced in this chapter, and I will not consider it. I'll confine myself to observing that one might maintain, at most, that in the exposition of his system (in the first part of *Grundgesetze*), Frege makes use of meta-variables in the current sense, denoted by capital Greek letters such as 'Γ', 'Δ', 'Θ', 'Λ', . . . for objects, and 'Φ', 'Ψ' and 'Υ', . . . for functions.

3 That is, a one-argument and a two-argument function, respectively, whose values are truth-values.

denote the same truth-value as '$\Phi(\Gamma)$' and '$\Psi(\Gamma, \Delta)$', they have different senses, in that the former expressions denote their truth-value as the value of the function $\xi = \zeta$, while the latter denote it as the value of the functions Φ and Ψ, respectively.

For whatever object Λ, let $\Lambda_{1,n}^{F}$ be the constant n-argument function ($n = 1$, 2, . . .) defined as follows:

$$\Lambda_{1,n}^{F}(\Gamma_1,...,\Gamma_n) = \Lambda,$$

for whatever n-tuple of objects $\Gamma_1, \ldots, \Gamma_n$—that is, the n-argument function that takes the value Λ for whatever n-tuple of arguments.

A function extensionally coinciding with such a constant function has a perfectly defined place in Frege's system. It is the concept $[x : x \neq x]$ (in the usual notation in predicate logic), extensionally coinciding with $F_{1,1}^{F}$. Let's denote its value-range by '$\emptyset$'. This results in stipulating that:

$$\emptyset = \mathcal{E}F_{1,1}^{F} = \mathcal{E}[x : x \neq x].$$

Other constant functions are easy to define in this system. An obvious example is the concept $[x : x = x]$ (keeping the previous notation), which extensionally coincides with $T_{1,1}^{F}$. Let us denote its value-range with '@', which results in stipulating that:

$$@ = \mathcal{E}T_{1,1}^{F} = \mathcal{E}[x : x = x].$$

Other functions such as $\Lambda_{1,n}^{F}$, in which Λ is not a truth-value, do not become visible at first sight within this system. But, as I will show in the following, their existence must be admitted if we admit that there are objects other than value-ranges. Whether these objects are admitted will play a crucial role in what follows. It is, therefore, important to clarify the nature and the implications of such acceptance.

First, let us observe that within the framework of *Grundgesetze*, any value-range is the value-range of a first-level, one-argument function. On the one hand, the value-range of a first-level function with several arguments is defined (in §I.36) as the value-range of a function with one argument which is associated with it in a standard way. On the other hand, the value-ranges of higher-level functions are simply not defined,[4] even though they could be thought of as the value-ranges of the first-level functions representing the former functions. Although the process of allowing first-level functions to represent higher-level functions depends on the function $\xi \frown \zeta$—to which I'll largely come back in what follows—nothing I will say depends on whether it is possible to think about (or even deal with) the value-ranges of higher-level functions as value-ranges of first-level functions. Here it will be sufficient to focus on the first point.

---

4 See footnote 2 above.

For whatever two-argument function $\Psi = \Psi(\xi, \zeta)$, let $\mathcal{E}_\varepsilon\Psi(\varepsilon, \zeta)$ [also denoted by '$\mathcal{E}_\alpha\Psi(\alpha, \zeta)$'] be the one-argument function resulting from the value-range of $\Psi(\xi, \zeta)$ relative to the ξ argument alone and $\mathcal{E}_\varepsilon\Psi(\xi, \varepsilon)$ [also denoted by '$\mathcal{E}_\alpha\Psi(\xi, \alpha)$'] the one-argument function resulting from the value-range of $\Psi(\xi, \zeta)$ relative to the ζ argument alone. The value-range $\mathcal{E}\Psi$ of $\Psi(\xi, \zeta)$ is the value-range of $\mathcal{E}_\varepsilon\Psi(\varepsilon, \zeta)$, that is, $\mathcal{E}_\alpha[\mathcal{E}_\varepsilon\Psi(\varepsilon, \alpha)]$ (also denoted by '$\mathcal{E}_\varepsilon[\mathcal{E}_\alpha\Psi(\alpha, \varepsilon)]$'), and it is thus a double value-range.[5]

Let us also note that Frege conceives of any function as being a total function, which means that, for any function, we can quantify on their values by quantifying on (all) the objects. In other words, if Φ is whatever one-argument function and Ψ is whatever two-argument function, then to say that (all) their values are so-and-so is equivalent to saying that, for whatever object Γ, $\Phi(\Gamma)$ is so-and-so, and for whatever pair of objects (Γ, Δ), $\Psi(\Gamma, \Delta)$ is so-and-so. For Frege, any value of any function is also an object, regardless of the function's level, which means that what is true for all objects is also true for all values of all functions.

Let us consider now the following statements concerning the existence of objects other than value-ranges (where 'x', 'z', 'f', and 'g' are used, for the sake of simplicity, as variables in the current sense, and 'x' and 'z' vary over objects, while 'f' and 'g' vary over one-argument functions):

A.   'Every object is a value-range', or '$\forall x \exists g[x = \mathcal{E}g]$';

　　¬A. 'Some objects are not value-ranges', or '$\exists x \forall g[x \neq \mathcal{E}g]$';

B.   'Any value of any one-argument function is a value-range', or '$\forall g \forall x \exists f[g(x) = \mathcal{E}f]$';

　　¬B. 'Some values of some one-argument function are not value-ranges', or '$\exists g \exists x \forall f[g(x) \neq \mathcal{E}f]$';

C.   'Some values of any one-argument function are value-ranges', or '$\forall g \exists x \exists f[g(x) = \mathcal{E}f]$';

　　¬C. 'No value of some one-argument function is a value-range', or '$\exists g \forall x \forall f[g(x) \neq \mathcal{E}f]$';

---

5 For the sake of simplicity, I use Frege's terminology here, calling 'double' the value-range of a two-argument function—that is also, as I am trying to explain, the value-range of another one-argument function—and 'simple' the value-range of a one-argument function that is not (or is not meant to be) the value-range of a two-argument function at the same time. This implies that any value-range is by definition the value-range of a first-level one-argument function, while some value-ranges, but not necessarily all of them, are also value-ranges of a first-level two (or more)-argument function. Therefore, in what follows, I will generally use 'value-range' to refer to the value-range of some one-argument first-level function—making it explicit only if required for clarity of exposition—regardless of whether this value-range is also the value-range of a two (or more)-argument function, whereas I will speak of the value-range of a two-argument function to refer to the value-range of a one-argument function that is also the value-range of a two-argument function.

D. 'Any one-argument function is such that all its values are value-ranges, or none is', or '$\forall g[\forall x \exists f[g(x) = \mathcal{E}f] \vee \forall x \forall f[g(x) \neq \mathcal{E}f]]$';

¬D. 'Some values of some one-argument functions are value-ranges and some are not', or '$\exists g[\exists x \exists f[g(x) = \mathcal{E}f] \wedge \exists x \forall f[g(x) \neq \mathcal{E}f]]$';

E. 'Any object is the value of a one-argument function', or '$\forall x \exists g \exists z[x = g(z)]$';

F. 'Some one-argument functions are such that any object is one of their values', or '$\exists g \forall x \exists z[x = g(z)]$';

G. 'There is a one-argument function such that, for any object as argument, the same object is the value of that function', or '$\exists g \forall x[x = g(x)]$'.

Obviously, A⇒B, and consequently ¬B⇒¬A. Moreover, E⇒(B⇒A), F⇒E, G⇒F; consequently, G⇒(B⇒A). It follows that, if we assume that G, then A⇔B.

This raises the question of whether G must be admitted. I think that the answer to this question depends on the unrestricted rule of substitution that Frege applies throughout the deductions within his formal system. This rule is equivalent to an axiom schema of unrestricted comprehension for functions. In the case of one-argument functions, the schema is the following:

$$\exists g \forall x[g(x) = \mathcal{A}(x)],$$

where '$\mathcal{A}(x)$' is any formula of the language of the system where 'x' occurs free. A simple instance of the schema is the following:

$$\exists g \forall x[g(x) = x],$$

which allows the introduction of the one-argument identity function defined by

$$\forall x[\mathbb{I}(x) = x].$$

It is then sufficient to admit, in agreement with Frege, that any well-formed name of function denotes a function, in order to derive a proof of G. If this argument is accepted, then one must conclude that, for Frege, A⇔B and ¬A⇔¬B. On the other hand, if it is assumed that there are objects and one-argument functions, which for Frege is certainly the case,[6] ¬C⇒¬B, and consequently ¬C⇒¬A and A⇒C.

---

6 As we have just seen in the case of objects, and as it follows for one-argument functions from the assumption that **T** and **F** are objects.

Moreover, it should be clear that A⇒D, and consequently ¬D⇒¬A, and that (C∧D)⇒B, so that, if it is assumed that there are objects and one-argument functions, D⇒(B⇔C); therefore, D⇒(A⇔C), (C∧¬A)⇒¬D, and (¬A∧D)⇒¬C.

This being said, the question arises whether Frege could have proved A (and thus B), C, and D or their negations. One could think that an answer comes from two well-known results obtained by Kai Wehmeier (1999). He showed that in Heck's system H (see Heck 1996), as well as in his system T_Δ, it is possible to prove that there is an object other than a value-range, that is, that:

$$\exists x \forall F[x \neq \mathcal{E}F] \tag{1}$$

where '*F*' varies over (first-level) concepts. In T_Δ, it is even possible to show more than this, namely that there is no concept of value-range, that is, that:

$$\neg \exists G \forall x [Gx \Leftrightarrow \exists F(x = \mathcal{E}F)] \tag{2}$$

Both H and T_Δ are consistent fragments of a second-order monadic logic system, enriched by an appropriate version of Frege's Basic Law V (or BLV), and can, as such, be considered consistent subsystems of a modern version of Frege's system, in which the second-order variables are restricted to monadic predicates.

In H, BLV is reformulated in such a way as to implicitly define '$\mathcal{E}$' as an operator acting on formulas (in particular, open formulas including one and only one free variable), and the comprehension axiom-schema is predicative. What Wehmeier proves there (by *reductio*) is that there is an $\mathcal{E}$-term (in particular, the $\mathcal{E}$-term of a Russell-like formula) such that there can be no (first-level) concept this term denotes the value-range of, which allows him to prove (1). In fact, one can also prove that there are infinitely many $\mathcal{E}$-terms like this, but this is not important for my present purpose.

In T_Δ, BLV is reformulated in such a way as to implicitly define '$\mathcal{E}$' as an operator acting on monadic predicates, and the comprehension axiom-schema is Δ_1^1. What Wehmeier proves there (again by *reductio*) is that there is no (first-level) concept such that an object falls under it if and only if it is the value-range of a (first-level) concept, which allows him to deduce (2) and then to show that from this it follows that (1) is provable (which does not, in turn, prevent us from writing the open formula '$\exists F[x = \mathcal{E}F]$').

Among others, there are two important differences between H and T_Δ, on the one hand, and Frege's original system, on the other. Both in H and in T_Δ, second-order variables vary only on concepts, whereas in Frege's system, it is more generally a question of functions (with one or more arguments), which can be reduced to concepts or relations with the help of the horizontal function, which is absent from H and T_Δ. Both in H and T_Δ, one distinguishes between open formulas and predicates for concepts and connects the two by means of comprehension axioms, whereas in Frege's system, concepts result immediately from appropriate expressions or, to be more precise, from their desaturation. These two differences make it far from clear whether and how Wehmeier's proofs can be reproduced in the latter system. Moreover, what Wehmeier proves about (1) in T_A is not the statement of the theorem itself but its provability, and it requires a meta-theoretical argument that cannot be easily transformed into a proof in Frege's system. Again, even if one could reproduce Wehmeier's proofs and theorems in Frege's system, it would remain that these proofs make essential use of Russellian devices that Frege would not have been able to envisage at the time of writing the *Grundgesetze*.

This is why I will here address a problem similar (but not identical) to the one that motivated Wehmeier's results in a completely different way: instead of reasoning deductively in modernized, consistent versions of Frege's system, I will ask whether, in his state of knowledge (at the time of writing the *Grundgesetze*), Frege would have had to admit objects other than value-ranges in order to be able to assign to his definition of *Anzahlen* the meaning he seems to assign to it. In trying to answer this question, I shall take A (and therefore B), C, and D, as well as their negations, as undecided hypotheses and reason conditionally, assuming in a first case either A (and therefore B, C, and D) or ¬A (and therefore ¬B)[7] and, in a second case, either C (and therefore ¬D) or ¬C and assuming either D (and therefore ¬C) or ¬D.

2.   The Functions $\xi \frown \zeta$ and $\xi \frown \Theta$ and the Concept $\_\xi \frown \Theta$

Among the primitive functions of Frege's system, there is the function $\backslash\xi$. This function is informally defined in §I.11 by stipulating that, for whatever object Γ and Δ,

$$\exists z \, [\mathcal{E}[x : x = z] = \Gamma \wedge z = \Delta] \Rightarrow \backslash\Gamma = \Delta, \tag{3}$$
$$\neg\exists z \, [\mathcal{E}[x : x = z] = \Gamma] \Rightarrow \backslash\Gamma = \Gamma.$$

The first clause of this stipulation is unambiguous in that, in accordance with Basic Law V,

$$\exists z \, [\mathcal{E}[x : x = z] = \Gamma] \Leftrightarrow \exists!z \, [\mathcal{E}[x : x = z] = \Gamma].$$

This means that the two clauses taken together set the value of the function $\backslash\xi$ for whatever argument.

---

7 It is clear that if G is rejected, then the equivalence of A and B is no longer proven, which leads to an increase in the number of possible cases. One should not, however, have difficulties in dealing with the new cases that would then be generated in analogy to those I am about to consider.

This allows Frege to use this function to define the most fundamental of his non-primitive functions, that is, the (two-argument) function $\xi \frown \zeta$. This is done in §I.34, where it is established that:

$$\forall x,y \; [x \frown y = \backslash\!\mathscr{E} \; [z : \exists g[y = \mathscr{E}g \wedge g(x) = z]]] \tag{4}$$

where 'x', 'y', and 'z' vary over objects and 'g' over one-argument functions.

For whatever one-argument function Φ, and for whatever object Γ, it follows from (3) and (4) that:

$$\Gamma \frown \mathscr{E}\Phi = \backslash\!\mathscr{E} \; [x : \exists g[\mathscr{E}\Phi = \mathscr{E}g \wedge g(\Gamma) = x]] = \backslash\!\mathscr{E} \; [x : x = \Phi(\Gamma)] = \Phi(\Gamma). \tag{5}$$

Under assumption A, this exhausts all possible cases.

On the other hand, under the assumption ¬A, one can imagine that the ζ-argument of the function $\xi \frown \zeta$ is given by an object other than a value-range, let us say Θ°. From (3)[8] and (4), it thus follows that:[9]

$$\Gamma \frown \Theta^{\circ} = \backslash\!\mathscr{E} \; [x : \exists g[\Theta^{\circ} = \mathscr{E}g \wedge g(\Gamma) = x]] = \backslash(\mathscr{E}\mathrm{F}^F_{1,1}) = \backslash\varnothing = \varnothing. \tag{6}$$

On this basis, the following conclusions can be drawn.

1. $\xi \frown \Theta$ is a one-argument function, namely:

$$\backslash\!\mathscr{E} \; [x : \exists g[\Theta = \mathscr{E}g \wedge g(\xi) = x]].$$

As a result:

i.   If, for some one-argument function Φ, $\Theta = \mathscr{E}\Phi$, then $\xi \frown \Theta$ is the same function as Φ, which makes $\xi \frown \mathscr{E}\Phi$ coincide with Φ (extensionally at least, if not also intensionally).

---

8 Clearly, the derivation depends on the second clause of (3). In the following, I will consider the possibility of eliminating this second clause while keeping the first, which would lead to a different conclusion than (6). In order to avoid any confusion between the cases where this clause is admitted and those where it is not, it is important to make clear that from now on, I will reason by default under the assumption that this clause is admitted, and I will explicitly point out when I reason under the opposite assumption.

9 It is easy to observe that from (3), it follows that:

$$\backslash\!\mathscr{E} \; [x : \exists g[\Theta^{\circ} = \mathscr{E}g \wedge g(\Gamma) = x]] = \mathscr{E} \; [x : \exists g[\Theta^{\circ} = \mathscr{E}g \wedge g(\Gamma) = x]]$$

but not that

$$\backslash\!\mathscr{E} \; [x : \exists g[\mathscr{E}\Phi = \mathscr{E}g \wedge g(\Gamma) = x]] = \mathscr{E} \; [x : \exists g[\mathscr{E}\Phi = \mathscr{E}g \wedge g(\Gamma) = x]],$$

the latter being not valid unless one accepts that $\mathscr{E} \; [x : x = \Phi(\Gamma)] = \Phi(\Gamma)$. This is something that Frege has to deny if $\Phi(\Gamma)$ is the value-range of a function other than a concept under which only one object falls, because, as he himself shows in §I.10 (second footnote), this would contradict BLV. In general, one is thus not allowed to identify $\Gamma \frown \Theta$ and $\mathscr{E} \; [x : \exists g[\Theta = \mathscr{E}g \wedge g(\Gamma) = x]]$, even though this could be allowed in some specific cases.

Under assumption A, there are no further cases to consider. On the contrary, under the assumption ¬A, another case is possible:

ii. It may be that $\Theta = \Theta^\circledast$, for whatever object $\Theta^\circledast$ other than a value-range, which means that $\xi \frown \Theta$ would be reduced to the constant function $\emptyset^F_{1,1}$.

It follows that, considering $\xi \frown \Theta$ as a well-formed function for whatever object Θ and assuming ¬A, one is forced to admit a constant function like $\emptyset^F_{1,1}$. (Note that it would be inappropriate to denote such a function with '$\xi \frown \Theta^\circledast$', since it is completely independent from $\Theta^\circledast$, provided the latter object is not a value-range.)

2. $\underline{\quad} \xi \frown \Theta$ (where $\underline{\quad} \xi$ is the horizontal function defined by Frege in §I.4) is a concept, namely:

$$\underline{\quad}\backslash\mathcal{E}\,[x : \exists g[\Theta = \mathcal{E}g \wedge g(\xi) = x]].$$

Therefore:

i. If, for whatever one-argument function Φ, $\Theta = \mathcal{E}\Phi$, then $\underline{\quad} \xi \frown \Theta$ is the same concept as $\underline{\quad} \Phi(\xi)$, which means that $\xi \frown \mathcal{E}\Phi$ coincides with Φ (extensionally at least, if not also intensionally).

Under assumption A, there are no further cases to consider. Under assumption ¬A, another case is possible:

ii. It may be that $\Theta = \Theta^\circledast$ for whatever object $\Theta^\circledast$ other than a value-range, which entails that $\underline{\quad} \xi \frown \Theta$ would be reduced to the constant function $\underline{\quad} \emptyset^F_{1,1}$ or $\mathbf{F}^F_{1,1}$.

It follows that, under assumption A as well as under assumption ¬A:

— for an argument Γ, the concept $\underline{\quad} \xi \frown \Theta$ takes the value T if and only if Θ is the value-range of a one-argument function Φ such that $\Phi(\Gamma) = \mathbf{T}$.

In both cases, this concept is thus the same as

$$[x : \exists g[\Theta = \mathcal{E}g \wedge \mathbf{C}_g x]]. \tag{7}$$

For brevity, let's denote this concept by $\mathfrak{T}_\Theta$.[10]

---

10 The difference between the two cases (the cases in which A and ¬A are respectively assumed) does not lie in the nature that has to be assigned to this concept but in the conditions that make an object fall under it or, more specifically, in the conditions under which Θ is a value-range.

With this in mind, let's go back to stipulations (3) and (4) to observe a crucial difference between the two. While the second is an (explicit) definition, integrated into Frege's formal system [i.e., definition (A) in §I.34 and in section 2 of the *Anhäng* of Volume I], the first is an informal (still explicit) definition that only appears in the preliminary 'exposition [*Darlegung*]' of the system and that within this system is rendered by Basic Law VI,

$$\vdash a = \backslash \overset{\text{2}}{\varepsilon}\, (a = \varepsilon)$$

or, in the language adopted here

$$\forall x[x = \backslash \mathcal{E}[z : z = x]],$$

which, however, only renders the first of its two clauses.

The problem is that this clause, and thus this law, only concerns the case in which the argument of the function $\backslash \xi$ is the value-range of a concept under which only one object falls. It follows that, even if A is admitted, Frege's formal system does not display any instruction on how to understand the function $\backslash \xi$ in its generality unless it is also assumed that any value-range is somehow reduced to the value-range of a concept under which only one object falls—an assumption that seems difficult to maintain in light of the second of the two footnotes in §I.10.[11] Nevertheless, under assumption A, this indetermination does not concern the function $\xi \frown \Theta$, since, as shown by (5), identifying the values of this function for $\Theta = \mathcal{E}\Phi$ (regardless of whether Φ is a concept under which only one object falls) only depends on the first clause of stipulation (3). Under assumption ¬A, on the other hand, the values of this function remain indeterminate within the system, unless it is established that Θ is a value-range. Indeed, as shown by (6), if the latter object were not a value-range, the reduction of this function to the constant function $\emptyset^F_{1,1}$ would depend on the second clause of (3); in the absence of the latter, it can only be said that $\xi \frown \Theta^{\circledcirc}$ reduces to $\backslash \emptyset^F_{1,1}$, with no way of establishing the constant value that the function takes for any argument: all we can say is that this value is the one taken by the function $\backslash \xi$ for the argument $\emptyset$.

Things are not going any better for the concept $\_\_ \xi \frown \Theta$. Indeed, the reduction of this concept to the concept $\mathfrak{X}_\Theta$ depends on the second clause of stipulation (3). Without such a clause, and if it is assumed, under ¬A, that Θ is not a value-range, it can only be concluded that this concept reduces to $\_\_\backslash \emptyset^F_{1,1}$, again with no way of establishing the constant value that the function takes for any argument: all we can say is (again) that this value is the one taken by the function $\_\_\,\xi$ for the argument $\backslash\emptyset$. It would thus be enough to accept that $\backslash\emptyset = T$ to conclude that

$$\_\_\Gamma \frown \Theta^\circ = \_\_\backslash\emptyset = \_\_T = T;$$

without being able to conclude that, for whatever objects Θ, the concept $\_\_\xi \frown \Theta$ takes the value T for a given argument Γ if and only if Θ is the value-range of a given one-argument function Φ, such as $\Phi(\Gamma) = T$, and reduce this concept to the concept $\mathfrak{X}_\Theta$. Therefore, by itself, Frege's formal system does not include any unequivocal instruction on how to understand the function $\_\_\xi \frown \Theta$.[12]

Whether this concept is to be understood as the same concept as $\mathfrak{X}_\Theta$ thus depends:

i.   either on the adoption of the second clause of stipulation (3);
ii. or on the acceptance of A;
iii. or, alternatively, on the rejection of the possibility that $\backslash\emptyset = T$.

Although Frege would agree with the first and last of these options—if not with the second—the fact remains that his formal system leaves the issue undecided.[13]

One could argue that this is not problematic for Frege. Indeed, it would be implausible to think that Frege was aiming at a complete system—in the sense of a system capable of proving all the statements formulated in its language which are true in accordance with the stipulations adopted in the preliminary 'exposition' concerning its fundamental functions. He seems only to have sought to provide the

---

12 Is this implied from what Frege says in commenting on his definition (A) in the *Anhäng* of the first volume? In the comment, Frege presents definition (A) as the definition of the 'relation of an object that falls under the extension of a concept'. But in the footnote, he observes: 'These short hints in words which I add to the *Begriffsschrift* definitions are not exhaustive and make no claim to be of the strictest precision [*machen keinen Anspruch auf strengste Genauigkeit*]' (G. F. Frege 2013, p. 240, with a slight change in the translation).

13 One could hope that there is a way to rule out the possibility that $\backslash\emptyset = T$ by showing that this leads to a contradiction (independent of the Russellian one, of course). A clever argument to show this has been suggested to me by Joan Bertran-San Millán. In accordance with the stipulation that Frege makes in §I.10, $T = \mathcal{E}[z : z = T]$, and, given BLV, $T \neq \emptyset$. It follows that, if $\backslash\emptyset = T$, then $\backslash\emptyset \neq \emptyset$. The value of $\backslash\emptyset$ is thus not established in accordance with the second clause of (3); consequently, it will be established in accordance with the first clause, so that

$$\exists z\, [\mathcal{E}[x : x = z] = \emptyset \wedge z = \backslash\emptyset = T]$$

and then

$$\mathcal{E}[x : x = T] = \emptyset,$$

which contradicts BLV. The argument is correct, but it is based on two stipulations that are external to Frege's formal system: the one exposed in §I.10 concerning T and stipulation (3) taken as a whole.

system with the minimum required to establish what he intended to prove in accordance with his foundational purposes. Thus, one could say, if the second clause of (3) is not rendered within the system, it is simply because there was no need to derive theorems in this system by appealing to a formal correlate of it.

This argument is convincing. But it is not conclusive, as such. For the absence of such a correlate not only prevents us from deriving in the system the statements which are true by virtue of such a clause, but, under the assumption $\neg A$, it also prevents us from identifying the meaning that must be assigned to the functions $\xi \frown \Theta$ and $\_\xi \frown \Theta$ (and thus to all the definitions and theorems involving them) while remaining within the system; that is, if I am allowed to express myself in this way, it prevents us from fixing the informal semantics intrinsic to such a system. Now, it is precisely this semantics that I am concerned with here. The problem is therefore whether such an absence has significant implications.

Since the question cannot be decided *a priori*, in the following, I will consider the case in which the second clause of (3) is adopted, as well as the case in which it is not, leaving to the end of the chapter the task of determining the consequences to be drawn in both cases.

3.   The Functions $\xi \frown (\zeta \frown \theta)$ and $\xi \frown (\zeta \frown \Theta)$ and the Relation $\_\xi \frown (\zeta \frown \theta)$

As Frege explains in §I.36, it also follows from (4) that for whatever objects Γ, Δ, and Θ,

$$\Gamma \frown (\Delta \frown \Theta) = \backslash\!\mathcal{E}\,[x : \exists g[\Delta \frown \Theta = \mathcal{E}g \wedge g(\Gamma) = x]]$$
$$= \backslash\!\mathcal{E}\,[x : \exists g\,[\backslash\!\mathcal{E}\,[z : \exists f[\Theta = \mathcal{E}f \wedge f(\Delta) = z]]$$
$$= \mathcal{E}g \wedge g(\Gamma) = x]] \tag{8}$$

where 'f' and 'g' vary over one-argument functions.

Therefore, for whatever two-argument function Ψ, from (3), it follows that:

$$\Gamma \frown (\Delta \frown \mathcal{E}\Psi) = \Gamma \frown (\Delta \frown \mathcal{E}_\alpha[\mathcal{E}_\varepsilon\Psi(\varepsilon,\alpha)])$$
$$= \backslash\!\mathcal{E}\,[x : \exists g\,[\backslash\!\mathcal{E}\,[z : \exists f[\mathcal{E}[\mathcal{E}_\varepsilon\Psi(\varepsilon, \zeta)] = \mathcal{E}f \wedge f(\Delta) = z]] = \mathcal{E}g \wedge g(\Gamma) = x]]$$
$$= \backslash\!\mathcal{E}\,[x : \exists g\,[\backslash\!\mathcal{E}\,[z : z = \mathcal{E}\Psi(\xi, \Delta)] = \mathcal{E}g \wedge g(\Gamma) = x]]$$
$$= \backslash\!\mathcal{E}\,[x : \exists g\,[\mathcal{E}\Psi(\xi, \Delta) = \mathcal{E}g \wedge g(\Gamma) = x]]$$
$$= \backslash\!\mathcal{E}\,[x : x = \Psi(\Gamma, \Delta)]$$
$$= \Psi(\Gamma, \Delta). \tag{9}$$

If Φ is whatever one-argument function, from (3) it follows that, on the other hand:

$$\Gamma \frown (\Delta \frown \mathcal{E}\Phi) = \backslash\!\mathcal{E}\,[x : \exists g\,[\mathcal{E}\,[z : \exists f[\mathcal{E}\Phi = \mathcal{E}f \wedge f(\Delta) = z]] = \mathcal{E}g \wedge g(\Gamma) = x]]$$
$$= \backslash\!\mathcal{E}\,[x : \exists g[\mathcal{E}[z : z = \Phi(\Delta)] = \mathcal{E}g \wedge g(\Gamma) = x]]$$
$$= \backslash\!\mathcal{E}\,[x : \exists g[\Phi(\Delta) = \mathcal{E}g \wedge g(\Gamma) = x]].$$

Once this has been established, the derivation can continue only after establishing whether $\Phi(\Delta)$ is or is not a value-range.

Under assumption A, it is certain that $\Phi(\Delta)$ is a value-range. Now let's suppose that $\Phi(\Delta)$ is the value-range of the (one-argument) function $\tilde{\Phi}_\Delta$. We should continue as follows:

$$\Gamma \frown (\Delta \frown \mathcal{E}\Phi) = \backslash\mathcal{E}\left[x : \exists g\left[\mathcal{E}\tilde{\Phi}_\Delta = \mathcal{E}g \wedge g(\Gamma) = x\right]\right]$$
$$= \backslash\mathcal{E}\left[x : x = \tilde{\Phi}_\Delta(\Gamma)\right] \qquad (10)$$
$$= \tilde{\Phi}_\Delta(\Gamma).$$

Now, if assumption A is made, given the function Φ, for whatever object Δ, we will have a corresponding one-argument function $\tilde{\Phi}_\Delta$ which depends on Δ, such as $\Phi(\Delta) = \mathcal{E}\tilde{\Phi}_\Delta$. If Λ were an object other than Δ, the function $\tilde{\Phi}_\Delta$ might as well be a function other than $\tilde{\Phi}_\Delta$ and have, in particular, a different value-range than the latter. Therefore, assuming that $\Phi(\Delta) = \mathcal{E}\tilde{\Phi}_\Delta = \mathcal{E}\tilde{\Phi}_\Delta(\alpha)$ is equivalent to assuming that $\Phi(\Delta) = \mathcal{E}_\alpha\Psi_{[\Phi]}(\alpha, \Delta)$—where $\Psi_{[\Phi]}$ is a two-argument function such that $\Phi(\zeta)$ is the same function (or at least it is a function that takes the same values for the same arguments) as $\mathcal{E}_\alpha\Psi_{[\Phi]}(\alpha, \Delta)$—and thus $\mathcal{E}\Phi = \mathcal{E}_\varepsilon[\mathcal{E}_\alpha\Psi_{[\Phi]}(\alpha, \varepsilon)]$.[14] This first case is thus reduced to the case assumed in (9) under the condition of the identification of Ψ and $\Psi_{[\Phi]}$.[15]

Under assumption ¬A, there is no reason to admit that, given the function Φ, for whatever object Δ, there is a one-argument function $\tilde{\Phi}_\Delta$ such that $\Phi(\Delta) = \mathcal{E}\tilde{\Phi}_\Delta$.

If assumption ¬C is accepted, it may be that Φ is such that none of its values is a value-range. Therefore, for whatever Δ, the derivation will continue as follows:

$$\Gamma \frown (\Delta \frown \mathcal{E}\Phi) = \backslash\left(\mathcal{E}F_{1,1}^F\right)$$
$$= \backslash\varnothing \qquad (11)$$
$$= \varnothing$$

14 The argument can of course be reiterated: under assumption A, and given the function $\Psi_{[\Phi]}$, for whatever pair of objects (Γ, Δ), there will be a corresponding two-argument function $\tilde{\Psi}_{[\Phi]}\big|_{(\Gamma,\Delta)}$ (depending on Γ and Δ) such that $\Psi_{[\Phi]}(\Gamma, \Delta) = \mathcal{E}\tilde{\Psi}_{[\Phi]\big|_{(\Gamma,\Delta)}} = \mathcal{E}\Upsilon_{\Psi_{[\Phi]}}(\alpha, \Gamma, \Delta)$ is a three-argument function such that $\Psi_0(\zeta, \xi)$ is the same function as $\mathcal{E}_\alpha\Upsilon_{\Psi_{[\Phi]}}(\alpha, \zeta, \xi)$, and so $\mathcal{E}\Psi_{[\Phi]} = \mathcal{E}_\eta\big[\mathcal{E}_\varepsilon\big[\mathcal{E}_\alpha\Upsilon_{\Psi_{[\Phi]}}(\alpha, \varepsilon, \eta)\big]\big]$. The same goes for the function $\Upsilon_{\Psi_{[\Phi]}}$, and so forth. It follows that, under assumption A, any value-range is the value-range of a one-, two-, . . . n-argument function, for any natural number n. The reader will judge whether this is a reason to discredit A.

15 For:

$$\Gamma \frown (\Delta \frown \mathcal{E}_\varepsilon[\mathcal{E}_\alpha\Psi_{[\Phi]}(\alpha, \varepsilon)]) = \backslash\mathcal{E}[x : \exists g[\backslash\mathcal{E}[z : \exists f[\mathcal{E}_\varepsilon[\mathcal{E}_\alpha\Psi_{[\Phi]}(\alpha, \varepsilon)] = \mathcal{E}f \wedge f(\Delta) = z]] = \mathcal{E}g \wedge g(\Gamma) = x]]$$
$$= \backslash\mathcal{E}[x : \exists g [\backslash\mathcal{E}[z : z = [\mathcal{E}_\alpha\Psi_{[\Phi]}(\alpha, \Delta)]] = \mathcal{E}g \wedge g(\Gamma) = x]]$$
$$= \backslash\mathcal{E}[x : \exists g[\mathcal{E}_\alpha\Psi_{[\Phi]}(\alpha, \Delta) = \mathcal{E}g \wedge g(\Gamma) = x]]$$
$$= \backslash\mathcal{E}[x : x = \Psi_{[\Phi]}(\Gamma, \Delta)]$$
$$= \Psi_{[\Phi]}(\Gamma, \Delta).$$

which indeed corresponds to what Frege himself observes at the end of §I.36.[16]

If, on the other hand, one assumes C, and therefore ¬D, or even only the latter, it may be that Φ is such that some of its values are value-ranges and others are not. Let Φ° be such a function, and suppose that $\Theta = \mathcal{E}\Phi^\circ$. We will then have two possible sub-cases. If Δ is such that $\Phi^\circ(\Delta)$ is a value-range, we will move back to the first of the two previous cases, and we will have that $\Phi^\circ(\Delta) = \mathcal{E}\tilde{\Phi}^\circ_\Delta = \mathcal{E}_\alpha \tilde{\Phi}^\circ_\Delta(\alpha) = \mathcal{E}_\alpha \Psi_{[\Phi^\circ]}(\alpha, \Delta)$, where $\Psi_{[\Phi^\circ]}$ is a two-argument function such that $\Phi^\circ(\zeta)$ is the same function as $\mathcal{E}_\alpha \Psi_{[\Phi^\circ]}(\alpha, \zeta)$, and so $\mathcal{E}\Phi^\circ = \mathcal{E}_\varepsilon \left[\mathcal{E}_\alpha \Psi_{[\Phi^\circ]}(\alpha, \varepsilon) \right]$, which brings us back to case (9) under the condition of the identification of Ψ and $\Psi_{[\Phi^\circ]}$. On the other hand, if Δ is such that $\Phi^\circ(\Delta)$ is not a value-range, we will move back the second of the two previous cases, which brings us back to case (11).[17]

This duality arising in the case where $\Theta = \mathcal{E}\Phi^\circ$ makes this case difficult to deal with using Frege's notion of a function, since from the description that we just outlined, it should be clear that, even though Φ° is a total function—as required by Frege for any function—this cannot, by definition, be the case for the functions $\Psi_{[\Phi^\circ]}(\xi, \zeta)$ and $\mathcal{E}_\alpha \Psi_{[\Phi^\circ]}(\alpha, \zeta)$. Since the former is defined by requiring that $\Phi^\circ(\zeta)$ be the same function as the latter, that is, under the condition that the values of the latter be value-ranges, it is only defined for the ζ-arguments for which the value of the function Φ° is a value-range. By using the modern notion of a variable,[18] one could say that $\Psi_{[\Phi^\circ]}(y, x)$ and $\mathcal{E}_\alpha \Psi_{[\Phi^\circ]}(\alpha, x)$ are defined by the condition $\Phi^\circ(x) = \mathcal{E}_\alpha \Psi_{[\Phi^\circ]}(\alpha, x)$ and so that they are only defined for the value

---

16 To be more precise, Frege limits himself to observe that, if Θ is a simple (in opposition to double) value-range, then $\Delta \frown \Theta$ is not a value-range, which implies that $\Gamma \frown (\Delta \frown \Theta) = \emptyset$. More precisely, if $\Phi^\circledast$ is a function no value of which is a value-range, and $\Lambda^\circledast_{[\Phi,\Delta]}$ is an object other than a value-range that provides its value for the argument Δ (which is equivalent to supposing that $\Phi^\circledast(\Delta) = \Lambda^\circledast_{[\Phi,\Delta]}$), then we have that:

$$\Gamma \frown (\Delta \frown \mathcal{E}\Phi^\circledast) = \backslash\mathcal{E} \left[x : \exists g \left[\backslash\mathcal{E} \left[z : \exists f [\mathcal{E}\Phi^\circledast = \mathcal{E}f \wedge f(\Delta) = z] \right] = \mathcal{E}g \wedge g(\Gamma) = x \right] \right]$$
$$= \backslash\mathcal{E} \left[x : \exists g \left[\backslash\mathcal{E} \left[z : z = \Lambda^\circledast_{[\Phi,\Delta]} \right] = \mathcal{E}g \wedge g(\Gamma) = x \right] \right]$$
$$= \backslash\mathcal{E} \left[x : \exists g \left[\Lambda^\circledast_{[\Phi,\Delta]} = \mathcal{E}g \wedge g(\Gamma) = x \right] \right]$$
$$= \backslash(\mathcal{E} \, \mathbf{F}^f_{1,1})$$
$$= \backslash\emptyset$$
$$= \emptyset.$$

17 This brings us back to the derivation outlined in footnote 16 by replacing the object $\Lambda^\circledast_{\Phi,\Delta}$ (which is an object other than a value-range) with $\Lambda^\circledast_{\Phi^\circ,\Delta}$ (which is also an object other than a value-range).

18 See footnote 2.

of x for which $\Phi^\circ(x)$ is a value-range. Given that, following Frege (as pointed out in §1), the value-range of whatever two-variable function $\Psi(\xi, \zeta)$ (or double value-range) is the value-range of a one-argument function $\mathcal{E}_\alpha\Psi(\alpha, \zeta)$ associated with $\Psi(\xi, \zeta)$, it would follow that Θ (i.e., $\mathcal{E}\Phi^\circ$) should be considered at the same time as the double value-range of a (two-argument) function only defined, with respect to its second argument, on certain objects (i.e., the objects giving the argument for which the value of Φ° is a value-range in turn) and as a simple value-range of another (one-argument) function only defined for the other objects. However, for Frege, a value-range is an object, and any object is fully saturated; that is, in current terms, it does not depend on any variable and therefore cannot change its nature according to the values taken by any such variable—or, in Frege's terms, according to the arguments of the function of which it is the value-range.

A way to account for this case without abandoning this fundamental idea—which would radically call into question the distinction between objects and functions, which for Frege is absolutely crucial and even constitutive of all his logical setting—is to distinguish between the two subcases, that is, either to argue that Θ can be two distinct objects depending on the choice of the object Δ or to argue that the function Φ° splits into two separate partial functions: the function $\mathcal{E}_\alpha\Psi_{[\Phi^\circ]}(\alpha, \zeta)$, which might be designated by 'Φ°_ε', defined on some objects, in particular on those objects for which one would say, by abuse of language, that the value of the function Φ° taking those objects as arguments is a value-range, and another function, which might be designated by 'Φ°_ε', defined on the other objects, those for which one might say, by abuse of language again, that the value of the function Φ° taking those objects as arguments is not a value-range. In the first case, Θ, that is, $\mathcal{E}\Phi^\circ$, or $\mathcal{E}\Phi^\circ_\varepsilon$, will be a double value-range (of a partial, two-argument function); in the second, Θ, that is, $\mathcal{E}\Phi^\circ$, or $\mathcal{E}\Phi^\circ_\varepsilon$, will be a simple value-range—of a partial, one-argument function. It would thus be nothing but a convenient misuse of language to affirm that Θ is the value-range of Φ° or to write the identity '$\Theta = \mathcal{E}\Phi^\circ$', as well as to affirm that the value of Φ° is a value-range for some arguments and not for others.

Although this option goes against Frege's conceptions, it seems to me to do so in a less radical way than the previous one, so to speak. This is why I choose this option here.

Even if this choice is allowed, it remains that assuming ¬A and ¬D (or ¬A and C) forces one to go against at least one of two fundamental ideas on which Frege bases the very conception of his logical system by making it impossible to keep the two at the same time. One might think that this is, on its own, an argument suggesting that Frege could not have accepted the statement ¬D or, even more radically, the statement ¬A. However, this would be too quick: after all, Frege tells us nothing about the relations between these statements and his ideas and often expresses himself (for example, at the end of §I.36) in such a way as to make us think that he accepts ¬A—if not ¬C. That is why I keep open the possibility of accepting ¬A and ¬D (or ¬A and C), at least for the time being.

This having been noticed, it is also important to remark that, under assumption ¬A alone (i.e., regardless of whether C, ¬C, D, and ¬D are accepted), nothing prevents Θ from being an object other than a value-range, for example, $\Theta^\circledcirc$. In this case, from (3), it follows that:

$$
\begin{aligned}
\Gamma \frown (\Delta \frown \Theta^\circledcirc) &= \grave{\mathcal{E}} \left[x : \exists g \left[\grave{\mathcal{E}} [z : \exists f [\Theta^\circledcirc = \mathcal{E}f \wedge f(\Delta) = z]] = \mathcal{E}g \wedge g(\Gamma) = x] \right] \right] \\
&= \grave{\mathcal{E}} \left[x : \exists g \left[\backslash(\mathcal{E}\, F_{1,1}^F) = \mathcal{E}g \wedge g(\Gamma) = x] \grave{A} \right] \right] \\
&= \grave{\mathcal{E}} \left[x : \exists g \left[\backslash \varnothing = \mathcal{E}g \wedge g(\Gamma) = x] \grave{A} \right] \right] \\
&= \grave{\mathcal{E}} \left[x : \exists g \left[\varnothing = \mathcal{E}g \wedge g(\Gamma) = x] \grave{A} \right] \right] \\
&= \grave{\mathcal{E}} \left[x : x = F \right] \\
&= F,
\end{aligned}
\tag{12}
$$

as Frege himself points out at the end of §I.36.

On this basis, the following conclusions can be drawn.

1. $\xi \frown (\zeta \frown \Theta)$ is a two-argument function, and in particular, it is the function:

$$
\grave{\mathcal{E}} \left[x : \exists g \left[\grave{\mathcal{E}} [z : \exists f [\Theta = \mathcal{E}f \wedge f(\zeta) = z]] = \mathcal{E}g \wedge g(\xi) = x] \right] \right].
$$

Therefore:

  i.   If $\Theta = \mathcal{E}\Psi$, for whatever two-argument function Ψ, or if $\Theta = \mathcal{E}\Phi$, for whatever one-argument function Φ° all values of which are value-ranges in turn, then $\xi \frown (\zeta \frown \Theta)$ is either the same function as Ψ or the same function as the (two-argument) function $\Psi_{[\Phi^\circ]}$ such that Φ° is the same function as $\mathcal{E}_\alpha \Psi_{[\Phi^\circ]}(\alpha, \zeta)$, which makes $\xi \frown (\zeta \frown \mathcal{E}\Psi)$ coincide (extensionally at least, if not also intensionally) with Ψ, or $\xi \frown (\zeta \frown \mathcal{E}\Phi^\circ)$ with $\Psi_{[\Phi^\circ]}$.

Under assumption A, no other case has to be considered. Under assumption ¬A, there are three other possible cases. Assuming ¬C, it may be the case that:

  ii. $\Theta = \mathcal{E}\Phi^\circ$, for whatever one-argument function Φ°, any value of which is a value-range, which implies that $\xi \frown (\zeta \frown \Theta)$ reduces to the constant function $\varnothing_{1,2}^F$.

Assuming C, hence assuming ¬D, or even only the latter, it may be the case that:

  iii. Θ is the value-range $\mathcal{E}\Phi^\circ$ of the one-argument function Φ°, some values of which are value-ranges and others are not; if this is the case, the behavior of the function $\xi \frown (\zeta \frown \Theta)$ depends on its ζ-argument; that is, the one-argument function $\xi \frown (\Delta \frown \Theta)$ behaves differently depending on whether the value of the function of which Θ is the value-range, for Δ as argument, is or is not a value-range in turn. In particular, if the function Φ° is such that $\Phi^\circ(\Delta)$ is a value-range—that is, $\Theta = \mathcal{E}\Phi^\circ_\varepsilon$ is the (double) value-range of the partial (two-argument) function $\Psi_{[\Phi^\circ_\varepsilon]}(\alpha, \zeta)$, such that $\Phi^\circ_\varepsilon(\zeta)$ is the same function as $\mathcal{E}_\alpha\Psi_{[\Phi^\circ_\varepsilon]}(\alpha, \zeta)$, only defined, regarding its ζ-argument, on the objects supplying the arguments for which the value of Φ° is a value-range in turn—then the function $\xi \frown (\Delta \frown \Theta)$ reduces to the function $\Psi_{[\Phi^\circ_\varepsilon]}(\xi, \Delta)$. If the function Φ° is such that $\Phi^\circ(\Delta)$ is not a value-range—that is, $\Theta = \mathcal{E}\Phi^\circ_\varepsilon$ is the simple value-range of a partial (one-argument) function, only defined on the objects for which the value of Φ° is not a value-range—then the function $\xi \frown (\Delta \frown \Theta)$ reduces to the constant function $\emptyset^F_{1,1}$.

Under assumption ¬A alone (i.e., regardless of whether C, ¬C, D, and ¬D are accepted), it may be the case that:

iv. $\Theta = \Theta^\circledcirc$ for whatever object $\Theta^\circledcirc$ other than a value-range, which makes $\xi \frown (\zeta \frown \Theta)$ reduce to the constant function $F^F_{1,2}$.

It follows that, if it is admitted that $\xi \frown (\zeta \frown \Theta)$ is a well-formed function for whatever object Θ, then:

- to accept ¬A forces us to accept that the constant function $F^F_{1,2}$ can be an instance of the function $\xi \frown (\zeta \frown \Theta)$;
- to accept ¬A and ¬C forces us to admit a constant function such as $\emptyset^F_{1,2}$;
- to accept ¬A and C or ¬A and ¬D forces us both to admit, again, a constant function such as $\emptyset^F_{1,1}$ and to accept that Θ can be the value-range of a partial function.

2. $\_\xi \frown (\zeta \frown \Theta)$ is a binary relation, and in particular, it is the function:

$$\_\backslash\!\mathcal{E}\,[x : \exists g\,[\backslash\!\mathcal{E}\,[z : \exists f[\Theta = \mathcal{E}f \wedge f(\zeta) = z]] = \mathcal{E}g \wedge g(\xi) = x]].$$

Therefore:

i. If $\Theta = \mathcal{E}\Psi$ for whatever two-argument function Ψ, or if $\Theta = \mathcal{E}\Phi$ for whatever one-argument function Φ°, all values of which are value-ranges, then $\_\xi \frown (\zeta \frown \Theta)$ is either the same relation as $\_\Psi$ or the same (two-argument) relation $\_\Psi_{[\Phi^*]}$ such that Φ° is the same function as $\mathcal{E}_\alpha\Psi_{[\Phi^*]}(\alpha, \zeta)$, which makes $\_\xi \frown (\zeta \frown \mathcal{E}\Psi)$ coincide with $\Re_\Psi$ (extensionally at least, if not intensionally), or $\_\xi \frown (\zeta \frown \mathcal{E}\Phi^\circ)$ with $\Re\Psi_{[\Phi^*]}$.

Under assumption A, no other case has to be considered. Under assumption ¬A, there are three other possible cases. Under assumption ¬C, it may be that:

ii. $\Theta = \mathcal{E}\Phi^{\circledast}$, for whatever one-argument function $\Phi^{\circledcirc}$, any value of which is a value-range, which makes $\_\xi \frown (\zeta \frown \Theta)$ reduce to the constant concept $\_\varnothing_{1,2}^{F}$, which is nothing but $\mathbf{F}_{1,2}^{F}$.

Assuming C, and hence ¬D, or even only assuming ¬D, it may also be that:

iii. Θ is the value-range $\mathcal{E}\Phi^{\circledcirc}$ of a one-argument function $\Phi^{\circledcirc}$, some values of which are value-ranges while others are not; if this is the case, the behavior of the concept $\_\xi \frown (\Delta \frown \Theta)$ depends on the choice of Δ. In particular, if the function $\Phi^{\circledcirc}$ is such that $\Phi^{\circledcirc}(\Delta)$ is a value-range—that is, $\Theta = \mathcal{E}\Phi_{\varepsilon}^{\circledast}$ is the (double) value-range of the partial (two-argument) function $\Psi_{\left[\Phi^{\circledast}\right]}(\xi,\zeta)$, such that $\Phi_{\varepsilon}^{\circledast}(\zeta)$ is the same function as $\mathcal{E}_{\alpha}\Psi_{\left[\Phi^{\circledast}\right]}(\alpha,\zeta)$—then the concept $\_\xi \frown (\Delta \frown \Theta)$ reduces to the concept $\Psi_{\left[\Phi^{\circledast}\right]}(\xi,\Delta)$. If the function $\Phi^{\circledcirc}$ is such that $\Phi^{\circledcirc}(\Delta)$ is not a value-range—that is, $\Theta = \mathcal{E}\Phi_{\overline{\varepsilon}}^{\circledcirc}$ is the simple value-range of another partial (one-argument) function—the concept $\xi \frown (\Delta \frown \Theta)$ reduces to the constant concept $\_\varnothing_{1,1}^{F}$, which is nothing but $\mathbf{F}_{1,1}^{F}$.

Under condition ¬A alone (i.e., regardless of whether C, ¬C, D, and ¬D are accepted), it may also be the case that:

iv. $\Theta = \Theta^{\circledcirc}$, for whatever object $\Theta^{\circledcirc}$ other than a value-range, which makes $\_\xi \frown (\zeta \frown \Theta)$ reduce to the constant concept $\_\mathbf{F}_{1,2}^{F}$, which is nothing but $\mathbf{F}_{1,2}^{F}$.

It follows that, both under assumption A and under assumption ¬A (and regardless of whether C, ¬C, D, and ¬D are accepted):

- if Θ is the value-range of a two-argument function Ψ, or of a one-argument function $\Phi^{\circledcirc}$, all values of which are value-ranges, then the relation $\_\xi \frown (\zeta \frown \Theta)$ takes the value **T** for Γ and Δ as arguments if and only if $\Psi(\Gamma, \Delta) = \mathbf{T}$ or $\Psi_{\left[\Phi^{\circledast}\right]}(\Gamma,\Delta) = \mathbf{T}$, respectively;
- if Θ is not a value-range, or if it is the (simple) value-range of a one-argument function $\Phi^{\circledcirc}$, any value of which is a value-range, then the relation $\_\xi \frown (\zeta \frown \Theta)$ never takes the value **T**;
- if Θ is the value-range of a one-argument function $\Phi^{\circledcirc}$, some values of which are value-ranges and others are not, the concept $\_\xi \frown (\Delta \frown \Theta)$ takes the value **T** for an argument Γ if and only if Δ is such that $\Phi^{\circledcirc}(\Delta)$ is a value-range in turn—that is, $\Theta = \varepsilon\Phi_{\varepsilon}^{\circledcirc}$

is the (double) value-range of the partial (two-argument) function $\Psi_{[\Phi_\varepsilon^\circ]}(\xi,\zeta)$, such that $\Phi_\varepsilon^\circ(\zeta)$ is the same function as $\mathcal{E}_\alpha \Psi_{[\Phi_\varepsilon^\circ]}(\alpha,\zeta)$ —and $\Psi_{[\Phi_\varepsilon^\circ]}(\Gamma,\Delta)=\mathbf{T}$.

Let us consider the two first cases. Given that the value-range of a one-argument function, no value of which is a value-range, is not a double value-range, both if A and ¬A are accepted (and regardless of whether C, ¬C, D, and ¬D are so), the relation $\underline{\quad}\xi \frown (\zeta \frown \Theta)$ will be the same as

$$[x,y : \exists h[\Theta = \mathcal{E}h \wedge x\mathfrak{R}_h y]], \tag{13}$$

where 'h' varies over two-argument functions.

In the third case, the concept $\underline{\quad}\xi \frown (\Delta \frown \Theta)$ is the same as

$$\left[x : \exists h \exists g\left[\Theta = \mathcal{E}g \wedge g(\Delta) = \mathcal{E}_\alpha h(\alpha,\Delta) \wedge x\mathfrak{R}_h\Delta\right]\right],$$

as long as it is assumed that, for any value of x, the open formula '$\Theta = \mathcal{E}g \wedge g(\Delta) = \mathcal{E}_\alpha h(\alpha,\Delta) \wedge x\mathfrak{R}_h\Delta$' can be satisfied by two partial functions (a one-argument function, for the variable 'g', and a two-argument one, for the variable 'h'), which makes the relation $\underline{\quad}\xi \frown (\zeta \frown \Theta)$ the same as

$$\left[x,y : \exists g,h\left[\Theta = \mathcal{E}g \wedge g(y) = \mathcal{E}_\alpha h(\alpha,y) \wedge x\mathfrak{R}_h y\right]\right]$$

It is, then, enough to notice that

$$\forall y\ [\exists h[\Theta = \mathcal{E}h] \Leftrightarrow \exists g,\ k[\Theta = \mathcal{E}g \wedge g(y) = \mathcal{E}_\alpha k(\alpha,\ y)]]$$

(where 'k' and 'h' vary over two-argument functions) to conclude that the third case reduces to the first two, as long as it is assumed that for any value of x and y, the open formula '$\Theta = \mathcal{E}g \wedge g(y) = \mathcal{E}_\alpha h(\alpha,y) \wedge x\mathfrak{R}_h y$' can be satisfied by two partial functions (a one-argument and a two-argument function, respectively).

It follows that, regardless of which assumptions are made among A, ¬A, C, ¬C, D, or ¬D, the relation $\underline{\quad}\xi \frown (\zeta \frown \Theta)$ will be the same as the relation (13).[19] Let's denote this relation by '$\mathfrak{S}_\Theta$'.

This being said, let us note that, if assumption A is admitted, the fact that only the first clause of stipulation (3) is rendered in Frege's formal system does not make the system incapable of always deciding which values the functions $\xi \frown (\zeta \frown \Theta)$ and $\underline{\quad}\xi \frown (\zeta \frown \Theta)$ take. Indeed—as shown in

---

19 The difference between the case where A is assumed and the one where ¬A is assumed (and with them C or ¬C and D or ¬D) does not lie in the nature to be assigned to the relation but in the conditions that elicit that a pair of objects fall under this relation, and thus, as mentioned previously, in its behavior.

derivations (9) and (10)—if A is assumed, the decision does not depend on the second clause of the stipulation. But, if ¬A is assumed, and if it is not assumed that Θ is the value-range of a two-argument function or of a one-argument function, all values of which are value-ranges, then this is no longer the case, since derivations (11) and (12) rely on the second clause. This implies that the argument on the basis of which the function $\_\xi \frown (\zeta \frown \Theta)$ reduces to $\mathfrak{S}_\Theta$ also rests on this second clause.

Let's focus on this latter point by assuming ¬A.

If ¬C is also assumed, then it could be that $\Theta = \mathcal{E}\Phi^\circ$, where Φ° is as previously. If coupled with stipulation (4), the first clause of stipulation (3) is thus sufficient to reduce the function $\xi \frown (\zeta \frown \Theta)$ to $\backslash\emptyset_{1,2}^F$. But the second clause is required to reduce this latter function to $\emptyset_{1,2}^F$, and consequently to reduce the function $\_\xi \frown (\zeta \frown \Theta)$ to $\mathbf{F}_{1,2}^F$. It would therefore be sufficient to reject this clause and to admit, once again, that $\backslash\emptyset = \mathbf{T}$ in order to conclude that $\_\xi \frown (\zeta \frown \mathcal{E}\Phi^\circ)$ reduces to $\mathbf{T}_{1,2}^F$.

If C and ¬D, or even only ¬D, are assumed, it may be that $\Theta = \mathcal{E}\Phi^\circ$, where Φ° is as previously. If Δ is such that $\Phi^\circ(\Delta)$ is not a value-range, and $\Theta = \mathcal{E}\Phi_{\bar{\varepsilon}}^\circ$ is, then, the simple value-range of a partial (one-argument) function, then the first clause of stipulation (3) is sufficient, if coupled with stipulation (4), to reduce the function $\xi \frown (\Delta \frown \Theta)$ to $\backslash\emptyset_{1,1}^F$. But the second clause is necessary to reduce the latter function to $\emptyset_{1,1}^F$ and consequently to reduce $\_\xi \frown (\Delta \frown \Theta)$ to $\mathbf{F}_{1,1}^F$. It would therefore be sufficient to reject this clause and to admit, once again, that $\backslash\emptyset = \mathbf{T}$ in order to conclude that $\_\xi \frown (\Delta \frown \mathcal{E}\Phi^\circ)$ reduces to $\mathbf{T}_{1,1}^F$.

Again, under assumption ¬A alone, it may be that $\Theta = \Theta^\circ$, where Θ° is as previously. The first clause of stipulation (3), together with stipulation (4), is therefore sufficient to reduce the function $\xi \frown (\zeta \frown \Theta)$ to $\backslash\mathcal{E}\left[x : \exists g\left[\backslash\emptyset_{1,1}^F = \mathcal{E}g \wedge g(\xi) = x\right]\right]$. But the second clause is required to reduce both the latter function and $\_\xi \frown (\zeta \frown \Theta)$ to $\mathbf{F}_{1,2}^F$. It would therefore be sufficient to allow that $\backslash\emptyset = @$ in order to conclude that these functions reduce to $\mathbf{T}_{1,2}^F$.[20] And for whatever object Γ, in order to reduce the function $\_\Gamma \frown (\zeta \frown \Phi^\circ)$ to $\mathbf{T}_{1,1}^F$, it would also be sufficient to take $\backslash\emptyset$ as the value-range of any function which takes $\mathbf{T}$ as a value for Γ as argument.[21]

---

20 Indeed, if $\backslash\emptyset = @$, then:

$$\begin{aligned}
\Gamma \frown (\Delta \frown \Phi^\circ) &= \backslash\mathcal{E}\,[x : \exists g[\backslash\emptyset = \mathcal{E}g \wedge g(\Gamma) = x]] \\
&= \backslash\mathcal{E}\,[x : \exists g[@ = \mathcal{E}g \wedge g(\Gamma) = x]] \\
&= \backslash\mathcal{E}\,[x : x = \mathbf{T}] \\
&= \mathbf{T}.
\end{aligned}$$

21 If $\Gamma = \emptyset$, it would be sufficient to allow, for instance, that $\backslash\emptyset = \mathcal{E}[z : z = \emptyset]$ in order to conclude that:

$$\begin{aligned}
\emptyset \frown (\Delta \frown \Phi^\circ) &= \backslash\mathcal{E}\,[x : \exists g[\backslash\emptyset = \mathcal{E}g \wedge g(\emptyset) = x]] \\
&= \backslash\mathcal{E}\,[x : \exists g[\backslash\mathcal{E}[z : z = \emptyset] = \mathcal{E}g \wedge g(\emptyset) = x]] \\
&= \backslash\mathcal{E}\,[x : x = (\emptyset = \emptyset)] \\
&= \backslash\mathcal{E}\,[x : x = \mathbf{T}] \\
&= \mathbf{T}.
\end{aligned}$$

It is clear that, in all these cases, one cannot conclude that the function $\_\xi \frown (\zeta \frown \Theta)$ takes the value **T** for Γ and Δ as arguments if and only if Θ is the value-range of a two-argument function Ψ such as $\Psi(\Gamma, \Delta) = \mathbf{T}$ and that this function reduces, then, to $\mathfrak{S}_\Theta$. It follows that Frege's formal system does not come with any explicit instructions on how to understand the relation $\_\xi \frown (\zeta \frown \Theta)$. The choice of conceiving of this relation as the same as $\mathfrak{S}_\Theta$ depends on the following assumptions:

i. either on the adoption of the second clause of stipulation (3);
ii. or on the acceptance of A;
iii. or, alternatively, on the rejection of both the possibility for $\backslash\emptyset$ to be the value-range of a one-argument function which may have the value **T** (and, *a fortiori*, of the possibility that $\backslash\emptyset = @$) and the possibility that $\backslash\emptyset = \mathbf{T}$.

Even though Frege could accept the first and the last of these options, the fact remains that his formal system does not decide in favor of any of them.

4. The Concept $I\xi$ and the Functions $\rangle\xi$ and $⦂\xi$

Having established all this, let us now follow Frege in his definition of the *Anzahlen*.

Let's start with the definition of the concept $I\xi$, stated in §I.37. Following this definition, the concept $I\xi$ is such that

$$\forall v \, [Iv = \forall x, y \, [\_x \frown (y \frown v) \Rightarrow \forall z[\_x \frown (z \frown v) \Rightarrow y = z]]]. \quad (14)$$

This amounts to stipulating that, for any object v, the value of Iv is the same truth-value as the one of the formula on the right. Therefore, if we assume that the relation $\_\xi \frown (\zeta \frown \Theta)$ is to be understood as (13), this definition is equivalent to the stipulation that:

$$\forall v \, [Iv \Leftrightarrow \forall x, y, z \, [(x\mathfrak{S}_v y \wedge x\mathfrak{S}_v z) \Rightarrow y = z]].$$

This implies that, for whatever object Θ,

$$I\Theta \Leftrightarrow \forall x, y, z \, [(x\mathfrak{S}_\Theta y \wedge x\mathfrak{S}_\Theta z) \Rightarrow y = z]. \quad (15)$$

This holds whether or not one assumes A, and, if one does not admit A, it holds whether or not Θ is a value-range (the simple or double value-range of a total or a partial function). But the formula on the right-hand side of the double implication does not have the same meaning in those different cases.

Under assumption A, Θ can only be the value-range of a two-argument function Ψ or the value-range of a one-argument function $\Phi^\circledast$ whose values are all value-ranges. It follows that this formula clearly expresses the fact that the relation $\mathfrak{S}_\Theta$ is functional, that is, for any object x there is at most one object y such that $x\mathfrak{S}_\Theta y$. But if Θ is $\mathcal{E}\Psi$ or $\mathcal{E}\Phi^\circledcirc$, then the relation $\mathfrak{S}_\Theta$ reduces to $x\mathfrak{R}_\Psi y$ or $x\mathfrak{R}_{\Psi_{[\Phi^\circledcirc]}}y$. From (15), it thus follows that

$$\mathbf{I}\mathcal{E}\Psi \Leftrightarrow \forall x,y,z\big[\big(x\mathfrak{R}_\Psi y \wedge x\mathfrak{R}_\Psi z\big)\Rightarrow y=z\big] \tag{16}$$

and

$$\mathbf{I}\mathcal{E}\Phi^\circledcirc \Leftrightarrow \forall x,y,z\left[\left(x\mathfrak{R}_{\Psi_{[\Phi^\circledcirc]}}y \wedge x\mathfrak{R}_{\Psi_{[\Phi^\circledcirc]}}z\right)\Rightarrow y=z\right] \tag{17}$$

which respectively state that $\mathbf{I}\mathcal{E}\Psi$ and $\mathbf{I}\mathcal{E}\Phi^\circledcirc$ if and only if the relations $\mathfrak{R}_\Psi$ and $\mathfrak{R}_{\Psi_{[\Phi^\circledcirc]}}$ are functional. If Ψ and $\Psi_{[\Phi^\circledcirc]}$ are relations in their turn, this amounts to asserting that $\mathbf{I}\mathcal{E}\Psi$ and $\mathbf{I}\mathcal{E}\Psi^\circledcirc$ if and only if these relations are themselves functional.[22]

On the other hand, under assumption ¬A, Θ may be an object $\Theta^\circledcirc$ other than the value-range, and, if ¬C is also assumed, Θ may be the range-value $\mathcal{E}\Phi^\circledcirc$ of a one-argument function, no value of which is a value-range. Finally, if C and ¬D are assumed, or if ¬D is assumed, it may be that Θ is the value-range of a one-argument function $\Phi^\circledcirc$, some values of which are value-ranges, while others are not.

In both the first and the second case, the relation $\mathfrak{S}_\Theta$ reduces to $\mathrm{F}_{1,2}^F$, which means that from (15), it follows that

$$\mathbf{I}\Theta^\circledcirc \Leftrightarrow \forall y,z[y\mathrm{F}_{1,2}^F z \Rightarrow y=z] \text{ and } \mathbf{I}\mathcal{E}\Theta^\circledcirc \Leftrightarrow \forall y,z[y\mathrm{F}_{1,2}^F z \Rightarrow y=z]. \tag{18}$$

As for the third case, it involves two sub-cases: the first one, where $\Theta=\mathcal{E}\Phi_\varepsilon^\circledcirc$, is analogous to the case to which the double implication (17) applies. The second sub-case, where $\Theta=\mathcal{E}\Phi_{\overline{z}}^\circledcirc$, is analogous to the case to which the double implication (18) applies.

In the first of these sub-cases, it follows from (15) that

$$\mathbf{I}\mathcal{E}\Phi_\varepsilon^\circledcirc \Leftrightarrow \forall_{\mathcal{D}_{\Phi^\circledcirc}} y,z\forall x[(x\mathfrak{R}_{\Psi_{[\Phi^\circledcirc]}}y \wedge x\mathfrak{R}_{\Psi_{[\Phi^\circledcirc]}}z)\Rightarrow y=z], \tag{19}$$

where the index of the first universal quantifier expresses the restriction of this quantifier to the domain $\mathcal{D}_{\Phi^\circledcirc}$ of the objects on which the partial

---

22 Note that if $\Phi^\circledcirc(\varsigma)$ is a concept, then it will be the same for $\mathcal{E}_\alpha\Psi_{[\Phi^\circledcirc]}(\alpha,\varsigma)$, so that for whatever object Δ, $\mathcal{E}_\alpha\Psi_{[\Phi^\circledcirc]}(\alpha,\Delta)=\mathbf{T}$ or $\mathcal{E}_\alpha\Psi_{[\Phi^\circledcirc]}(\alpha,\Delta)=\mathbf{F}$. This means that if $\mathbf{T}$ and $\mathbf{F}$ are as Frege (informally) states in §I.10 (i.e., if $\mathbf{T}=\mathcal{E}[z:z=\mathbf{T}]$ and $\mathbf{F}=\mathcal{E}[z:z=\mathbf{F}]$: see footnote 13), then the one-argument function $\Psi_{[\Phi^\circledcirc]}(\xi,\Delta)$ reduces either to the concept $[z:z=\mathbf{T}]$ or to the concept $[z:z=\mathbf{F}]$, which means that the two-argument function $\Psi_{[\Phi^\circledcirc]}(\xi,\varsigma)$ is a relation. Hence, if the stipulation in §I.10 is admitted, the fact that $\Phi^\circledcirc$ is a concept is a sufficient condition for $\Psi_{[\Phi^\circledcirc]}$ to be a relation.

function $\Phi_{\bar{\varepsilon}}^{\circ}$ is defined. This double implication thus tells us that, if Δ and Δ' are whatever two objects on which this function is defined [i.e., they are such that $\Phi^{\circ}(\Delta)$ and $\Phi^{\circ}(\Delta')$ are value-ranges], then for any object x, it follows from $x\Re_{\Psi[\Phi_{\bar{\varepsilon}}^{\circ}]}\Delta$ and $x\Re_{\Psi[\Phi_{\bar{\varepsilon}}^{\circ}]}\Delta'$ that $\Delta = \Delta'$, that is, that the relation $\Re_{\Psi[\Phi_{\bar{\varepsilon}}^{\circ}]}$ is functional on its domain of definition (which includes every ordered pair of objects such that the first object is any object whatsoever, while the second is an object on which $\Phi_{\bar{\varepsilon}}^{\circ}$ is defined).

In the second sub-case, it follows from (15) that

$$I\mathcal{E}\Phi_{\bar{\varepsilon}}^{\circ} \Leftrightarrow \forall_{\mathcal{D}_{\Phi_{\bar{\varepsilon}}^{\circ}}} y, z[yF_{1,2}^{F} z \Rightarrow y = z], \tag{20}$$

where the index of the universal quantifier expresses the restriction of this quantifier to the domain $\mathcal{D}_{\Phi_{\bar{\varepsilon}}^{\circ}}$ of the objects on which the partial function $\Phi_{\bar{\varepsilon}}^{\circ}$ is defined. This means that, if Δ and Δ' are whatever two objects on which this function is defined [i.e., they are such that $\Phi^{\circ}(\Delta)$ and $\Phi^{\circ}(\Delta')$ are not value-ranges], then it follows from $\Delta F_{1,2}^{F}\Delta'$ that $\Delta = \Delta'$.

While the double implications (16), (17), and (19) correspond to the meaning that Frege seems to assign to his definition—apart from the fact that $\Phi_{\bar{\varepsilon}}^{\circ}$ is a partial function—the double implications (18) and (20) convey a very different meaning. Indeed, given that the implications on their right-hand sides are tautological, it follows that

$$I\Theta^{\circledcirc}; I\mathcal{E}\Phi^{\circledcirc}; I\mathcal{E}\Phi_{\bar{\varepsilon}}^{\circ}$$

for whatever object $\Theta^{\circledcirc}$ which is not a value-range; whatever value-range $\mathcal{E}\Phi^{\circledcirc}$ of a total function $\Phi^{\circledcirc}$, no value of which is a value-range; and whatever value-range $\mathcal{E}\Phi_{\bar{\varepsilon}}^{\circ}$ of a partial function $\Phi_{\bar{\varepsilon}}^{\circ}$, no value of which is a value-range.

But that's not the end of the story yet. As shown previously, expressing the relation $\_\xi \frown (\zeta \frown \Theta)$ by (13) comes at a price. One could thus suppose that the conditions making this possible do not obtain. If one admitted, for example, not only ¬A but also that the second clause of stipulation (3) is not valid and that $\backslash\emptyset = @$, from (14), it would follow that, for whatever object $\Theta^{\circledcirc}$ other than a value-range:[23]

$$I\Theta^{\circledcirc} \Leftrightarrow \forall y, z\left[T_{1,2}^{F}(y,z) \Rightarrow y = z\right].$$

This implies that $I\Theta^{\circledcirc}$ if and only if any object is identical to any other object, that is, if and only if there is one object at most. But, given that we assumed that $\Theta^{\circledcirc}$ is an object other than a value-range, this implies that $I\Theta^{\circledcirc}$ if and only if there is only one object and this object is not a

---

23 See footnote 20.

value-range. The existence of two different objects such as **T** and **F** would thus be sufficient to grant that $\neg I\Theta^\circledcirc$.

Of course, this observation does not allow us to conclude that this is what Frege had in mind when advancing definition (14). It only shows that taking definition (14) to have this consequence is coherent with the formal requirements imposed by Frege's system.[24] However, taking definition (14) to be equivalent to (15) is also coherent with Frege's formal requirements; moreover, this would be in line with the second clause of stipulation (3). This is even mandatory, if one assumes A. Under this assumption, all that definition (14) says is that one object (or value-range) Θ falls under the concept $I\xi$ if and only if the relation $\Re_\Psi$ (if $\Theta = \mathcal{E}\Psi$) or $\Re_{\Psi_{[\Phi]}}$ (if $\Theta = \mathcal{E}\Phi$) is functional, for whatever two-argument function Ψ and whatever one-argument function Φ.

All this having been noticed, let's move on to the definition of the function $\rangle\xi$, presented in §I.38:

$$\forall v \,[\rangle v = \mathcal{E}\,[x,y : Iv \wedge \forall z\,[\_z \frown x \Rightarrow \exists w[\_w \frown y \wedge \_z \frown (w \frown v)]]]] \quad (21)$$

This definition states that for any object v, the value of $\rangle v$ is the double value-range denoted by the formula on the right, that is, the extension of a relation. If the concept $\_\xi \frown \Theta$ and the relation $\_\xi \frown (\zeta \frown \Theta)$ are understood in agreement with (7) and (13), to adopt this definition is equivalent to stating that

$$\forall v\,[\rangle v = \mathcal{E}\,[x, y : Iv \wedge \forall z\,[\mathfrak{T}_x z \Rightarrow \exists w\,[\mathfrak{T}_y w \wedge z\mathfrak{S}_v w]]]],$$

which implies that, for whatever object Θ,

$$\rangle\Theta = \mathcal{E}\,[x, y : I\Theta \wedge \forall z\,[\mathfrak{T}_x z \Rightarrow \exists w\,[\mathfrak{T}_y w \wedge z\mathfrak{S}_\Theta w]]]. \quad (22)$$

---

24 In order to deny this, one could, for example, try to prove, within this system, that there are at least one value-range and one object $\Theta^\circledcirc$ other than a value-range which has the property I. As for the first task, one may observe that '$\_[(\mathcal{E}[x : x \neq x] = \mathcal{E}[x : x \neq x]) = \_\forall z[(z \neq z) = (z \neq z)]]$' is an occurrence of BLV and that '$\_\forall z[(z \neq z) = (z \neq z)]$' is an occurrence of a (propositional) theorem that could be proved within Frege's system, which implies that within this system, we can also prove that $\mathcal{E}[x : x \neq x] = \mathcal{E}[x : x \neq x]$ and consequently that $\exists x[x = \mathcal{E}[z : z \neq z]]$ and then that $\exists g \exists x[x = \mathcal{E}g]$. This argument is questionable, however. It would be valid if the rule of existential generalization were admitted. But this is not a rule that Frege admits explicitly (what he explicitly admits, in §§I.17 and I.48.5, is the rule of universal generalization, which implies the rule of existential instantiation, which is just the reciprocal rule of existential generalization). Moreover, even if this conclusion were accepted (which would, by the way, make it possible to easily prove, in a similar way, within the system, that there are countably many *Anzahlen*), in order to conclude that it is contradictory to take definition (14) as having the aforementioned consequence—that is, that one cannot reject the second clause of stipulation (3) and at the same time accept the existence of objects other than value-ranges and that $\emptyset = @$—one would also have to prove that, under the same conditions, there is at least one object $\Theta^\circledcirc$ other than a value-range, such as $I\Theta^\circledcirc$. But this is something we certainly cannot prove in Frege's system.

Again, this would hold under any of the assumptions mentioned previously, but to admit or not to admit either of these assumptions leads to identifying Θ and, consequently, $\rangle\Theta$ with objects of different nature.

Under assumption A, Θ can only be the value-range of a two-argument function Ψ or of a one-argument function $\Phi^{\circledast}$, all values of which are value-ranges. It follows that $\rangle\Theta$ can only be the extension of the relation ⌜pair of objects (x, y) such that the objects falling under the concept $\mathfrak{T}_x$ stand with all (respectively, some) of the objects falling under the concept $\mathfrak{T}_y$ in the relation $x\mathfrak{R}_\psi y$ (respectively, $\mathfrak{R}_{\psi[\Phi^{\circledast}]}$), provided that this relation is functional⌝. It is evident that, if the latter relation is not functional, no pair of objects falls under the relation whose extension is $\rangle\Theta$. Hence, any pair of objects (Γ, Δ) whatsoever falls under a relation whose extension is $\rangle\Theta$, for some object Θ, if and only if for any object falling under $\mathfrak{T}_\Gamma$, there is one (and only one) object falling under $\mathfrak{T}_\Delta$ that is in a functional relation with the first one. Given that, following Frege, this relation must be taken as total, and given that, under assumption A, Γ and Δ can only be value-ranges, it means that any objects falling under $\mathbf{C}_{\Phi_\Gamma}$ corresponds to one (and only one) object falling under $\mathbf{C}_{\Phi_\Delta}$, where Φ_Γ and Φ_Δ are the one-argument functions whose value-ranges are Γ and Δ. If Φ_Γ and Φ_Δ are concepts in their turn, then Γ and Δ fall under a relation whose extension is $\rangle\Theta$, for some object Θ, if and only if any object falling under Φ_Γ corresponds to one (and only one) object falling under Φ_Δ.

Under assumption ¬A, however, things change. Θ can now be an object Θ° different from a value-range or, under assumption ¬C, the value-range $\mathcal{E}\Phi^{\circ}$ of a one-argument function, no value of which is a value-range, or else, under assumption ¬D (and thus under assumption C), the value-range $\mathcal{E}\Phi^{\circ}$ of a one-argument function, some values of which are value-ranges, while others are not.

In the first two cases, from (18), (22), and points 2.*ii* and 2.*iv* of §3 previously, it follows that

$$\rangle\Theta = \mathcal{E}[x, y : \forall z\, [\mathfrak{T}_x z \Rightarrow \exists w[\mathfrak{T}_y w \wedge \mathbf{F}^F_{1,2}(z,w)]]]$$
$$= \mathcal{E}[x, y : \forall z\, [\neg\mathfrak{T}_x z]]$$
$$= \mathcal{E}[x, y : \forall g \forall z [x \neq \mathcal{E}v \vee \neg\mathbf{C}_g z]].$$

This implies that $\rangle\Theta$ is the value-range of the relation that holds between any object that is not a value-range or that it is the value-range of a function that never takes the value T and any other object. Let us denote this relation by '$\wp$'. We thus have that

$$\forall x, y\, [x\wp y \Leftrightarrow \forall g\, [x = \mathcal{E}g \Rightarrow \neg\exists z\, [\mathbf{C}_g z]]]$$

and

$$\rangle\Theta^{\circ} = \rangle\, \mathcal{E}\Theta^{\circ} = \mathcal{E}\wp.$$

The third case splits into two different sub-cases: the one in which $\Theta = \varepsilon \Phi_\varepsilon^\circ$ and the one in which $\Theta = \varepsilon \Phi_{\underset{\varepsilon}{}}^\circ$, where Φ° and Φ_ε° are two partial functions, the former being defined on the objects on which the latter is not defined, and vice versa. The first of these two sub-cases corresponds to the second case considered under assumption A. In this case, from (22), it follows that

$$\rangle \mathcal{E}\Phi_{\underset{\varepsilon}{}}^\circ = \mathcal{E}\left[x, y : \mathrm{I}\mathcal{E}\Phi_\varepsilon^\circ \wedge \forall z \left[\mathfrak{T}_x z \Rightarrow \exists_{D_{\oplus\odot}} w [\mathfrak{T}_y w \wedge z \mathfrak{S}_\Theta w]\right]\right]$$

which tells us that, if the (partial) relation $\Re_{\psi[\Phi_\varepsilon^\circ]}$ is functional on its domain of definition, whatever pair of objects (Γ, Δ) falls under the relation whose extension is $\rangle\Theta$ if and only if, for any objects falling under $\mathfrak{T}_\Gamma$, within the same domain of definition, there is one (and only one) object falling under $\mathfrak{T}_\Delta$ that stands in the relation $\Re_{\psi[\Phi_\varepsilon^\circ]}$ with the former. The second case reduces to the case in which $\Theta = \mathcal{E}\Phi^\circ$, which implies that

$$\rangle \mathcal{E}\Phi_{\underset{\varepsilon}{}}^\circ = \mathcal{E}\wp.$$

However, in this case, too, expressing the concept $\_\xi \frown \Theta$ by (7) and the relation $\_\xi \frown (\zeta \frown \Theta)$ by (13) comes at a price. If one assumes that the conditions which make that possible do not obtain and, in particular, not only that a is assumed but also that the second clause of stipulation (3) is not valid and that $\backslash\emptyset = @$, from (21) together with the assumption that there exist at least two distinct objects, one may conclude that, for whatever object Θ° other than a value-range,

$$\rangle\Theta^\circ = \mathcal{E}[x, y : \forall z, [z = w] \wedge \forall z [\_z \frown x \Rightarrow \exists w [\_w \frown y \wedge \mathbf{T}_{1,2}^F (z, w)]]]$$
$$= \mathcal{E}[x, y : \forall z, [z = w] \wedge \forall z [\_z \frown x \Rightarrow \exists w [\_w \frown y]]]$$
$$= \mathcal{E}\mathbf{F}_{1,2}^F.$$

Once again, this remark only shows that taking definition (21) as having this consequence is coherent with the formal requirements imposed by Frege's system. But the same holds if one takes this definition as equivalent to (22), and this would also be in line with the second clause of stipulation (3); moreover, this must be the case if one accepts assumption a. Therefore, under this assumption, definition (21) only states that a pair of objects (or value-ranges) (Γ, Δ) falls under a relation whose extension is $\rangle\Theta$, for some object Θ, if and only if any objects falling under $\mathfrak{C}_{\Phi_\Gamma}$ correspond to one and only one object falling under $\mathfrak{C}_{\Phi_\Delta}$, as long as Φ_Γ and Φ_Δ are the one-argument functions of which Γ and Δ are the value-ranges.

Let's consider, now, the definition of the one-argument function $\maltese\xi$, presented in §I.39:

$$\forall v [\maltese v = \mathcal{E}_\alpha \mathcal{E}_\varepsilon [\alpha \frown (\varepsilon \frown v)]]. \tag{23}$$

Under assumption A, and by (9) and (10), this amounts to stipulating that, for whatever two-argument function Ψ and for whatever one-argument function Φ (all values of which must be value-ranges),

$$\text{🙼}\mathcal{E}\Psi = \mathcal{E}_\alpha\mathcal{E}_\varepsilon\,[\alpha \frown (\varepsilon \frown \mathcal{E}\Psi)] = \mathcal{E}_\alpha\mathcal{E}_\varepsilon\,\Psi(\alpha,\varepsilon) = \mathcal{E}\Psi^\sim$$

and

$$\text{🙼}\mathcal{E}\Phi = \mathcal{E}_\alpha\mathcal{E}_\varepsilon\,[\alpha \frown (\varepsilon \frown \mathcal{E}\Psi_{[\Phi]})] = \mathcal{E}_\alpha\mathcal{E}_\varepsilon\Psi_{[\Phi]}\,(\alpha,\varepsilon) = \mathcal{E}\Psi^\sim_{[\Phi]}$$

where $\Psi^\sim$ is the function obtained by Ψ by inverting the places of its arguments, that is, the inverse relation of Ψ if the latter is a relation (and the same, of course) for $\Psi^\sim_{[\Phi]}$ and $\Psi_{[\Phi]}$).

Under assumption A, this is the only possible understanding of definition (23). On the other hand, if ¬A is assumed, one not only has to specify that the second of these identities does not apply to whatever one-argument function Φ but rather to whatever one-argument function Φ^*, all values of which are value-ranges, but also, above all, to account for other possibilities: first of all, the possibility to apply (23) to an object Φ° other than a value-range; then, under assumption ¬C, to the value-range $\mathcal{E}\Phi^\circ$ of a one-argument function, no value of which is a value-range; and finally, under assumption ¬D (or C), to the value-range $\mathcal{E}\Phi^\odot$ of a one-argument function, some values of which are value-ranges and others are not. From (23), it will thus follow that:

$$\text{🙼}\Theta^\circ = \mathcal{E}_\alpha\mathcal{E}_\varepsilon\,[\alpha \frown (\varepsilon \frown \Theta^\circ)] \quad = \mathcal{E}_\alpha\mathcal{E}_\varepsilon\,[\mathbf{F}^F_{1,2}(\alpha,\varepsilon)]$$
$$= \mathcal{E}_\alpha\mathcal{E}_\varepsilon\,[(\mathbf{F}^F_{1,2})^\sim(\varepsilon,\alpha)]$$
$$= \mathcal{E}[(\mathbf{F}^F_{1,2})^\sim] = \mathcal{E}\mathbf{F}^F_{1,2}$$

$$\text{🙼}\mathcal{E}\Phi^\circ = \mathcal{E}_\alpha\mathcal{E}_\varepsilon\,[\alpha \frown (\varepsilon \frown \Phi^\circ)] \quad = \mathcal{E}_\alpha\mathcal{E}_\varepsilon\,[\emptyset^F_{1,2}(\alpha,\varepsilon)]$$
$$= \mathcal{E}_\alpha\mathcal{E}_\varepsilon\,[(\emptyset^F_{1,2})^\sim(\varepsilon,\alpha)]$$
$$= \mathcal{E}[(\emptyset^F_{1,2})^\sim] = \mathcal{E}\emptyset^F_{1,2}$$

$$\text{🙼}\mathcal{E}\Phi^\circ_\varepsilon = \mathcal{E}_\alpha\mathcal{E}_\varepsilon\,[\alpha \frown (\varepsilon \frown \Phi^\circ_\varepsilon)] \quad = \mathcal{E}_\alpha\mathcal{E}_\varepsilon\,[\Psi_{[\Phi^\circ_\varepsilon]}(\alpha,\varepsilon)]$$
$$= \mathcal{E}_\alpha\mathcal{E}_\varepsilon\,[\Psi^\sim_{[\Phi^\circ_\varepsilon]}(\varepsilon,\alpha)]$$
$$= \mathcal{E}\Psi^\sim_{[\Phi^\circ_\varepsilon]}$$

$$\text{🙼}\mathcal{E}\Phi^\odot_\varepsilon = \mathcal{E}_\alpha\mathcal{E}_\varepsilon\,[\alpha \frown (\varepsilon \frown \Phi^\odot_\varepsilon)] \quad = \mathcal{E}_\alpha\mathcal{E}_\varepsilon\,[\emptyset^F_{1,2}(\alpha,\varepsilon)]$$
$$= \mathcal{E}_\alpha\mathcal{E}_\varepsilon\,[(\emptyset^F_{1,2})^\sim(\varepsilon,\alpha)]$$
$$= \mathcal{E}[(\emptyset^F_{1,2})^\sim] = \mathcal{E}\emptyset^F_{1,2}$$

It follows that, for whatever object Θ and independently of which among A, ¬A, C, ¬C, D, or ¬D one accepts, $\maltese\Theta$ is the value-range of the two-argument function resulting from $\xi \frown (\zeta \frown \Theta)$ by inverting the places of its arguments, that is:[25]

$$\maltese\Theta = \mathcal{E}[x, y : y \frown (x \frown \Theta)],$$

which entails that, under any of these assumptions,[26]

$$\forall x, y \, [x\mathfrak{S}_{\maltese\Theta}y \Leftrightarrow y\mathfrak{S}_{\maltese\Theta}x]. \tag{24}$$

Unlike that of the concept $\mathbf{I}\xi$ and of the function $\rangle\xi$, the definition of the function $\maltese\xi$ has no unexpected consequences under assumption A, even if ¬C or ¬D are assumed. It is easy to see that this is so even if, in addition to ¬A, it is also assumed that the conditions which make it possible to understand the concept $\_\_\xi \frown \Theta$ in agreement with (7) and the relation

---

25 The fact that any two-argument constant function is invariant under the exchange of its arguments, so that $\mathcal{E}\left[\left(\mathbf{F}_{1,2}^{F}\right)^{\sim}\right] = \mathcal{E}\mathbf{F}_{1,2}^{F}$ and $\mathcal{E}\left[\left(\varnothing_{1,2}^{F}\right)^{\sim}\right] = \mathcal{E}\varnothing_{1,2}^{F}$, has no influence on this conclusion.

26 Indeed:

$$\forall x, y \begin{bmatrix} x\mathfrak{S}_{\maltese\mathcal{E}\Psi}y & \Leftrightarrow & \exists h[\maltese\mathcal{E}\Psi = \mathcal{E}h \wedge x\mathfrak{R}_h y] \\ & \Leftrightarrow & \exists h[\mathcal{E}\Psi^{\sim} = \mathcal{E}h \wedge x\mathfrak{R}_h y] \\ & \Leftrightarrow & x\mathfrak{R}_{\Psi^{\sim}}y \\ & \Leftrightarrow & \Psi^{\sim}(x, y) = \mathbf{T} \\ & \Leftrightarrow & \Psi(y, x) = \mathbf{T} \\ & \Leftrightarrow & y\mathfrak{R}_\Psi x \\ & \Leftrightarrow & \exists h[\mathcal{E}\Psi = \mathcal{E}h \wedge y\mathfrak{R}_h x] \\ & \Leftrightarrow & y\mathfrak{S}_{\mathcal{E}\Psi}x \end{bmatrix}$$

(and in a similar way in the case that $\Theta = \mathcal{E}\Theta^{\odot}$ and $\Theta = \mathcal{E}\Theta_\varepsilon^{\circ}$),

$$\forall x, y \begin{bmatrix} x\mathfrak{S}_{\maltese\Theta^{\odot}}y & \Leftrightarrow & \exists h[\maltese\Theta^{\odot} = \mathcal{E}h \wedge x\mathfrak{R}_h y] \\ & \Leftrightarrow & \exists h[\mathcal{E}\mathbf{F}_{1,2}^{F} = \mathcal{E}h \wedge x\mathfrak{R}_h y] \\ & \Leftrightarrow & \mathbf{F}_{1,2}^{F}(x, y) \\ & \Leftrightarrow & \mathbf{F}_{1,2}^{F}(y, x) \\ & \Leftrightarrow & y\mathfrak{S}_{\Theta^{\odot}}x \end{bmatrix}$$

and

$$\forall x, y \begin{bmatrix} x\mathfrak{S}_{\maltese\Phi^{\odot}}y & \Leftrightarrow & \exists h[\maltese\Phi^{\odot} = \mathcal{E}h \wedge x\mathfrak{R}_h y] \\ & \Leftrightarrow & \exists h[\mathcal{E}\varnothing_{1,2}^{F} = \mathcal{E}h \wedge x\mathfrak{R}_h y] \\ & \Leftrightarrow & \varnothing_{1,2}^{F}(x, y) \\ & \Leftrightarrow & \varnothing_{1,2}^{F}(y, x) \\ & \Leftrightarrow & y\mathfrak{S}_{\Phi^{\odot}}x \end{bmatrix}$$

(and in a similar way in the case that $\Theta = \mathcal{E}\Theta_\varepsilon^{\circ}$).

$\_\xi \frown (\zeta \frown \Theta)$ in agreement with (13) do not obtain. If one rejects, for example, the second clause of stipulation (3) and assumes that $\backslash\emptyset = @$, $\maltese\Theta°$ is determined, *mutatis mutandis*, as previously:

$$\begin{aligned}
\maltese\Theta° = \mathcal{E}_\alpha\,\mathcal{E}_\varepsilon[\alpha\frown(\varepsilon\frown\Theta°)] \;\; &= \mathcal{E}_\alpha\,\mathcal{E}_\varepsilon[\mathbf{T}^F_{1,2}(\alpha,\varepsilon)] \\
&= \mathcal{E}_\alpha\,\mathcal{E}_\varepsilon[\mathbf{T}^F_{1,2})\widetilde{}(\varepsilon,\alpha)] \\
&= \mathcal{E}[\mathbf{T}^F_{1,2})\widetilde{}] = \mathcal{E}\,\mathbf{T}^F_{1,2}.
\end{aligned}$$

This does not prevent the consequences of this definition under the assumption ¬A from having some unpleasant effects on the definition of the *Anzahlen*.

5.   *Anzahlen*

The definition is presented just after the one of the function $\maltese\xi$ in §I.40. Here it is:

$$\forall u\,[\mathscr{A}u = \mathcal{E}\,[x : \exists y[\_x\frown(u\frown\rangle y) \wedge \_u\frown(x\frown\rangle\maltese y)]]] \tag{25}$$

It states that for any object u, the value of $\mathscr{A}u$ is the value-range denoted by the formula on the right, that is, the extension of a concept. If the relation $\_\xi\frown(\zeta\frown\Theta)$ is understood in agreement with (13)—which means that the concepts $\mathbf{I}\xi$ and $\rangle\xi$ are understood in agreement with (15) and (22)—to adopt this definition amounts to stating that

$$\forall u\,[\mathscr{A}u = \mathcal{E}\,[x : \exists y[x\mathfrak{S}_{yy}u \wedge u\mathfrak{S}_{\rangle\maltese y}x]]] \tag{26}$$

This implies that, for whatever object Γ,

$$\begin{aligned}
\mathscr{A}\Gamma = \mathcal{E}[x : &\exists y[x\mathfrak{C}_{yy}\Gamma \wedge \Gamma\mathfrak{C}_{\rangle\maltese y}x]] \\
= &\mathcal{E}[x : \exists y[\exists k[\rangle y = \mathcal{E}k \wedge x\mathfrak{R}_k\Gamma] \wedge \exists h[\rangle\maltese y = \mathcal{E}h \wedge \Gamma\mathfrak{R}_h x]]] \\
= &\mathcal{E}[x : \exists y[\mathbf{I}y \wedge \forall z\,[\mathfrak{I}_x z \Rightarrow \exists w[\mathfrak{I}_\Gamma w \wedge z\mathfrak{S}_y w]] \wedge \mathbf{I}\maltese y \wedge \forall z \\
&[\mathfrak{I}_\Gamma z \Rightarrow \exists w\,[\mathfrak{I}_x w \wedge z\mathfrak{S}_{\maltese y} w]]]].
\end{aligned} \tag{27}$$

The problem is that, under assumption ¬A, to adopt this definition does not warrant that, for whatever object Γ, $\mathscr{A}\Gamma$ is what Frege seems to want it to be, that is, the value-range of the concept of being an object x such that the objects falling under the concept $\_\xi\frown\Gamma$ are in a bijective correspondence with the objects falling under the concept $\_\xi\frown x$. In other words, under the assumption ¬A, adopting this definition does not guarantee that

$$\forall u\,[\mathscr{A}u = \mathcal{E}[x : \mathfrak{I}_x \approx \mathfrak{I}_u]], \tag{28}$$

where '$\mathfrak{T}_x \approx \mathfrak{T}_u$' is, as usual, the shorthand for the second-order formula expressing the condition that the concepts $\mathfrak{T}_x$ and $\mathfrak{T}_u$ are equinumerous. And this is no more the case even if one understands the concept $\_\xi \frown \Theta$ and the relation $\_\xi \frown (\zeta \frown \Theta)$ in agreement with (7) and (4), which is what makes it possible to understand (25) in agreement with (27).

In agreement with (15) and (24), the assumption that, for whatever objects Γ and Δ, the object y satisfying the condition

$$\mathbf{I}y \wedge \forall z[\mathfrak{T}_\Delta z \Rightarrow \exists w[\mathfrak{T}_\Gamma w \wedge z\mathfrak{S}_y w]] \wedge \mathbf{I}\!\!\!/_{\!\!y} y \wedge$$
$$\forall z[\mathfrak{T}_\Gamma z \Rightarrow \exists w [\mathfrak{T}_\Delta w \wedge z\mathfrak{S}_{\!\!/_{\!\!y}} w]] \tag{29}$$

is the value-range $\mathcal{E}\Psi$ of a two-argument function makes this function meet the condition:

$$\forall x, y, z[(x\mathfrak{R}_\psi y \wedge x\mathfrak{R}_\psi z) \Rightarrow y = z] \wedge \forall z[\acute{A}\ \mathfrak{T}_\Delta z \Rightarrow \exists w[\mathfrak{T}_\Gamma w \wedge z\mathfrak{R}_\psi w]]\wedge$$
$$\forall x, y, z[(y\mathfrak{R}_\psi x \wedge z\mathfrak{R}_\psi x) \Rightarrow y = z] \wedge \forall z[\acute{A}\ \mathfrak{T}_\Gamma z \Rightarrow \exists w[\mathfrak{T}_\Delta w \wedge w\mathfrak{R}_\psi z]],$$

and the situation is similar for the assumption that y is the value-range $\mathcal{E}\Phi^\circ$ of a one-argument function, all values of which are value-ranges, with $\Psi_{[\Phi^\circ]}$ replacing Ψ. Therefore, it is sufficient to observe that, for whatever Ψ and Φ°, $\mathfrak{R}_\psi$ and $\mathfrak{R}_{\psi_{[\Phi^\circ]}}$ are binary relations (i.e., they are *ipso facto* so, in Frege's sense, or they can be reduced to such relations in the modern sense by Σ_1^1-comprehension), to conclude that requiring that this condition be satisfied by a value-range such as $\mathcal{E}\Psi$ or $\mathcal{E}\Phi^\circ$ is equivalent to requiring the concepts $\mathfrak{T}_\Delta$ and $\mathfrak{T}_\Gamma$ to be equinumerous. This will reduce the identity (27) to the identity (28), as expected.

However, under assumption $\neg$A, to require the existence of an object y satisfying condition (29) is not the same thing as requiring the latter condition to be satisfied for a value-range such as $\mathcal{E}\Psi$ or $\mathcal{E}\Phi^\circ$. This is because, under assumption $\neg$A, the former request is equally satisfied if condition (29) is met only by objects such as Θ° other than value-ranges or, under assumption $\neg$C, only by value-ranges $\mathcal{E}\Phi^\circ$ of a one-argument function, no value of which is a value-range, or, alternatively, under assumption $\neg$D (and so under assumption C), only by the value-range $\mathcal{E}\Phi^\circ$ of a one-argument function, some values of which are value-ranges, while others are not. If this is so, meeting condition (29) comes down to something other than granting that the concepts $\mathfrak{T}_\Delta$ and $\mathfrak{T}_\Gamma$ are equinumerous. In particular, from (13), (15), and (the consequences of), it follows that to assume condition (29) as satisfied by $\Theta^\circledast$, $\mathcal{E}\Phi^\circledast$, or $\mathcal{E}\Phi_{\bar{z}}^\circ$ reduces to the assumption that

$$\forall z[\mathfrak{T}_\Delta z \Rightarrow \exists w [\mathfrak{T}_\Gamma w \wedge F_{1,2}^F (z, w)]] \wedge \forall z [\mathfrak{T}_\Gamma z \Rightarrow \exists w [\mathfrak{T}_\Delta w \wedge F_{1,2}^F (z, w)]],$$

that is,

$$\forall z[\mathfrak{T}_\Delta z \Rightarrow F_{1,1}^F (z)] \wedge \forall z [\mathfrak{T}_\Gamma z \Rightarrow F_{1,1}^F (z)],$$

that is,

$$\forall z[\neg \mathfrak{I}_\Delta z \wedge \neg \mathfrak{I}_\Gamma z].$$

Hence, if condition (29) were satisfied only by such object, one would have that

$$\forall u[\,\mathcal{V}u = \mathcal{E}\,[x : \forall z[\neg \mathfrak{I}_x z \wedge \neg \mathfrak{I}_u z]]]. \tag{30}$$

Of course, this does not mean that admitting assumption ¬A, as well as either assumption ¬C or ¬D, or both, forces us to consider definition (25) equivalent to (30). In fact, it's even worse than that: we have just shown that, under these assumptions, one cannot decide if definition (25) is equivalent to (28) or to (30), or in some cases to one and in some cases to the other. To make sure that it is equivalent to (28), it should be established that the existentially quantified formula in (27) is satisfied only by such value-ranges as $\mathcal{E}\Psi$ and $\mathcal{E}\Phi^\circ$. On the other hand, to ensure that it is equivalent to (30), it should be proved that this formula is satisfied only by objects such as Θ° or value-ranges such as $\mathcal{E}\Phi^\circ$. Finally, if it is the case only for some value-ranges such as $\mathcal{E}\Phi^\circ$ (or, more precisely, such as $\mathcal{E}\Phi^\circ_\varepsilon$ or $\mathcal{E}\Phi^\circ_{\bar\varepsilon}$), it should be assessed on a case-by-case basis if under the concept $\mathfrak{I}_u$ some objects fall on which $\mathcal{E}\Phi^\circ_\varepsilon$ is not defined (i.e., on which $\varepsilon\Phi^\circ_{\bar\varepsilon}$ is defined): if it were so, even if this formula were satisfied only by value-ranges such as $\mathcal{E}\Phi^\circ_\varepsilon$, it would not follow that the instance of definition (25) that one considers is equivalent to the corresponding instance of (28).

The difficulty lies in the fact that (28) and (30) are not at all equivalent, or, more precisely, their equivalent instances are only those where $u = \Theta^\circ$ or $u = \mathcal{E}\Phi^\Theta$, where Φ^Θ is whatever one-argument function such as $\neg\exists x\,[\mathbf{C}_{\Phi^\Theta}(x)]$, that is, $\neg\exists x[\Phi^\Theta(x) = \mathsf{T}]$—which is the case, among others, for the function $\mathbf{F}^F_{1,1}$, so that $\emptyset$ is an occurrence of $\mathcal{E}\Phi^\Theta$.

Indeed, on the one hand, from (28) it follows that

$$\mathcal{V}\Theta^\circ = \mathcal{V}\mathcal{E}\Phi^\Theta = \mathcal{E}\,[x : \mathfrak{I}_x \approx \mathfrak{I}_{\Theta^\circ}] = \mathcal{E}\,[x : \mathfrak{I}_x \approx \mathfrak{I}_{\mathcal{E}\Phi^\Theta}] = \mathcal{E}\,[x : \mathfrak{I}_x \approx \mathbf{F}^F_{1,1}],$$

because

$$\forall z\,[\mathfrak{I}_{\Theta^\circ} z \Leftrightarrow \exists g[\Theta^\circ = \mathcal{E}g \wedge \mathbf{C}_g z] \Leftrightarrow \mathbf{F}^F_{1,1}\,z]$$

and

$$\forall z\,[\mathfrak{I}_{\mathcal{E}\Phi^\Theta} z \Leftrightarrow \exists g[\mathcal{E}\Phi^\Theta = \mathcal{E}g \wedge \mathbf{C}_g z] \Leftrightarrow \mathbf{C}_\Phi z \Leftrightarrow [\Phi^\Theta(z) = \mathsf{T}] \Leftrightarrow \mathbf{F}^F_{1,1} z].$$

On the other hand, from (30), it follows that

$$\begin{aligned}
\mathcal{V}\Theta^\circ &= \mathcal{V}\mathcal{E}\Phi^\Theta = \mathcal{E}\,[x : \forall z\,[\neg\mathfrak{I}_x z \wedge \neg\mathfrak{I}_{\Theta^\circ} z]] = \mathcal{E}\,[x : \forall z[\neg\mathfrak{I}_x z \wedge \neg\mathfrak{I}_{\mathcal{E}\Phi^\Theta} z] \\
&= \mathcal{E}\,[x : \neg\exists z[\mathfrak{I}_x z]] = \mathcal{E}\,[x : \mathfrak{I}_x \approx \mathbf{F}^F_{1,1}]
\end{aligned}$$

because

$$\forall z \,[\neg \mathfrak{T}_{\Theta^\circ} z \Leftrightarrow \neg \exists g \,[\Theta^\circ = \mathcal{E}g \wedge \mathbf{C}_g z] \Leftrightarrow \neg F^F_{1,1} \, z \Leftrightarrow T^F_{1,1} \, z]$$

and

$$\forall z \,[\neg \mathfrak{T}_{\mathcal{E}\Phi^\Theta} z \Leftrightarrow \neg \exists g \,[\mathcal{E}\Phi^\Theta = \mathcal{E}g \wedge \mathbf{C}_g z] \Leftrightarrow \neg \mathbf{C}_{\Phi^\Theta} z \Leftrightarrow \neg (\Phi^\Theta(z) = \mathbf{T}) \Leftrightarrow T^F_{1,1} \, z]$$

In any other case—that is, for whatever value-range $\mathcal{E}\Phi^\oplus$, where $\Phi^\oplus$ is any one-argument function such that $\exists x \,[\mathbf{C}_{\Phi^\oplus}(x)]$, that is, $\exists x [\Phi^\oplus(x) = \mathbf{T}]$ (so that $\exists x [\Phi^\oplus(x) \neq \mathbf{F}]$)—it follows from (28) that

$$\mathfrak{N}\mathcal{E}\Phi^\oplus = \mathcal{E}\,[x : \mathfrak{T}_x \approx \mathfrak{T}_{\mathcal{E}\Phi^\oplus}] = \mathcal{E}\,[x : \mathfrak{T}_x \approx \mathbf{C}_{\Phi^\oplus}] \neq \mathcal{E}\,[x : F^F_{1,1}(x)] = \mathcal{E}F^F_{1,1} = \emptyset$$

while from (30), it follows that

$$\mathfrak{N}\mathcal{E}\Phi^\oplus = \mathcal{E}\,[x : \forall z \,[\neg \mathfrak{T}_{\mathcal{E}\Phi^\oplus} z \wedge \neg \mathfrak{T}_x z]] = \mathcal{E}\,[x : F^F_{1,1}] = \mathcal{E}F^F_{1,1} = \emptyset]$$

Clearly, this does not imply that definition (25) is defective or mathematically inappropriate (besides the inconsistency of Frege's system): for none of this implies that Frege's proofs, based on this definition, of several theorems about the *Anzahlen*—among which are his reformulations of the fundamental theorems of arithmetic (and in particular of second-order Peano axioms) and other theorems about the *Anzahl* of the extension of the concept of being a natural number, or *Endlos* (his version of $\aleph_0$)—are incorrect. Moreover, for whatever objects Γ and Δ, be it under the assumption A or $\neg$A, the definition allows us to decide whether Δ does or does not fall under the concept whose extension is $\mathfrak{N}\Gamma$, that is, if there is an object y satisfying the formula (29).[27] The point at issue here is not the

---

27 For, let's suppose that we assume $\neg$A and that Γ is not a value-range. It will follow that $\forall z \,[\neg \mathfrak{T}_\Gamma z]$. Therefore, there is an object y which satisfies formula (28) if and only if this object is such that $\mathbf{I} y \wedge \forall z \,[\neg \mathfrak{T}_\Delta z] \wedge \mathbf{I}^y_x y$. But if Δ is not a value-range in its turn, this is the case if and only if there is an object y such that $\mathbf{I} y \wedge \mathbf{I}^y_x y$—which is certainly the case, since it is the case for both the extension of the relation $[x, z : x = y]$ and for Δ itself. On the other hand, if Δ is a value-range, for example, $\mathcal{E}\Phi$, it would be necessary and sufficient for that to hold that Δ be such that $\forall z \,[\neg \mathbf{C}_\Phi x]$. If, still under assumption $\neg$A, it is assumed that Δ is not a value-range, one can therefore reason as follows, reversing the role of Γ and Δ. One is thus left with the case that both Γ and Δ are value-ranges. Under $\neg$A, any object other than a value-range will do the job. On the other hand, if one assumes A, then there will be an object y which satisfies (28) if and only if this object satisfies (29)—and that will depend, of course, on the concepts $\mathfrak{T}_\Gamma z$ and $\mathfrak{T}_\Delta z$. In addition, note that the case in which Δ is not a value-range, while Γ is, leads to an important consequence. Indeed, if $\Gamma = \mathcal{E}\Phi$, then the extension of the relation $[x, z : x = y]$ satisfies (28) if and only if $\forall z [\neg \mathbf{C}_\Phi x]$, that is, under assumption $\neg$A whatever object other than a value-range falls under the concept whose extension is $\mathfrak{N}\mathcal{E}\Phi$ if and only if no object falls under the concept $\mathbf{C}_\Phi$. But this would mean that $\mathcal{E}\mathbf{C}_\Phi = \emptyset$. Under this assumption, whatever object other than a value-range thus falls under the concept whose extension is $\mathfrak{N}\mathcal{E}\Phi$ if and

formal correction of Frege's approach but the meaning to be assigned to his definitions, that is, as said previously, the informal semantics intrinsic to his system. The previous considerations only show that this semantics is unambiguous only if assumption A is made, namely if all objects other than value-ranges are banned.

It should also be noticed that, even if, in order to understand the concept $\_\xi \frown \Theta$ and the relation $\_\xi \frown (\zeta \frown \Theta)$ in agreement with (7) and (13) [and, consequently, the concepts $I\xi$ and $)\xi$ in agreement with (15) and (22)], it is enough to accept the second clause of stipulation (3) or to deny the possibility for $\emptyset$ to be either the value-range of a one-argument function that can take the value **T** or **T** itself, this is not sufficient to make definition (25) unambiguous within this semantics. In order to do so, the only option is to ban all objects other than value-ranges.

6.   Zero

A less radical option might be to appropriately revise Frege's definition: instead of banning any object other than value-ranges, one may simply exclude all such objects from the domain of quantification of (25) and hence of (26) and (27). This would lead to the replacement of these latter two identities by the following ones:

$$\forall u\Big[\mathfrak{y}u = \mathcal{E}\big[x : \exists f \exists y\big[y = \mathcal{E}f \wedge x\mathfrak{S}_{yy}u \wedge u\mathfrak{S}_{\mathfrak{Y}y}x\big]\big]\Big]$$

and

$$\mathfrak{y}\Gamma \;\; = \mathcal{E}\big[x : \exists f \exists y\big[y = \mathcal{E}f \wedge x\mathfrak{S}_{yy}\Gamma \wedge \Gamma\mathfrak{S}_{\mathfrak{Y}y}x\big]\big]$$

$$= \mathcal{E}\big[x : \exists f \exists y\big[y = \mathcal{E}f \wedge \exists g[)y = \mathcal{E}g \wedge x\mathfrak{R}_{g}\Gamma\big] \wedge \exists h[)\mathfrak{Y}y = \mathcal{E}h \wedge \Gamma\mathfrak{R}_{h}x\big]\big]$$

$$= \mathcal{E}\left[x : \exists f \exists y \begin{array}{|l|} y = \mathcal{E}f \wedge Iy \wedge I\mathfrak{Y}y \wedge \\ \forall z\big[\mathfrak{I}_{x}z \Rightarrow \exists w\big[\mathfrak{I}_{\Gamma}w \wedge z\mathfrak{S}_{y}w\big]\big] \wedge \\ \forall z\big[\mathfrak{I}_{\Gamma}z \Rightarrow \exists w\big[\mathfrak{I}_{x}w \wedge z\mathfrak{S}_{\mathfrak{Y}y}w\big]\big] \end{array}\right],$$

only if $\mathcal{E}\mathfrak{C}_{\Phi} = \emptyset$. It follows that, also according to assumption ¬A, the behavior of this concept is easily determinable on the basis of considerations relating to the cardinality of concepts, in particular the concept $\mathfrak{C}_{\Phi}$. This is important because Frege's reformulations of the fundamental theorems of arithmetic and of his theorems about *Endlos* only concern the *Anzahlen* of the extensions of concepts (so that in all these cases, $\mathfrak{C}_{\Phi}$ coincides with Φ), and $\mathfrak{y}\emptyset$ corresponds, in his reformulation, to 0, whereas all natural numbers are the *Anzahlen* of the extensions of concepts that can be defined recursively from 0, and *Endlos* is, as we have just said, the *Anzahl* of the extension of the concept of being a natural number. I would like to thank Joan Bertran-San Millán for suggesting this argument to me.

which one might more simply write, respectively, as follows

$$\forall u\left[\mathfrak{N}u = \mathcal{E}\left[x : \exists f\left[x\mathfrak{S}_{\rangle\mathcal{E}f}u \wedge u\mathfrak{S}_{\rangle\mathfrak{L}\mathcal{E}f}x\right]\right]\right] \tag{31}$$

and

$$
\begin{aligned}
\mathfrak{N}\Gamma &= \mathcal{E}\left[x : \exists f[x\mathfrak{S}_{\rangle\mathcal{E}f}\Gamma \wedge \Gamma\mathfrak{S}_{\rangle\mathfrak{L}\mathcal{E}f}x]\right] \\
&= \mathcal{E}\left[x : \exists f \exists g\left[\rangle\mathcal{E}f = \mathcal{E}g \wedge x\mathfrak{R}_g\Gamma\right] \wedge \exists h\left[\rangle\mathfrak{L}\mathcal{E}f = \mathcal{E}h \wedge \Gamma\mathfrak{R}_h x\right]\right] \\
&= \mathcal{E}\left[x : \exists f \left[\begin{array}{l} \mathrm{I}\mathcal{E}f \wedge \mathrm{I}\mathfrak{L}\mathcal{E}f \wedge \\ \forall z\left[\mathfrak{T}_x z \Rightarrow \exists w\left[\mathfrak{T}_\Gamma w \wedge z\mathfrak{S}_{\mathcal{E}f}w\right]\right] \wedge \\ \forall z\left[\mathfrak{T}_\Gamma z \Rightarrow \exists w\left[\mathfrak{T}_x w \wedge z\mathfrak{S}_{\mathfrak{L}\mathcal{E}f}w\right]\right] \end{array}\right]\right].
\end{aligned} \tag{32}
$$

This move would make Frege's definition also unambiguously equivalent to (28) in the absence of the second clause of definition (3). But it does not solve all the problems. Under assumption ¬A, another difficulty arises from the fact that, from (28), it follows that $\mathfrak{N}\Theta^\circ = \mathfrak{N}\emptyset$ or, more precisely, from the fact that, from the latter statement and from the definition of 0 that Frege presents in §I.41, it follows that 0—here defined as the same object as $\mathfrak{N}\emptyset$—is the same *Anzahl* as $\mathfrak{N}\Theta^\circ$ for whatever object Θ° that is not a value-range.[28]

The problem here is that, even without contradicting Hume's Principle (which is the most important consequence that Frege draws from his definition of the *Anzahlen*, both on the deductive level and on the interpretative or, if one prefers, philosophical level).[29] This fact still undermines the idea that the *Anzahlen* belong to (or are) concepts. Although there is no reason $\mathfrak{N}\Gamma$ should not be taken, for whatever object Γ, not only literally, as the *Anzahl* of Γ, as I have said so far, but also as the *Anzahl* of $\mathfrak{T}_\Gamma$ (which makes any *Anzahl* both the *Anzahl* of an object and of a concept at the same time), as Frege himself seems to do,[30] the fact remains that what

---

28 Note that (28) also implies that $\mathfrak{N}\emptyset = \mathfrak{N}\mathcal{E}\Phi^\circ$, for whatever one-argument function that never takes T as value. As there is no guarantee that such a function is a concept, there is also no guarantee that $\mathcal{E}\Phi^\circ = \emptyset$ and that 0, as defined by Frege, is the *Anzahl* of a single value-range. This does not depend, however, on assumption ¬A, for the same is also true if the latter is not assumed. Moreover, this is also the case for all the *Anzahlen*, as defined by (28), since, from the fact that $\mathfrak{T}_\Theta \approx \mathfrak{T}_\Lambda$, for two objects Θ and Λ, it does not follow at all that $\Theta = \Lambda$, even if these objects are value-ranges.

29 The reason is that $\mathfrak{T}_\Theta \approx \mathfrak{T}_{\Theta^\circ}$ and that, in *Grundsetze* (§I.65, th. 32, and §I.69, th. 49), this principle is reduced to: $\forall x, z[\mathfrak{N}x = \mathfrak{N}z \Leftrightarrow \mathfrak{T}_x \approx \mathfrak{T}_z]$.

30 Indeed, in §I.40, just before presenting definition (25), Frege observes that, in line with his definition of *Grundlagen* (§68), the extension occurring in the right-hand side of this definition, with 'Δ' replacing '*u*', is "the *Anzahl* which belongs to the concept $\_\xi \frown \Delta$" or, more briefly, "the *Anzahl* of the Δ-concept".

makes it the case that the concept $\mathfrak{T}_{\Theta^\circ}$ is empty is only the fact that Θ° is not a value-range rather than the way in which objects are distributed among the concepts under which they fall.

That is not all, yet. For this means that the nature of 0 changes drastically depending on whether one assumes A or ¬A. If 0 is identified with $\eta\varnothing$, then it is the extension of the concept $\left[x : \mathfrak{T}_x \approx \mathrm{F}_{1,1}^{F}\right]$ of being the value-range of a one-argument function that does not take the value T for any argument. Now, while under assumption ¬A, all the objects other than value-ranges fall under this concept so that 0 is the extension of a concept under which all these objects fall, under assumption A, 0 is just the extension of a concept under which only the extensions of the functions that never take the value T fall. Therefore, if one cannot decide between assumptions A and ¬A, one cannot decide the nature of 0 either.

Even though Frege's definition of the *Anzahlen* was amended in line with (31) and (32), in order to make it unambiguous under assumption ¬A, it would still remain that 0 would be a different object depending on whether the latter assumption is admitted, and, in case it is admitted, its nature would not be coherent with the fundamental idea according to which the *Anzahlen* belong to concepts.

7.   Conclusions

The latter observation, together with all the difficulties mentioned concerning the functions $\xi \frown \Theta$, $\_\xi \frown \Theta$, $\xi \frown (\zeta \frown \Theta)$, $\_\xi \frown (\zeta \frown \Theta)$, I$\xi$, $\rangle\xi$, $\nparallel\xi$ and $\eta\xi$, suggests that, despite some of its formulations, Frege implicitly assumed that any object is a value-range.

This assumption can take at least two different forms.

1. On the one hand, one can imagine that Frege took his system to admit (in a sense to be clarified) several interpretations and implicitly worked, to put it in modern terms, within the intended model: a minimal model including nothing but value-ranges as zero-level elements. These elements would, in particular, be given by T and F—identified, in accordance with the stipulation presented in §I.10, with the extensions of $[x : \_x]$ and $[x : x = \neg\forall z[z = z]]$, respectively— and by the other value-ranges of the functions that are definable within the language of this system. Indeed, it easy to see that, if the second clause of stipulation (3) is accepted, and T and F are supposed to be value-ranges, then the values of these functions could be objects other than value-ranges only if their arguments are so. This is because these functions result, by composition, from the eight primitive functions of the system—horizontal, negation, implication, identity, the first- and second-order universal quantifiers, the value-range function or $\acute{\epsilon}\phi(\epsilon)$, and the function $\backslash\xi$—the first six of which are concepts or relations and can have, then, only T and F as values, while the seventh only has, by definition, value-ranges as values, and, in agreement with the second clause of stipulation (3),

the eighth cannot but take as values either value-ranges or the same objects providing its arguments.

2. On the other hand, one can imagine that Frege believed that the objects in his system are nothing but the objects that exist independently of it, so that this system admits, strictly speaking, and at least as far as its objects are concerned, only one interpretation (if we can speak of interpretation in this case), given by the world as it is, independently of the system itself. To grant this, one would also have to admit that this world includes, among its objects, all the value-ranges of the functions that can be defined within the language of the system, which is, however, something that Frege would have been ready to accept.[31] To ensure that every object is a value-range, it would be necessary, moreover, to associate with any object that is not immediately defined as a value-range a first-level function (possibly definable only by extending the language and adding a name for this object) whose value-range is, by definition, nothing more than such an object.

A way to do that is suggested by the stipulation discussed by Frege in the second footnote of §I.10.[32] After suggesting, in the *corpus* of this section, that, as we said previously,

$$\mathbf{T} = \mathcal{E}[x : \_\_x] \text{ and } \mathbf{F} = \mathcal{E}[x : x = \neg \forall z[z = z]],$$

which is equivalent to stating that

$$\mathbf{T} = \mathcal{E}[x : x = \mathbf{T}] \text{ and } \mathbf{F} = \mathcal{E}[x : x = \mathbf{F}], \tag{33}$$

Frege considers the possibility to generalize this stipulation by stating that, for whatever object Θ,

$$\Theta = \mathcal{E}[x : x = \Theta]. \tag{34}$$

He then observes that, if Θ were the value-range $\mathcal{E}\Phi$ of a (first-level) function Φ other than a concept under which only one object falls, then this stipulation would conflict with BLV. This is because, from

$$\mathcal{E}\Phi = \mathcal{E}[x : x = \mathcal{E}\Phi],$$

it would then follow, according to BLV, that

$$\forall z[\Phi(z) = (z = \mathcal{E}\Phi)],$$

---

31 Note that this assumption would go hand in hand with the one that the functions within this system are only those definable in the system's language, which would allow repeating, also in this context, *mutatis mutandis*, the previous argument about the values of these functions.

32 See footnote (9).

which can be the case only if Φ is a concept under which only the object $\mathcal{E}\Phi$ falls, something that would contradict our assumption about Φ. This leads Frege to revise his stipulation (34).

Yet it is not difficult to see that for whatever object Θ, if Φ is the concept $[x : x = \Theta]$ under which only Θ falls, it follows from (34) that

$$\mathcal{E}[x : x = \Theta] = \mathcal{E}[x : x = \mathcal{E}[x : x = \Theta]],$$

and so, according to BLV,

$$\forall z[(z = \Theta) \Leftrightarrow (z = \mathcal{E}[x : x = \Theta])],$$

which is perfectly coherent with (34). It seems, then, at first glance, that nothing prevents adopting stipulation (34) under the restriction that Θ should not be the value-range of a (first-level) function other than a concept under which a single object falls.[33] It would follow that any object would be either the value-range of a function other than a concept under which a single object falls or the value-range of the concept under which it is the only object to fall.

But let us consider the value-range $\mathcal{E}\Phi^<$ of a (first-level) function $\Phi^<$ other than a concept under which a single object falls, for example, $\emptyset$ or @. What will be the extension of the concept $[x : x = \mathcal{E}\Phi^<]$, for instance, that of $[x : x = \emptyset]$ or of $[x : x = @]$? As Frege's argument shows, it could certainly not be, respectively, $\mathcal{E}\Phi^<$, $\emptyset$, or @, as this would contradict BLV. It should then be identified with an object distinct from the latter, which could be denoted by '$\{\mathcal{E}\Phi^<\}$', '$\{\emptyset\}$', or '$\{@\}$'. But then what will be the extension of $[x : x = \{\mathcal{E}\Phi^<\}]$, $[x : x = \{\emptyset\}]$, or $[x : x = \{@\}]$? If stipulation (34) were restricted only as just said, then one should conclude that it would be nothing more than $\{\mathcal{E}\Phi^<\}$, $\{\emptyset\}$, or $\{@\}$, in that these objects are the value-ranges of a (first-level) concept under which a single object falls, which would contradict BLV anew. Hence, if we wanted to adopt stipulation (34), we would have to restrict it in a different way: we should clarify that it should apply only to whatever object Θ that is different from $\{\mathcal{E}\Phi^<\}^n (n = 0, 1, \ldots)$, provided that

$$\{\mathcal{E}\Phi^<\}^0 = \mathcal{E}\Phi^< \quad \text{and} \quad \{\mathcal{E}\Phi^<\}^{n+1} = \mathcal{E}[x : x = \{\mathcal{E}\Phi^<\}^n]$$

and that $\mathcal{E}\Phi^<$ be a function other than a (first-level) concept under which a single object falls.

---

33 In discussing stipulation (34), Frege rejects the possibility of adopting it 'only for such objects which are not given to us as value-ranges', because 'the way an object is given must not be regarded as its immutable property'. Whatever one thinks of this remark, when it applies to abstract objects, the fact remains that the limitation in question here does not fall under this criticism, because it does not concern the way in which Θ is presented, but it concerns what Θ is. If this limitation makes it possible, indeed, to apply this stipulation to any object that is not immediately defined as a value-range, it is only because, within the framework of what has been described as the informal semantics of Frege's system, to define an object as a value-range amounts to determining its intrinsic nature.

Let's adopt, then, the stipulation under this restriction, and let us imagine that $\Theta^\triangleright$ is an object other than $\{\mathcal{E}\Phi^\triangleleft\}^n$, for example, Julius Caesar, provided that he is such an object.[34] It will follow that $\Theta^\triangleright$ is the same object as $\mathcal{E}[x : x = \Theta^\triangleright]$ and Julius Caesar the same object as $\mathcal{E}[x : x = \text{Julius Caesar}]$ and that they will thus be value-ranges. For any statement $\mathcal{A}$, let's denote the truth-value of $\mathcal{A}$ as $\mathfrak{M}(\mathcal{A})$. According to (4), (8), and the first clause of (3), for whatever objects Γ and Δ, we will have that

$$
\begin{aligned}
\Gamma \frown \Theta^\triangleright &= \backslash\mathcal{E}\left[x : \exists g\left[\Theta^\triangleright = \mathcal{E}g \wedge g(\Gamma) = x\right]\right] \\
&= \backslash\mathcal{E}\left[x : \exists g\left[\mathcal{E}[z : z = \Theta^\triangleright] = \mathcal{E}g \wedge g(\Gamma) = x\right]\right] \\
&= \backslash\mathcal{E}\left[x : x = \mathfrak{M}\left(\Gamma = \Theta^\triangleright\right)\right] \\
&= \mathfrak{M}\left(\Gamma = \Theta^\triangleright\right)
\end{aligned}
\tag{35}
$$

and

$$
\begin{aligned}
\Delta \frown \left(\Gamma \frown \Theta^\triangleright\right) &= \backslash\mathcal{E}\left[x : \exists g\left[\Gamma \frown \Theta^\triangleright = \mathcal{E}g \wedge g(\Delta) = x\right]\right] \\
&= \backslash\mathcal{E}\left[x : \exists g\left[\mathfrak{M}\left(\Gamma \frown \Theta^\triangleright\right) = \mathcal{E}g \wedge g(\Delta) = x\right]\right],
\end{aligned}
\tag{36}
$$

From (35), one will, then, immediately infer, on the one hand, that

$$
\forall x[(x = \Theta^\triangleright \Rightarrow x \frown \Theta^\triangleright = \mathbf{T}) \wedge (x \neq \Theta^\triangleright \Rightarrow x \frown \Theta^\triangleright = \mathbf{F})],
\tag{37}
$$

and from (36) and (33), one we will infer, on the other hand, that

$$
\forall x, y \left| \begin{aligned} &\left[x = \Theta^\triangleright \Rightarrow y \frown \left(x \frown \Theta^\triangleright\right) = \backslash\mathcal{E}[z : (y = \mathbf{T}) = z]\right] \wedge \\ &\left[x \neq \Theta^\triangleright \Rightarrow y \frown \left(x \frown \Theta^\triangleright\right) = \backslash\mathcal{E}[z : (y = \mathbf{F}) = z]\right] \end{aligned} \right|
$$

and, consequently, according to the first clause of (3):

$$
\forall x, y \left| \begin{aligned} &\left[\left(x = \Theta^\triangleright \wedge y = \mathbf{T}\right) \Rightarrow y \frown \left(x \frown \Theta^\triangleright\right) = \backslash\mathcal{E}[z : z = \mathbf{T}] = \mathbf{T}\right] \wedge \\ &\left[\left(x = \Theta^\triangleright \wedge y \neq \mathbf{T}\right) \Rightarrow y \frown \left(x \frown \Theta^\triangleright\right) = \backslash\mathcal{E}[z : z = \mathbf{F}] = \mathbf{F}\right] \wedge \\ &\left[\left(x \neq \Theta^\triangleright \wedge y = \mathbf{F}\right) \Rightarrow y \frown \left(x \frown \Theta^\triangleright\right) = \backslash\mathcal{E}[z : z = \mathbf{T}] = \mathbf{T}\right] \wedge \\ &\left[\left(x \neq \Theta^\triangleright \wedge y \neq \mathbf{T}\right) \Rightarrow y \frown \left(x \frown \Theta^\triangleright\right) = \backslash\mathcal{E}[z : z = \mathbf{F}] = \mathbf{F}\right] \end{aligned} \right|
\tag{38}
$$

---

34 Of course, the supposition that Julius Caesar is an object other than $\{\mathcal{E}\Phi^\triangleleft\}^n$ cannot but be, in the context of the hypothesis that we are discussing, an assumption about the world as it is. Even if one wanted to reject it, one would in any case have to assume that this world includes some objects other than $\{\mathcal{E}\Phi^\triangleleft\}^n$.

For whatever object $\Theta^{\triangleright}$ other than $\{\mathcal{E}\Phi^{\triangleleft}\}^{n}$, the first clause of (3) would thus be sufficient, if associated with stipulation (33), to make sure that the functions $\xi \frown \Theta^{\triangleright}$ and $\xi \frown (\zeta \frown \Theta^{\triangleright})$ are reduced to an unambiguous concept and an unambiguous relation, respectively, in perfect agreement with (5) and (9).[35]

These latter identities will, on the other hand, apply as such to functions $\xi \frown \{\mathcal{E}\Phi^{\triangleleft}\}^{n}$ and $\xi \frown (\zeta \frown \{\mathcal{E}\Phi^{\triangleleft}\}^{n})$.

Within Frege's system, there would thus be two kinds of objects: those like $\Theta^{\triangleright}$ that one could qualify as primitive, counting as value-ranges of concepts such as $[x : x = \Theta^{\triangleright}]$, and among which one would find **T** and **F**, and those like $\{\mathcal{E}\Phi^{\triangleleft}\}^{n}$ that one could qualify as defined, among which there would be Ø, @, and many others. The two kinds of objects would exist in the world as it is, but, while the latter would be the elements of a multiplicity of recursive hierarchies, the former would remain isolated, admitting only a recursive infinity of names, since:

$$\Theta^{\triangleright} = \mathcal{E}[x : x = \Theta^{\triangleright}]$$
$$= \mathcal{E}[x : x = \mathcal{E}[x : x = \Theta^{\triangleright}]]$$
$$= \mathcal{E}[x : x = \mathcal{E}[x : x = \mathcal{E}[x : x = \Theta^{\triangleright}]]]$$
$$= \dots$$

Whatever the respective merits of these two hypotheses, it is clear that choosing the first would not make the objects of the second kind disappear, and even less so for the two value-ranges **T** and **F**, which, according to the stipulation made in §I.10, would in any case remain outside the multiplicity of recursive hierarchies formed by those objects. In order to better understand the informal semantics of Frege's system, it would

---

35 By letting $\Theta^{\triangleright} = \mathcal{E}[z : z = \Theta^{\triangleright}]$, from (5), it follows, indeed, that

$$\Gamma \frown \Theta^{\triangleright} = \Gamma \frown \mathcal{E}\Big[z : z = \Theta^{\triangleright}\Big] = \backslash\mathcal{E}\Big[x : \exists g\Big[\mathcal{E}\big[z : z = \Theta^{\triangleright}\big] = \mathcal{E}g \wedge g(\Gamma) = x\Big]\Big]$$
$$= \backslash\mathcal{E}\Big[x : x = (\Gamma = \Theta^{\triangleright})\Big] = \mathfrak{M}\big(\Gamma = \Theta^{\triangleright}\big),$$

according to (37). To show that (38) fits with (9), let's suppose that $\Psi^{\triangleright}$ is a two-argument function such as

$$\mathcal{E}_{\varepsilon}\Psi^{\triangleright}(\varepsilon, \Gamma) = \begin{cases} \mathcal{E}[x : x = \mathbf{T}] = \mathbf{T} & \text{if} \quad \Gamma = \Theta^{\triangleright} \\ \mathcal{E}[x : x = \mathbf{F}] = \mathbf{F} & \text{if} \quad \Gamma \neq \Theta^{\triangleright}. \end{cases}$$

The functions $\mathcal{E}_{\varepsilon}\Psi^{\triangleright}(\varepsilon, \zeta)$ and $\Psi^{\triangleright}$ would thus coincide with the concept $[x : x = \Theta^{\triangleright}]$ and the relation $[x, y : (y = \Theta^{\triangleright} \Rightarrow x = \mathbf{T}) \wedge (y \neq \Theta^{\triangleright} \Rightarrow x = \mathbf{F})]$, respectively, and from (9), it would follow that

$$\Delta \frown (\Gamma \frown \Theta^{\triangleright}) = \Delta \frown \Big(\Gamma \frown \mathcal{E}\big[x : x = \Theta^{\triangleright}\big]\Big)$$
$$= \Delta \frown \Big(\Gamma \frown \mathcal{E}_{\alpha}\big[\mathcal{E}_{\varepsilon}\Psi^{\triangleright}(\varepsilon, \zeta)\big]\Big)$$
$$= \Psi^{\triangleright}(\Delta, \Gamma)$$
$$= \mathfrak{M}\Big(\big[\Gamma = \Theta^{\triangleright} \Rightarrow \Delta = \mathbf{T}\big] \wedge \big[\Gamma \neq \Theta^{\triangleright} \Rightarrow \Delta = \mathbf{F}\big]\Big)$$

in agreement with (38).

therefore be necessary, in any case, to investigate the structure of this multiplicity of hierarchies and what they could become in the different variants of this system that make it consistent. But this is not a task that can be accomplished here.

References

Frege, G. (1893–1903). *Die Grundgesetze der Arithmetick* (Vol. I–II). Jena: H. Pohle.

Frege, G. (1984). *Die Grundlagen der Arithmetik: eine logische mathematische Untersuchung über den Begriff der Zahl.* Breslau: Koebner.

Frege, G. F. (2013). *Basic Laws of Arithmetic.* Oxford: Oxford University Press. (Translated and edited by P. A. Ebert and M. Rossberg with C. Wright)

Heck, R. G. (1996). "The consistency of predicative fragments of Frege's *Grundgesetze der Arithmetik*". *History and Philosophy of Logic, 17*(1), 209–220.

Wehmeier, K. F. (1999). "Consistent fragments of *Grundgesetze* and the existence of non-logical objects". *Synthese, 121*(3), 309–328.

5 Dedekind's Logicism
A Reconsideration and Contextualization

Erich H. Reck

Logicism, primarily in Russell's and earlier in Frege's version, is often seen as one of the main attempts to provide a foundation for modern mathematics, with Hilbertian formalism and Brouwerian intuitionism as its two rivals. This, in any case, is how Rudolf Carnap described the situation in an influential 1930 lecture (1931). At that point, Carnap presented himself as a logicist, too, along Russellian lines. In the following decades, logicism fell more and more out of favor. It was only in the 1980s that Crispin Wright, Bob Hale, and others revived it in a Fregean form: Scottish neo-logicism. These four positions are the most widely known variants of logicism today. But there are more unorthodox forms as well. An early example is provided by Richard Dedekind's foundational work, as we will see in this chapter.[1]

While Dedekind is sometimes recognized as an early logicist (cf. Stein 1998; Demopoulos & Clark 2007, recently also Hellman & Shapiro 2019), this classification has been challenged. The challenges range from denying that he was a successful logicist to questioning, in several ways, whether he pursued a logicist project at all. Much of the present chapter has the goal of responding to those challenges. What will become evident, along the way, is that Dedekind's case raises interesting questions about what is, or should be, meant by "logicism" in the first place. Roughly, logicism is the thesis that all of mathematics, or at least core parts of it, can be reduced to logic. But that leads to questions about what is meant by "logic" and about the form the "reduction" is meant to take, in Dedekind and more generally.

---

1 The present chapter continues my discussion of Dedekind in several earlier publications, especially Reck (2013a, 2017, 2019a) and Reck and Keller (forthcoming), partly also Reck (2003, 2008, 2013b, 2018, 2019b) and Awodey and Reck (2002). The main way in which this chapter goes beyond them is by providing a more detailed response to challenges that Dedekind should be seen as a logicist. In defending a logicist reading of him, I see myself as in line with, and am partly influenced by, Ferreirós (forthcoming), Klev (2017), Demopoulos and Clark (2007), and Stein (1998).

DOI: 10.4324/9780429277894-6

To shed further light on Dedekind's logicism, including its relation to better-known variants like Frege's, Russell's, Carnap's, and Scottish neo-logicism, a second goal of the present chapter is to put his approach into historical context. By doing so, his position will reveal itself as not merely a response to philosophical claims about mathematics but as the continuation and extension of certain developments in nineteenth-century mathematics. This is relevant with respect to Dedekind but also for clarifying the significance of logicism itself, including its relation to mathematical practice. A third and related goal is to resist the identification of logicism with some versions, like Frege's, that add very strong philosophical assumptions to it, assumptions that are not part of Dedekind's version. Insisting on those assumptions obscures a proper evaluation of logicism, in Dedekind's case and more generally.

1.   Dedekind's Foundational Essays and His Logicist Remarks

The most important text for documenting Dedekind's adoption of logicism is his booklet, *Was sind und was sollen die Zahlen?* (1888). In it, he provides a novel characterization of the natural numbers in terms of his notion of a "simple infinity" (basically, the Dedekind-Peano axioms). As the background for this characterization and his further treatment, Dedekind uses a general theory of sets and functions. This is the first time in the history of mathematics that such a theory is used for elaborating the foundations of mathematics in a systematic way, as should be noted.[2] Moreover, Dedekind considers his basic notions of "thing", "set", and "function" to be "purely logical".[3] His project is, then, to develop "that part of logic which deals with the theory of numbers" (Dedekind 1963, p. 31).

Remarks like these seem to confirm immediately that Dedekind is pursuing a form of "logicism", although he does not use that term yet (nor does Frege). But how exactly should we conceive of "logic" here? Like Frege around the same time, and like Russell later, Dedekind does not provide a fully satisfactory answer to that question. We can get a basic sense of his relevant views from a remark he makes about the fundamentality of the notion of function ["Abbildung"]:

> If we scrutinize closely what is done in counting a set or a number of things, we are led to consider the ability of the mind to relate things

---

2 Frege only starts to do so slightly later, in Vol. 1 of *Basic Laws of Arithmetic* (1893), possibly under the influence of Dedekind; see Reck (2019b). And while Cantor develops transfinite set theory significantly further at the time, his perspective remains less foundational than Dedekind's.

3 In an 1887 draft of *Was sind und was sollen die Zahlen?*, Dedekind writes explicitly that a theory of sets, or of "systems of elements", simply "is" logic; see Ferreirós (1999), p. 225.

to things, to let a thing correspond to a thing, or to represent a thing by a thing, an ability *without which no thinking is possible*. Upon this *unique and therefore absolutely indispensable foundation* the whole science of number must, in my opinion, be established.

(Dedekind 1963, p. 32, my emphasis)

Apparently what "logic" encompasses for Dedekind are such "indispensable" and "foundational" notions, together with general results about them. With respect to sets ["Systeme"] Dedekind does not go quite as far, he only notes:

It *very frequently happens* that different things *a*, *b*, *c* . . . can be considered from a common point of view . . . and we say that they form a system *S*.

(*ibid.*, my emphasis)

Then again, sets are closely related to functions (as their domains and ranges), so that they can be seen as indispensable for thinking, too (or at least for scientific thinking). This explains why Dedekind considers both "logical".[4]

As the technical details of Dedekind's treatment of the natural numbers in his 1888 essay have been documented before (cf. Reck 2003, 2008), let me only provide a brief summary here. With the notions of function and set in the background, Dedekind defines the notion of "infinity" for sets, on which the notion of a "simple infinity" is then based. He also proves, or attempts to prove (in his "Theorem 66", more on which later), the existence of infinite sets and thus of simple infinities; he establishes that any two simple infinities are isomorphic (i.e., the Dedekind-Peano axioms are categorical), and this implies, as he observes, that exactly the same theorems hold for all of them (the axioms are semantically complete).[5] He then takes these results to justify the introduction of "the natural numbers" by "free creation" based on a kind of "abstraction" ("Definition 73", more again later). Finally, he provides explicit justifications for, among others, recursive definitions, inductive proofs, and the use of natural numbers as cardinal numbers. The overall suggestion is clearly that the rest of arithmetic can be developed on this basis as well.

---

4 This is not far from Frege and Russell, who both tie logic to scientific reasoning and mention its "generality" as crucial without turning this into a precise definition.

5 See Awodey and Reck (2002) for a further discussion of this result, including a historically more nuanced description of what exactly Dedekind established and how it was received.

In Dedekind's words again, the conclusion to draw is that arithmetic is "a part of logic". He clarifies the significance of this as follows:

> In speaking of arithmetic (algebra, analysis) as a part of logic I mean to imply that I consider the number concept entirely *independent of the notions of intuition of space and time*, that I consider it an *immediate result from the laws of thought*.
>
> (Dedekind 1963, p. 31, my emphasis)

If we take everything resulting immediately and exclusively from "laws of thought" (concerning "indispensable" and "foundational" notions) to constitute "logic", as seems implied here, this passage represents another straightforward endorsement of logicism, including the rejection of "intuition" as necessary for foundational purposes typical for it. However, Dedekind does not elaborate such suggestive passages in any substantive way. As a result, various questions can be raised about his approach, as we will see in what follows.

Before moving on to such questions, let me add two further observations. First, note Dedekind's parenthetical phrase in the quotation just given. Evidently his goal is not just to show that "arithmetic" in the narrow sense (the theory of the natural numbers) is "a part of logic" but also "analysis" (both real and complex, as is clear elsewhere) and "algebra" (the study of solutions to equations involving such numbers). What that means is that virtually all the "pure mathematics" of his time is meant to be covered.[6] In the background is Dedekind's earlier foundational booklet, *Stetigkeit und irrationale Zahlen* (1872), in which he treated the real numbers. The basic framework in that essay, too, is a general theory of sets and functions, although more implicitly. Dedekind shows, once again, that the crucial notion he defines—here that of a "continuous ordered field" (the "Dedekind-Hilbert axioms", as they should perhaps be called)—is satisfiable, namely by constructing the system of all cuts on the ordered field of rational numbers. Finally, "the real numbers" are introduced by an act of "free creation" on that basis.[7] Thus, his two foundational essays are part of a larger, comprehensive logicist project.[8]

---

6 See Ferreirós (2007b) on the notion and the range of "pure mathematics" at issue here.

7 Dedekind does not use the term "abstraction" in this earlier essay yet, nor does he prove that all continuous ordered fields are isomorphic (although this is true); see Reck (2003, 2008) for more.

8 Dedekind treats the natural numbers and the real numbers as the two crucial ingredients in the step-wise construction of the usual number systems. He already knows about how to construct the integers and the rational numbers out of the natural numbers as

As a second observation, note that at the end of the nineteenth century Dedekind's works were taken to aim at logicist conclusions by various interested thinkers. Three illustrations should suffice here. In his *Vorlesungen über die Algebra der Logik* (1890/1895), Ernst Schroeder writes that he is tempted to join "those who, like Dedekind, consider arithmetic a branch of logic". Similarly Frege, in Volume I of his *Basic Laws of Arithmetic* (1893), remarks: "Dedekind too is of the opinion that the theory of numbers is a part of logic". And David Hilbert, in lectures on geometry from 1899, writes: "As a given we take the laws of pure logic and in particular arithmetic. (On the relationship between logic and arithmetic see Dedekind, *Was sind und was sollen die Zahlen?*)".[9] To be sure, these thinkers do not use the term "logicism" either.[10] During Dedekind's time, projects like his are simply described as attempts to show that arithmetic is a "part" or "branch" of "pure logic". And that captures the general sense of "logicism" used in the present chapter.

2. First Challenge: Dedekind Was Not a Successful Logicist

Based on the evidence so far, it seems hard to deny that Dedekind pursued a form of "logicism", even if he did not use that label. Why, then, do some people resist classifying him as such? In what follows, we will consider several reasons for such resistance. A first reason concerns whether his contributions were at all promising for establishing logicist conclusions, that is, whether it was even potentially successful as a form of logicism. As we will see, there are indeed gaps in Dedekind's procedure that can make one wonder. But whether that implies he shouldn't be seen as a logicist is another question, since it is not inconceivable that these gaps could be

---

(equivalence classes) of pairs (cf. Sieg & Schlimm 2005); similarly for constructing the complex numbers out of the real numbers. Presumably, he views these number systems to be "purely logical" as well. What Dedekind thinks about geometry is less clear. The standard treatment of three-dimensional Euclidean geometry in terms of triples of real numbers would seem to provide him with the obvious means to treat it "purely logically" too. Moreover, he is familiar with Riemann's more general approach, including his treatment of various non-Euclidean geometries (also *n*-dimensional). But Dedekind does not remark on the status of geometry in any explicit and definite way anywhere, as far as I am aware.

9 Two further witnesses from the early twentieth century are C.S. Peirce and Ernst Cassirer; see Ferreirós (1999, 2009), Reck (2013a, 2013b), and Reck and Keller (forthcoming), also for further references. In Hilbert (1926), both Frege and Dedekind are still characterized as logicists.

10 That term seems to occur first, in the sense relevant for us, in texts by Carnap, Fraenkel, and so on from around 1930; see again Carnap (1931). Earlier "logicism" was sometimes used (as a term of abuse) for attempts to reduce all of metaphysics to logic, especially in the works of Leibniz, Wolff, and their followers. As such, it coincided with a strong form of rationalism; see Gabriel (1980) for more.

filled for him—or so I will argue in the present section. Later I will respond to other, more general challenges to seeing Dedekind as a logicist.

Several parts of Dedekind's procedure in his 1888 essay that were mentioned previously, in passing, will play an important role in this section. Thus, we need to be more explicit and consider them in some detail now. The first part consists of his putative proof that infinite sets, and thus also simply infinite sets, exist (Theorem 66). What Dedekind does in this context is rely on the following three surprising ingredients: (i) "the totality S of all things which can be objects of my thought" (as a kind of universal set), (ii) his own "self" (as a distinguished element s of S), and (iii) the function Φ that maps any element x of S onto the thought "x can be the object of my thought" (as a 1-1 successor function) (Dedekind 1963, p. 64). His argument is then that the set N constructed by closing $\{s\}$ under Φ in S (the corresponding "chain") is a simply infinite subset of S, one he then uses in his further procedure.

There are several problems with this argument. One issue is, of course, that Dedekind's assumption of a universal set S (together with a general separation principle implicit in his approach) leads to technical difficulties, most famously Russell's antinomy.[11] This raises the question: Is there any way to avoid these difficulties while still establishing the existence of simple infinities (and the other systems Dedekind needs) and, more specifically, while doing so in a "purely logical" way? Often the answer is assumed to be negative (more on why in the following). The first problem is, thus, whether a Dedekindian logicist approach can be developed in a consistent way. A second problem concerns Dedekind's appeal to his "self" ["Selbst"] and to "thoughts" ["Gedanken"] as elements of S. This appeal appears to make his approach problematically psychologistic. Or, more basically, it encourages the objection, first voiced by Russell, that Dedekind is relying on notions that "are not appropriate to mathematics" (1904, p. 258). A very forceful expression of that point, and a clear echo of Russell's objection, is presented by George Boolos, who calls Dedekind's attempted proof of Theorem 66 "one of the strangest pieces of argumentation in the history of logic", since it appeals to "as wildly non-mathematical an idea as his own ego" (Boolos 1999, p. 202).

With respect to this second issue, the main problem can be put this way: How can Dedekind be considered a logicist if he appeals, at a crucial step, to such non-mathematical, non-logical, and problematic entities as his "self" and "thoughts"? This seems a strong challenge, too. However, some initial, partial responses are possible right away. To begin with,

---

11 Dedekind was informed by Cantor, in the late 1890s, that assuming a universal set leads to inconsistencies. While his initial reaction was shock, later he expressed confidence that a solution could be found; see Reck (2013b, 2019a), also for references.

it is not clear that Dedekind conceives of "thoughts" in a problematic mentalistic way, that is, that he understands them as entities existing in someone's subjective consciousness. Instead, we can compare what he does to Frege's familiar appeal to "thoughts" in an objective sense, as well as to Bernard Bolzano's earlier appeal to "sentences in themselves".[12] And this might lead one to invoke intensional logic for Dedekind's logicist purposes.[13] Likewise, his appeal to the "self" may be less psychologistic than it appears at first, for example, if we read him as appealing to a "transcendental self" in a non-psychologistic Kantian sense, thus calling on transcendental logic.[14] But we can be charitable to Dedekind in a way more familiar to current logicians and mathematicians, too. Namely, it is not hard to replace the appeal to his "self" and his successor function Φ by, say, Zermelo's appeal to the empty set, $\varnothing$, and his successor function f, which maps x onto $\{x\}$.[15] Along such lines, one still needs to justify viewing their existence as "part of logic", but that may be a more tractable challenge.

So far, the two main problems are how to understand or replace Dedekind's appeal to certain basic entities in his proof of Theorem 66, namely his "self" and "thoughts", together with the simply infinite set N that contains them, along lines that are both consistent and *prima facie* logical. A third, less familiar problem concerns his appeal to "abstraction" and "free creation" in a later step ("Definition 73"). Here is the crucial passage, which we now also need to reconsider more:

> If in the consideration of a simply infinite system N set in order by a mapping Φ we entirely neglect the special character of the elements, merely retaining their distinguishability and taking into account only their relations to one another, then are these elements called *natural*

---

12 Both Frege and Russell suggested reading Dedekind's "thoughts" in such a non-psychologistic way; see Reck (2013b, 2019b). For Bolzano and the broader background, see Klev (2018).

13 Whether an appeal to intensional logic is the best way to save Dedekind's logicism is another question, among others because of the well-known thread of paradox for it. In addition, one may need "hyperintensional" logic for this purpose, with a very fine-grained conception of thoughts, as an anonymous reader pointed out to me. I will not pursue this issue further in the present chapter.

14 In Klev (2017), and earlier in McCarty (1995), the suggestion is to read Dedekind along such lines, that is, as appealing to a Kantian transcendental logic. I will again not explore that kind of appeal further in the present chapter, but see Reck and Keller (forthcoming) for a neo-Kantian variant.

15 This is exactly what Zermelo did in his axiomatization of set theory, and for that reason, he called his version of the Axiom of Infinity "Dedekind's axiom". Alternatively, we can follow von Neumann by again starting with $\varnothing$ but now using the successor function f^*: $x \rightarrow x \cup \{x\}$, together with the now-standard Axiom of Infinity. Yet Zermelo's procedure is particularly close to Dedekind's.

numbers. . . . With reference to this freeing the elements from every other content (*abstraction*) we are justified in calling numbers a *free creation of the human mind.*

(Dedekind 1963, p. 68, emphasis added,
translation slightly amended)

Dedekind's procedure in this passage has again been criticized as psychologistic, in this case as involving a mental operation supposed to result in entities that are "purely structural". As Michael Dummett puts the point: "[Dedekind] believed that the magical operation of abstraction can provide us with specific objects having only structural properties" (1991, p. 52). Dummett understands Dedekind as appealing to a kind of "creation" that has mental objects as the result. He brushes that idea aside as follows: "Frege devoted a lengthy section of *Grundlagen*, sections 29–44, to a detailed and conclusive criticism of this misbegotten theory".[16]

Similar to our second problem, some initial and partial responses on Dedekind's behalf are again possible. Most importantly, one can deny that his appeal to "abstraction" is meant in a psychologistic sense. Then the question becomes: How else should it be understood, and, especially, can we understand "Dedekind abstraction" as a logical operation?[17] Unfortunately, Dedekind himself again says very little about this topic. Partly for that reason, his appeal to "abstraction" is often ignored, or it is interpreted in a more minimal way, as simply "forgetting" all the non-arithmetic features of his original system N while still working with it as "the natural numbers" in a pragmatic manner (similar to what one does in Hilbertian formalism).[18] Along such lines, "Dedekind abstraction" does not create a new, "purely structural" system. However, what about his explicit invocation of "free creation", in the previous passage and elsewhere? Also, why not try to help him out in a different way, such as by formulating logical "abstraction principles" for him?

But even if we try the latter, a more general and deeper issue becomes apparent here. Namely, Dedekind does not formulate any principles, basic laws, or axioms—neither to back up his crucial existence claims (e.g.,

---

16 Against Dummett, it is worth noting that Frege's discussion in *Grundlagen der Arith-metik*, §§29–44, was not directed against Dedekind's 1888 essay, which only appeared in print four years later. Having said that, Frege did criticize Dedekind's appeal to "creation" in connection with the real numbers in his *Grundgesetze der Arithmetik*, Vol. 2 (1903). See Reck (2019b) for further details.

17 For the suggestion to understand "Dedekind abstraction" as a logical operation, *pace* Dummett and others, see Tait (1996). For further discussion, see Reck (2003), Yap (2017), and Reck (2018).

18 For a sophisticated interpretation of Dedekind along such lines, see Sieg and Morris (2018). I will come back to connections between Dedekind's logicism and Hilbert's formalism in the following.

Theorem 66) nor for the kind of abstraction at issue (Definition 73). In retrospect this is a significant lacuna, no doubt. But does it, by itself, establish that he should not be seen as a logicist? Well, there is a gap in his approach, and that gap makes it hard to be sure what is going on at bottom. This, as it turns out, is exactly Frege's main criticism of Dedekind, already in the 1890s.[19] It can be found in two passages from his *Basic Laws or Arithmetic, Vol. I*. The first starts with Frege acknowledging Dedekind to be a fellow logicist, as we already saw; then he adds a more critical twist:

> Mr. Dedekind too is of the opinion that the theory of numbers is a part of logic; but his essay barely contributes to the confirmation of this opinion since his use of the expression "system", "a thing belongs to a thing", are neither customary in logic nor reducible to something acknowledged as logical.
>
> (1893, viii)

Dedekind does use the notions of "system" and "belonging to a system" in his 1888 essay. Frege's response is: Why should they be considered "logical"? It is actually surprising that Frege appeals to what is "customary in logic" and "acknowledged as logical" in this context. Surely it would be better to have a more principled criterion at hand, and Frege does not provide one. Still, he clearly has a point.

The full force of Frege's challenge for Dedekind becomes clear later:

> [N]owhere in [Dedekind's] essay do we find a list of the logical or other laws he takes as basic; and even if it were there, one would have no chance to verify whether in fact no other laws were used, since, for this, the proofs would have to be not merely indicated but carried out gaplessly.
>
> (*ibid.*)

There are two weak spots in Dedekind's procedure, as Frege notes. First, his proofs are not spelled out in enough detail to see which laws are needed in them. Second, Dedekind never formulates his basic laws explicitly anywhere. Seen together, we have no clear sense which basic laws are at play for him and what their nature is, including whether they are "purely logical".[20] This criticism deepens the previous one as follows:

---

19 For Frege's general relationship to Dedekind, including this criticism, see Reck (2019b).

20 As one example, Dedekind is often taken to use a naïve comprehension principle. But that is not entirely clear, since he does not formulate it explicitly. (An alternative is to interpret him as using a "dichotomy conception"; cf. Ferreirós forthcoming.) There are also implicit uses of the axioms of choice and replacement in his works, at least according to some reconstructions; see Reck (2019a).

Maybe Dedekind intends to use "system" and "belonging to a system" as logical notions. But to check whether they really are, we would have to examine the laws governing them, and that is exactly what we cannot do.

Once again, does this Fregean argument establish that Dedekind should not be seen as a logicist? Not really (also not in Frege's eyes, I believe). What it establishes is that his procedure is incomplete in an important way. Assume one could provide a list of the needed laws for him and, in addition, argue convincingly that they are logical. Then he would be fully rehabilitated as a logicist, wouldn't he? Assume, in contrast, we could show that he needs assumptions which are undoubtedly not logical. Then the opposite conclusion would follow. Where does that leave us? For the purpose of defending an interpretation of Dedekind as a successful logicist, three things would be needed: First, explicit laws for the constructions he uses (especially constructions of a simple infinity and a complete ordered field); second, explicit laws for Dedekind abstraction (to introduce "the natural numbers" and "the real numbers"); and third, plausible arguments for taking both to be "purely logical".[21]

In light of later developments, it may appear quite implausible, or sheer wishful thinking, that these three desiderata could possibly be satisfied together. On the construction side, we have learned that Zermelo-Fraenkel set theory (ZFC) provides us with all Dedekind needs. In fact, the ZFC axioms were partly formulated with that goal in mind, that is, to be able to repeat Dedekind's constructions. But usually those axioms are not considered logical truths; thus a standard set-theoretic reduction of mathematics is not seen as a vindication of logicism. Does that settle the matter? Only if it is evident that there are no other alternatives. Maybe some restricted or otherwise modified version of ZFC, still sufficient for Dedekind's purposes, could be considered "purely logical".[22] Alternatively, suppose we use neo-Fregean "abstraction principles", such as Hume's Principle, to establish Dedekind's existence claims.[23] While not uncontroversial, there are arguments that these principles have a logical status, or at least a quasi-logical (definitional) status sufficient for logicism. Beyond that, suppose we introduce parallel "structure

---

21 The latter would, presumably, also give us assurance that the laws needed by Dedekind are consistent after all, perhaps relative to ZFC or a similar system.

22 For example, one might consider using a modified form of axiomatic set theory that satisfies Gödel's "$V=L$" and argue that it is "purely logical". How to do the latter is not obvious, of course. But note that Gödel's constructible universe L was derived explicitly from Russellian type theory, so from a logicist system.

23 This would mean using Hume's Principle for establishing the existence of a simple infinity (of cardinal numbers) and a parallel principle for the existence of a complete ordered field; see Ebert and Rossberg (2016) and the literature referred to in it. From Dedekind's point of view, one would, in a second step, apply "Dedekind abstraction" to arrive at "the natural numbers" as ordinals, and so on.

abstraction principles" to take care of "Dedekind abstraction", as has been explored in the literature already.[24] Finally, suppose we try to argue that they deserve to be considered "purely logical" too, or as much so as neo-logicist abstraction principles. Many difficult details remain to be worked out along such lines, to be sure, but the last word on them has not been spoken yet.[25] And insofar as that is the case, the arguments considered so far do not suffice to disqualify Dedekind as a logicist.

3. Second Challenge: Dedekind Did Not Pursue a Logicist Project at All

In the previous section, I went over a number of arguments to the effect that Dedekind's logicism, in its original form, was unsuccessful. I also considered ways of supplementing his project that might revive it as a candidate for logicism. None of that was meant as a conclusive defense of Dedekindian logicism but only as a counter-weight to dismissing its logicist nature too quickly. However, some interpreters claim that Dedekind should not be interpreted as pursuing a logicist project at all, and these people will surely find my supplements wrongheaded. I now want to consider three forms this more basic challenge to viewing Dedekind as a logicist has taken. According to the first, the suggestion is to see him as aiming at an early set-theoretic foundation for mathematics instead, which, in turn, should be understood along Hilbertian formalist lines.

3.1. First Variant: Dedekind Was a Hilbertian Formalist, Not a Logicist

There are certainly good reasons for seeing Dedekind as an important influence on axiomatic set theory and as a precursor to it.[26] Thus, he was in correspondence with Cantor when he, or both of them together, started to explore set-theoretic ideas. Dedekind is also the first thinker to use a general theory of sets (and functions) as a foundation for mathematics, as already mentioned. In addition, his foundational essays contain

---

24 See Linnebo and Pettigrew (2014) and Reck (2018). Interestingly, what we would end up with is a rapprochement between Fregean neo-logicism and Dedekindian structuralism.

25 As yet another alternative, one might try to reconstruct Dedekind's approach along category-theoretic lines, within a version of type theory and with some structure abstraction principle added. In fact, perhaps the Univalent Foundations project, building on Homotopy Type Theory, could be considered as relevant here. See Awodey (2018) on whether the result might count as a form of logicism.

26 See Ferreirós (2007a) or, for a more complete and very rich account, Ferreirós (1999).

various notions, techniques, and results that are built firmly into axiomatic set theory today, starting with his extensional conception of set and his definition of infinity. Indeed, Ernst Zermelo followed Dedekind's (and Cantor's) work carefully and self-consciously when he axiomatized set theory in the early twentieth century, even including an Axiom of Infinity modeled on Dedekind's "Theorem 66". Similarly, John von Neumann generalized Dedekind's treatment of induction and recursion to the transfinite, another core part of ZFC.

There are also good reasons for seeing Dedekind as an important influence on Hilbert. In Hilbert's early work, he built on Dedekind's foundational writings explicitly, with his axiom systems for geometry and for the real numbers. More generally, his early "structural axiomatics" is influenced by Dedekind, basically in terms of the conceptual approach to mathematics he had helped to establish.[27] In his later proof-theoretic period, Hilbert held on to the set-theoretic "paradise" created by Cantor, Dedekind, Zermelo, and others, now defended in a "formal axiomatic" way.[28] This led to the "formalist" conception of mathematics widely adopted from then on, including with respect to axiomatic set theory. Finally, all of this fits together well with conceiving of "Dedekind abstraction" in the minimalist, pragmatic sense mentioned previously. Hence interpreting Dedekind as a Hilbertian formalist undoubtedly has its attractions, as does the resulting position in itself.[29]

Nevertheless, there are reasons to resist a full identification of Dedekind's position with set-theoretic axiomatics or a related Hilbertian formalism. To begin with, Dedekind does not formulate axioms for his general framework, as we already saw (neither in the traditional sense of "axiom" nor in Hilbert's sense). Instead, he leaves his basic assumption implicit; or he introduces them in informal explanations, for example, when he adopts an extensional criterion of identity for sets. Why is that the case? It is hard to be sure, since Dedekind says little about this aspect. Perhaps he accepts the traditional view that only specific mathematical theories, like geometry, have axioms, while logic does not.[30] In any case, his basic framework of sets and functions is not treated as another mathematical theory by him, as Hilbertian formalism would suggest. Even with

---

27 See Sieg (2014). Compare Stein (1988) for the idea of a "conceptual" approach to mathematics. In Reck (2003, 2008), I explored these Dedekind-Hilbert connections myself.

28 See again Sieg (2014) for Hilbert's later "formal axiomatics" and its debt to Dedekind.

29 I take a Hilbertian formalist reading of Dedekind to be one of the main alternatives to my logicist reading, even though I disagree with it. Wilfried Sieg and his students have argued for it in detail; see Sieg (2010, 2014), Sieg and Schlimm (2005, 2017), Sieg and Morris (2018). Compare also Hallett (2003).

30 See Ferreirós (forthcoming) for the suggestion that Dedekind relies on such an older notion of "logic".

respect to the natural and real numbers, he does not start with axioms but with definitions (of a simple infinity and a continuous ordered field).[31] All of this points more in a logicist than a formalist direction.

The differences between Dedekind's approach and Hilbertian formalism go further. Hilbert was initially attracted to "logicism" himself, first in a Dedekindian form (cf. my quote from his 1899 lectures on geometry previously) and later in a Russell-Whitehead form (after the publication of *Principia Mathematica* in 1910–13).[32] But by the 1920s, Hilbert has given up on the latter so as to adopt formalism instead. In distinguishing his novel approach from Dedekind's, he characterizes the latter as an attempt to "ground the finite in the infinite", a suggestion he finds "dazzling and captivating" but no longer feasible (Hilbert 1922, p. 162).[33] He now rejects starting with an infinitary theory of sets and functions that is simply assumed in the background because of the set-theoretic antinomies. Instead, such an theory has to be "grounded in the finite", that is, formulated axiomatically and backed up by finitary consistency proofs, starting with concrete, intuitive symbols.

For the mature Hilbert of the 1920–30s, there is thus a strong distinction between the "finitary" basis for such consistency proofs, on the one hand, and the "ideal" superstructure in mathematics, on the other hand. Strictly speaking, only the former has "meaning", while the latter is "meaningless" (cf. Hilbert 1925). This goes along with focusing on formulas and proofs, treated as un-interpreted strings of symbols, for all parts of mathematics that go beyond the finitary, including full arithmetic, real analysis, and higher set theory. What we should do is study corresponding axiom systems but always from a formalist point of view in the end. Crucially, the axioms of set theory are not truths any more, especially not logical truths. Like the axioms of other parts of "ideal" mathematics, they are just the formal starting point for mathematical proofs. What matters about them is that they are syntactically consistent and possibly complete.[34] Once more, strictly speaking, only the finitist basis has "meaning" but not the "ideal" superstructure.

---

31 For textual support that Dedekind starts from definitions, see Klev (2011). I do not mean to deny that Dedekind's approach is "axiomatic" in an informal, practice-oriented sense, as spelled out nicely in Schlimm (2017); see also Sieg and Schlimm (2005, 2017). But that itself does not make it "formalist".

32 See Ferreirós (2008) for Hilbert's early logicist leanings, including their debt to Dedekind.

33 In the original German: "So glänzend und bestechend Dedekinds Idee, die endliche Zahl auf das Unendliche zu begründen, erschien, heute ist die Ungangbarkeit dieses Weges . . . außer Zweifel gesetzt" (Hilbert 1922, p. 162).

34 Dedekind is sometimes seen as a precursor of Hilbertian formalism because of remarks about the importance of proving consistency for arithmetic in Dedekind (1890); see Reck (2003) for references. He is not talking about syntactic consistency proofs, however, but semantic ones. In addition, I take my logicist reading to be compatible with seeing

I do not mean to deny that one can take core contributions by Dedekind and extend them in such a formalist direction. This is exactly what Hilbert did, in very fruitful ways, and many logicians and mathematicians have followed suit. But should we read Hilbertian formalism, in the full sense, back into Dedekind, thus taking him to pursue a formalist rather than a logicist project? That is the suggestion I am resisting. Hilbert was exactly right, I think, that Dedekind (like Frege) tried to "ground the finite in the infinite", contrary to formalism. Would it have been acceptable for Dedekind to say that full arithmetic and analysis, as well as his framework of sets and functions, are "meaningless"? That seems very doubtful to me too. Interpreting Dedekind as a formalist also requires explaining away, or downplaying drastically, his remarks about "abstraction" and "free creation". And, most basically, it requires ignoring his explicit endorsements of a logicist goal, as acknowledged by several of his contemporaries, including Hilbert. Hence, there are strong systematic and textual reasons against interpreting Dedekind along Hilbertian lines.

Finally, I take a formalist reading of Dedekind, in line with Hilbert's mature position, to be anachronistic. Unlike such formalism, which became prominent only in the 1920s, logicism was around during the last few decades of the nineteenth century, as acknowledged by Frege, Schröder, Hilbert, and others. The same is true for approaches to "abstraction" that treat it in more substantive ways than what we get along Hilbertian lines. The latter includes, for example, the use of "abstraction principles" (comparable to Scottish neo-logicism) in the Peano School.[35] My suggestion is to situate Dedekind's talk of "abstraction" and "free creation" in that historical context. If combined with taking seriously his expressed goal of showing that arithmetic and analysis are "part of logic", we are led to a logicist reading of him. Dedekind's project represents an informal, pre-Fregean form of logicism, to be sure, one in which "logic" does not yet have explicit axioms or laws. But, like Frege's project, it is meant to provide an "infinitary" foundation for "meaningful" mathematics.

3.2.   *Second Variant: Dedekind Was an Arithmetizer, Not a Logicist*

At the end of the previous section, I started to contextualize Dedekind's logicist position historically. I will do so more in the next section. But before turning to that task, let me respond to two other forms of denying

consistency as a core ingredient in mathematical existence. For the history of the notion of completeness involved here, see Awodey and Reck (2002).

35 See Mancosu (2016) for such "abstractionist" views in the late nineteenth century; earlier also Scholz and Schweitzer (1935), as an anonymous referee pointed out to me.

that Dedekind pursued a logicist project. So far I have concentrated on Dedekind's foundational writings, that is, his essays on the natural and the real numbers. However, it has been suggested (also in Reck 2008) that one should not read these two essays in isolation but as related to his other mathematical contributions, especially those in algebra and number theory. After all, Dedekind was a very productive and influential mathematician (more so than Frege and Russell). In fact, among historians of mathematics, he is known at least as much for several other contributions: his theory of ideals; his introduction of the notions of field, algebraic number, and so on; and his rethinking of Galois theory.[36]

Taking seriously Dedekind's contributions to algebra, number theory, and so on can lead to philosophical insights, too, especially for the "philosophy of mathematical practice". One example relevant for us is the following: An important feature of Dedekind's approach in those areas is that it involves a crucial, wide-ranging, but philosophically neglected form of "arithmetization". Namely, his introduction of ideals as new mathematical objects, together with operations of addition and multiplication on them, allows him to "calculate" with them like with numbers; and this is so not just in algebraic number theory (where Dedekind's ideals replace Kummer's "ideal divisors") but also in algebraic geometry (in his celebrated work with Heinrich Weber).[37] If one puts this together with Dedekind's more familiar contributions to the "arithmetization of analysis" (in his essays on the natural and real numbers) and to the "arithmetization of algebra" (his new approach to Galois theory), it might appear to be the essence of his whole approach. It is, then, only a small step to add that Dedekind is primarily an "arithmetizer", not a logicist.[38]

In my view, recognizing the importance of this side of Dedekind's methodology is indeed significant philosophically. However, why should adopting the technique of "arithmetizing", even in this broad sense, be seen as opposed to Dedekindian logicism? This step requires further justification. I can see two different arguments for taking that additional step. First, one might argue that Dedekind was primarily a mathematician, thus someone who was focused on making mathematical progress, and his "arithmetizing" approach clearly led to such progress. In contrast, his remarks about "logic", "abstraction", "free creation", and so on were

---

36 See Reck (2008) and Ferreirós and Reck (2020) for further details.
37 Compare Emmylou Haffner's recent writings on Dedekind, such as Haffner (2015, 2017), in which this aspect is worked out in insightful detail and emphasized as central to his methodology. Haffner's approach is influenced by Hourya Benis-Sinaceur's writings; compare the next few footnotes.
38 This last step is taken in important work by Hourya Benis-Sinaceur; see Benis-Sinaceur (2008, 2015, 2017). I take the resulting approach, including an "epistemological" conception of Dedekind abstraction, to constitute another main alternative to my logicist interpretation of Dedekind.

not only far and few between, they also remained inert and static. Especially from the perspective of a "philosophy of mathematical practice", attention to the former should thus replace attention to his supposed "logicism"; the latter misrepresents his position. Second, because of its centrality in Dedekind's re-conceptualization of analysis, algebra, number theory, and so on, one might argue that arithmetic, in a broad sense, is what is basic for Dedekind also in a philosophical sense. In other words, arithmetic is "rock bottom" for him, both methodologically and philosophically speaking, not some supposedly deeper "logic".[39]

I am again not convinced by this challenge to interpreting Dedekind as a logicist, and for several reasons. First, his algebraic number theory and related works rely directly on the construction of ideals as infinite sets; similarly for defining "arithmetical" operations on them in set-theoretic terms. But then, his "logic"—his infinitary theory of sets and functions—contributes to mathematical progress in those areas after all, and it does so in a way that is parallel to his treatment of the natural and real numbers. More generally, it seems to me a mistake to consider Dedekind's "logicism" mathematically inert, a mistake perhaps based on too narrow a conception of "logic" (e.g., as just having to do with formal inference). Second, had Dedekind considered arithmetic philosophically basic, one would expect him to say so and perhaps also to justify that claim. Compare here Henri Poincaré's explicit argument that mathematical induction, viewed as a non-logical and intuition-based arithmetic principle, is fundamental. But no such argument can be found in Dedekind. On the contrary, he bases mathematical induction on "logic" (in terms of his notion of "chain"), and he downplays intuition in connection with the natural and the real numbers, as we saw earlier.

Third, and most importantly, there is textual evidence that Dedekind does not take arithmetic to be "rock bottom" and, indeed, is interested in "digging deeper". As he writes in a well-known letter to the teacher Keferstein:

> [The central questions in my 1888 essay are:] What are the mutually independent fundamental properties of the sequence N, that is, those properties that are not derivable from one another but from which all others follow? And how should we *divest these properties of their*

---

39 In Benis-Sinaceur (2008, 2015, 2017), it is a generalized form of "arithmetic" that is presented as crucial, one that results from a structuralist kind of "abstraction". I find this suggestion helpful and intriguing (cf. also the next footnote). However, I don't see that it undermines my logicist reading of Dedekind in the end, assuming that "logicism" is understood in an appropriate way.

specifically arithmetic character so that they are *subsumed under more general notions and under activities of the understanding without which no thinking is possible at all* but with which a foundation is provided for the reliability and completeness of proofs and for the construction of consistent notions and definitions?

(Dedekind 1890, pp. 99–100, my emphasis)

Had Dedekind taken arithmetic to be basic, he could have stopped after the first question in this passage, couldn't he? But he wants to "divest the properties [of numbers] of their specific arithmetic character"; this is done by "subsuming them under more general notions and activities of the understanding", and without the latter, "no thinking is possible at all". This whole passage clearly echoes the remarks in his 1888 essays, quoted previously, in which he characterizes his "logical" notions.[40]

Finally, there is a more basic response to the anti-logicist challenge in this sub-section. If one looks at Dedekind's writings overall, they display a unity, systematicity, and depth that is remarkable and deserves further attention. The same "logical" tools are employed throughout. Take his notions of cut and of ideal as core examples. Both are introduced as infinite sets, with set-theoretic operations on them. Yes, those operations satisfy "arithmetic" laws, but they are defined in "logical" terms. From this perspective, what is most striking about Dedekind, also compared to other logicists, is the way in which he manages to combine pushing mathematics forward with re-thinking its foundations.[41] These two goals are not in conflict for him but two sides of the same coin. Consequently, his search for logical foundations does not take a static, inert, and "merely philosophical" form; it is part of a broad, forward-looking, and dynamic research program. Put briefly, Dedekind's approach is a form of "dynamic logicism", one in which "arithmetizing" in a broad sense is grounded in basic logical notions in such a way that it is mathematically fruitful. This point is significant beyond Dedekind, as it undercuts a common prejudice against "logicism", but it applies especially to his version of it.[42]

---

40 Should Dedekind's "more general notions and activities of the understanding" be understood as part of a "generalized form of arithmetic", as Benis-Sinaceur argues, or as part of "logic", as I suggest? The difference is subtle and I can see arguments on both sides. I will provide more support for my side in the following.

41 On this point, I am indebted to José Ferreirós; see also Ferreirós and Reck (2020).

42 Dedekind's dynamic way of looking at mathematics is apparent already in Dedekind (1854). In characterizing his approach as "dynamic logicism", I am indebted to Pierre Keller; see Reck and Keller forthcoming.

3.3. *Third Variant: Dedekind Was Too Different From Frege to Be a Logicist*

But should the position just described still be labeled "logicism"? This question leads to one more challenge for my interpretation of Dedekind. Its core is to point out how different Dedekind's approach is from Frege's while presenting the latter as the paradigm of logicism.[43] It is certainly true that Dedekind spends much less time than Frege on justifying his "logical" notions philosophically. In particular, while both take the notions of set and function to be fundamental, Frege devotes many pages to exploring their connection to notions from the philosophy of logic and language, such as those of object, concept, truth, sense, and reference (thereby taking the "linguistic turn"). Nothing of that sort can be found in Dedekind. Frege thus offers a "language-based" understanding of these notions, while Dedekind has a more mathematical understanding. Finally, the latter should not be seen as "logicist"—or so the challenge goes now.

The difference at issue is often assumed to be clearest with respect to their conceptions of function. As Frege is usually interpreted, functions have to be defined everywhere; that is, they take their arguments from a universal domain. In contrast, Dedekindian functions have restricted domains and ranges, as is usual in mathematics. These two views are quite different, arguably even incompatible. A first response to this challenge concerns Frege's side. Namely, the usual assumption about his views on functions is not uncontroversial; it has been challenged in the literature (cf. Blanchette 2012). More basically, Frege presents his conception of function also as an outgrowth of mathematical developments. But what about their conceptions of set or class; aren't they radically different? Clearly, for Frege, a class is always defined by, and thus parasitic to, a concept; it is "the extension of a concept". Dedekind's notion, on the other hand, is typically seen as a precursor to the iterative conception of set, where a set is constituted by, and thus parasitic to, its elements. But again, is the latter interpretation forced on us? Dedekind does, indeed, talk about sets as "consisting" of their elements occasionally; yet he also talks about them as determined by "laws" (Dedekind 1963, p. 45, footnote *), which may be more compatible with Frege's perspective. In addition, note that

---

43 In the recent literature, this challenge has been put forward most forcefully and insightfully in Benis-Sinaceur (2015, 2017). In what follows, I will not be able to do justice to all its details but only respond in a general way. Also again, I take Benis-Sinaceur's "epistemological" conception of Dedekind's abstraction to be subtle and intriguing, but it seems compatible with my "logical abstractionist" understanding of it. I plan to explore both aspects further in future work.

a clear distinction between a "logicist" and an "iterative" conceptions of class/set was not available until later; similarly for clearly different notions of function. Hence the danger of being anachronistic looms again here, by attributing "radically different" views to the two thinkers when the difference was far from clear at the time.

But assume it is true that Frege and Dedekind have significantly different conceptions of function and class/set. Assume also that Frege's "language-based" understanding of his basic logical notions is foreign to Dedekind. Does it follow, from those assumptions, that Dedekind is not a logicist? It only follows if Frege's position is seen as definitive for logicism. But should we give it that status? Note here that Frege does not have exactly the same conceptions of set and function as Russell, either. For Russell, functions are identified intensionally, unlike for Frege, and Russell's underlying ramified theory of types is distinctive. So is Russell not a logicist either? Going further, neither Frege nor Russell conceive of "logic" in the same way as Carnap. For Carnap, following Wittgenstein, logic does not have any "factual content"; it simply provides a "linguistic framework" for science. For the mature Carnap, there are also many different "logics" to consider, and the choice between them is pragmatic. Could either Frege or Russell agree with such views? I don't think so. But then, is Carnap not a logicist?

If one takes Frege's position, with its distinctive assumptions about functions, classes, and so on, to be definitive for "logicism", Dedekind is not a logicist, but nor are Russell and Carnap. Similarly, if one takes Carnap's position, with his deflationary view about logic, to be definitive, Dedekind is again not a logicist, but nor are Frege and Russell. Now, any such exclusionary view wouldn't just go against how these thinkers saw the situation themselves, it would also make using the "logicism" label largely pointless. To be able to talk about "logicism" with respect to all of these thinkers—as I think we want—we need a more inclusive conception of it, one that allows for interesting subdivisions. Which such conception should we use? My basic suggestion here is the following: What unites the positions of these "logicists" is the following: For all of them, "logic" encompasses a general framework of sets and functions; given this framework, core mathematical entities are introduced by some form of logical abstraction; the shared goal is to show that mathematics can be reconstructed on that basis, including seemingly non-logical principles like mathematical induction; and this is viewed as eliminating the need to appeal to intuition at the foundational level.[44] I suggest that

---

44 For Frege, Russell, Carnap, and the Scottish neo-logicists, "logic" explicitly covers propositional and quantificational logic as well, including deductive calculi for them (usually in the form of a theory of types, or at least second-order logic). Dedekind does not address these topics. Is he not a logicist for that reason? Note that it is the set/

this is a reasonably precise, historically useful conception of "logicism". My further claim is that it applies to Dedekind too, that is, that his approach is "logicist" in this general sense.

It is true that Dedekind is more of a "working mathematician" than Frege, Russell, or Carnap. As a consequence, he focuses more on applying his logical tools than on clarifying or justifying them philosophically. But this does not mean that he is not interested in philosophical questions. Some relevant remarks are sprinkled through his foundational essays, as we saw—indeed, rather striking remarks. In addition, we should not separate Dedekind's "foundational" writings too much from his more "mathematical" writings, as I argued already. Dedekind has an overall goal that is motivated both mathematically and philosophically, namely to re-think the "pure mathematics" of his time systematically. Given that goal, interest in methodological issues, and thus in mathematical practice, is inseparable from interest in its foundations for him. Should he therefore not be seen as a "logicist"? I don't see why. In fact, I see him as a particularly interesting example. Moreover, doing so can broaden and deepen our understanding of "logicism".

Let me close this section with one more related observation and argument. Often "logicism" is connected, almost definitionally, with a "platonist" view about mathematical objects (and contrasted as such with formalism and intuitionism). The paradigm is Frege's position as traditionally understood, partly also Russell and Whitehead's. This is then turned into another argument against interpreting Dedekind as a "logicist", since he is clearly not a "platonist"[45] and perhaps not interested in ontological questions at all.[46] My response to this final argument is threefold. First, Frege's views about mathematical objects have been interpreted in "thin" ways recently, which calls crude, stereotypical assumptions about his "platonism" into question.[47] Second, one of Dedekind's basic notions

---

function-theoretic part of "logic" that is crucial for a logicist reconstruction of mathematics; hence I do not see this lack as decisive. (It does make his position incomplete, however, as Frege noted.) See Ferreirós (forthcoming) on this point.

45 Because of his remarks about "abstraction" and "free creation", Dedekind is often viewed as holding psychologistic views, as already mentioned. See Reck (2013b, 2018, 2019b) and Yap (2017) for rebuttals. In fact, I take a crudely platonist reading of Frege and a psychologistic reading of Dedekind to be red herrings.

46 In Stein (1988), it is claimed that Dedekind, unlike Frege, is not "obsessed" with the nature of numbers as abstract objects. While I partly agree—nothing like Frege's long polemics against crude formalist and psychologistic views occurs in Dedekind's writings—this claim goes too far, it seems to me. Note, among others, part of the title of Dedekind (1888): What *are* numbers?

47 See Reck (1997, 2005). I take the most systematic development of a relevant "thin" view of mathematical objects, inspired by Frege's approach, to be Linnebo (2018). Note that abstraction principles play a central role in it, like in my interpretation of Dedekind.

in his 1888 essay is that of "object" too, and he goes on to treat sets and numbers as objects in such a way that their properties are determined objectively. This again blurs the line to Frege, at least to some degree. Third and more generally again, neither Russell, with his "no-classes theory of classes", nor Carnap, with his general anti-metaphysical position, subscribes to "platonism" in any straightforward sense. Consequently, treating it as definitional for "logicism" would again rule them out as well.[48]

4.   Dedekind's Logicism Further Contextualized

As I just argued, "logicism" should not be identified with Frege's position, with all its idiosyncracies, nor with that of any other single figure. Instead, we should use it as an umbrella term for a variety of positions held together by some shared commitments. The core of those commitments are, again, to take a theory of sets and functions to be part of "logic", to use a form of abstraction for the introduction of core entities, to reduce mathematics to logic on that basis, and to view this as eliminating the appeal to intuition for foundational purposes. At this point, it should be clear that my notion of "logicism" involves a conception of "logic" that is quite different from what is common today. After Wittgenstein, Tarski, Quine, and others, all "ontological commitments"—including those concerning sets and functions—have been excluded from "logic", which is thus often identified with first-order logic alone. But if such a conception is adopted, "logicism" is ruled out almost by definition. Moreover, this usage of the term was not the one prevalent in the nineteenth and early twentieth centuries. It is thus anachronistic to employ it in discussions of Dedekind (or Frege and Russell), at least if done uncritically.

To further clarify and defend my interpretation of Dedekind's logicism, together with my older conception of logic, let me place them more fully in the context of certain developments in nineteenth-century mathematics.

---

48 My threefold response in this paragraph is clearly broad brush and in need of refinement. I also admit that this last challenge can be developed in more subtle ways. As an example of the latter, an interesting variant of Stein's point about Dedekind's "non-ontological" perspective, in contrast to Frege's, occurs in Benis-Sinaceur (2008, 2015, 2017). According to Benis-Sinaceur, Frege is working with strict identity for mathematical objects in his approach, as appropriate for a "platonist" and "logicist", and Dedekind only with a form of equivalence or "analogy", along "epistemological" and "structuralist" lines. (This puts Frege in the tradition of Aristotle and Dedekind in that of Euclid, as she adds suggestively.) I agree that there is an interesting difference between the two at this level, but much depends on the kind of structuralism attributed to Dedekind. More generally, I don't see again that this rules Dedekind out as a "logicist". Still, I take this to be an intriguing aspect, one to which I have not done justice in the present chapter. I plan to come back to it in a future publication.

Logicism will reveal itself then as the natural continuation, and in some respects as the culmination, of them. Those developments include the radical broadening of geometry, in terms of various non-Euclidean geometries; the "rigorization" and "arithmetization" of analysis by Cauchy, Bolzano, Weierstrass, and others; the transformation of algebra, from contributions by British algebraists, like Peacock and Hamilton, to Galois' approach to the solvability of equations; the increasing use of "ideal" elements in various parts of mathematics, from complex numbers and Kummer's "ideal divisors" in number theory to "points at infinity" and points with complex coordinates in geometry; and, finally, the beginnings of a rethinking of "logic" by Boole, DeMorgan, Peirce, and so on. Overall, there was a revolutionary transformation of mathematics in this period, one that called for foundational investigations.[49]

Two mathematicians very close to Dedekind were involved in the rethinking of geometry at the time: Gauss, his dissertation advisor, and Riemann, his colleague and friend. While Gauss did not publish much on the subject, his push beyond Euclidean geometry led him to some well-known philosophical conclusions. In particular, he came to see arithmetic as grounded in the "laws of thought" alone, while he took geometry, understood as the theory of space, to be empirical. In Gauss, Riemann, and others, one can also find an increasing reluctance to take the notion of magnitude as basic, including seeing analysis as grounded in geometric magnitudes (a kind of "geometricism" increasingly called into question). At the same time, confidence grew that analysis could be reduced to the theory of the natural numbers plus "laws of thought". It became clear, for example, how to think about complex numbers as pairs of reals; likewise for parallel treatments of rational and negative numbers. The two central issues that remained at that point were to clarify the notions of real number and natural number, exactly the tasks Dedekind set himself in his foundational booklets from 1872 and 1888.

With respect to algebra, algebraic number theory, and related fields, some other developments influenced Dedekind, too. His teachers in Göttingen, in this case especially Dirichlet and Riemann, inspired him to rethink these areas of mathematics by proceeding in a new "conceptual" way, that is, by distilling out the basic concepts needed (field, group, algebraic number, etc.) and proving mathematical theorems solely based on those. (This pointed towards Peano's and Hilbert's later "axiomatic" approach, although it was not called that yet.) There were also attempts to clarify the use of "ideal" elements by constructing them out of more basic mathematical objects; and this usually involved using sets or classes, including by Dedekind (a form of "local" or "relative logicism";

---

49 For further details, see Reck (2013a), on which I draw heavily in this section.

cf. Wilson 1992). Beyond that, Cantor used sets fruitfully for other purposes as well, which led to the beginnings of transfinite set theory, as Dedekind was well aware. And in the Boolean school, among others, classes began to be treated as part of "logic". In other words, the "laws of thought", a term also used by Boole, explicitly covered results about classes.

If we now go back to Dedekind's main contributions—his treatment of the real numbers and the natural numbers, on the one hand, and his rethinking of algebra, algebraic number theory, and algebraic geometry, on the other—the following becomes evident: Not only did he develop a broad form of "arithmetization"; not only did he use sets or classes locally to clarify particular notions; he also actively pushed mathematics in a systematic foundational direction. That is to say, he aimed—quite self-consciously and systematically, even if this was only made explicit in a few philosophical asides in his publications—to "divest [various crucial notions] of their specifically arithmetic character so that they are subsumed under more general notions and under activities of the understanding without which no thinking is possible at all" (Dedekind 1963, pp. 99–100). This amounts to clarifying and making more substantive appeals to the "laws of thought", from Gauss to Boole. What we have arrived at, now in fuller context, is Dedekind's "logicism".

If we take arithmetic, analysis, and algebra together to constitute all the pure mathematics at the end of the nineteenth century, the ambitions driving Dedekind's logicism are evident. Again, his goal was to systematically rethink all the pure mathematics of his time, as well as to push it further. He proceeded through various forms of "arithmetization" all the way to the "laws of thought", and the latter included general assumptions about sets and functions, meant to replace traditional appeals to geometric magnitudes and to intuition. It is in this sense that Dedekind's logicism is the culmination and fruitful extension of various developments in nineteenth-century mathematics. Admittedly, he did not make his underlying "laws of thought" explicit enough. Frege is right in his criticism of that aspect. Hence Dedekind's approach remained an informal and incomplete version of logicism. Frege attempted to fill that gap by means of his own theory of functions and classes (or value ranges); and that attempt revealed, as Russell and others showed, that one has to be more careful than either Frege or Dedekind were aware initially.

When Dedekind found out about the set-theoretic antinomies, in the late 1890s, he did not attempt a solution himself, at least not in print. At the time, he was beyond his prime and already retired. We know that he was interested in Ernst Schröder's theory of classes in the 1890s, developed along Boolean lines. He also found out about Frege's work during that period. It would be fascinating to learn about his reaction to Zermelo's axiomatization of set theory from after the turn of the century, but as far as I know, there is no record of it. What we get as the next step in the foundational dynamic, in the 1910s–20s, is the three-part division we know from Carnap's survey: Russellian logicism, as an attempt to revive Frege's approach; Hilbertian formalism, as a novel response to the antinomies; and Brouwerian intuitionism, as a more radical alternative. At that point, Frege's and Dedekind's logicist approaches were put aside, since they were seen as inconsistent. Or, rather, Dedekind's approach was assimilated into Zermelo-Fraenkel set theory and Hilbertian formalism, while Frege's was revived in neo-logicism. This has made it hard to see them as two variants of logicisms originating in a shared context.

5. Conclusion: Dedekind and the Significance of Logicism

In this chapter, I reconsidered Dedekind's logicism, starting with his stated goal to show that arithmetic, understood in a broad sense, is "part of logic". Then I turned to four challenges for interpreting Dedekind as a logicist: that his project, because of some flaws, is not even potentially successful as a form of logicism; that he should be seen as an early set-theorist and a Hilbertian formalist instead; that he should at bottom be read as a mathematician "arithmetizing" large parts of mathematics; and that his position is too different from Frege's to count as logicism. In response, I acknowledged important gaps in Dedekind's approach; I granted that axiomatic set theory and Hilbert-style formalism are fruitful extensions of various Dedekindian ideas; I agreed that his approach involves a strikingly broad form of arithmetization; I also agreed that it does not coincide with Frege's in noteworthy ways. Yet none of these points rule out conceiving of Dedekind as a logicist, as I went on to argue.

As I understand Dedekind's project, it is a distinctive form of logicism, one of a variety of such forms and one worth reviving. It is incomplete, but in ways one can try to amend; for example, parts of it are plausibly reconstructed in terms of structural abstraction principles, which opens up interesting connections to neo-logicism. My historical contextualization of Dedekind's approach also revealed that it constitutes a dynamic form of logicism, one that is the natural continuation and extension of important developments in nineteenth-century mathematics, as opposed to a static and merely philosophical position. His concern for how to advance mathematics and his pursuit of philosophical goals are systematically intertwined in it. Finally, recognizing both ambitions as inherent in his project helps us see the full significance of logicism beyond Dedekind, including its origins in and often-neglected relationship to mathematical practice.[50]

50 An early version of this chapter was presented at the Workshop *Origins and Varieties of Logicism*, University of Pavia, Italy, March 2015. I would like to thank the organizers, Andrea Sereni and Francesca Boccuni, for inviting me to that workshop and later to contribute an chapter to this volume. I am also grateful to them and to an anonymous referee for helpful comments on a draft.

References

Awodey, S. (2018). "Univalence as a Principle of Logic". *Indagationes Mathematicae* 29: 1497–1510; https://doi.org/10.1016/j.indag.2018.01.011.

Awodey, S. & Reck, E. (2002). "Categoricity and Completeness, Part I: Nineteenth-Century Axiomatics to Twentieth-Century Metalogic". *History and Philosophy of Logic* 23: 1–30.

Benis-Sinaceur, H. (2008). *Richard Dedekind, la creation des nombres*, Paris: Vrin.

———— (2015). "Is Dedekind a Logicist? Why Does Such a Question Arise?". In H. Benis-Sinaceur et al., eds., *Functions and Generality of Logic: Reflections on Dedekind's and Frege's Logicism*, New York: Springer, pp. 1–50.

———— (2017)."Dedekind's and Frege's Views of Logic". In *In Memoriam Richard Dedekind (1831–1916)*, K. Scheel et al., eds., Münster: WTM Verlag, pp. 50–62.

Blanchette, P. (2012). *Frege's Conception of Logic*, Oxford: Oxford University Press.

Boolos, G. (1999). *Logic, Logic, and Logic*, Cambridge, MA: Harvard University Press.

Carnap, R. (1931). "The Logicist Foundations of Mathematics"; reprinted in *Philosophy of Mathematics*, 2nd ed., P. Benacerraf & H. Putnam, eds., Cambridge: Cambridge University Press, 1983, pp. 41–50.

Dedekind, R. (1854). "Über die Einführung neuer Funktionen in der Mathematik"; reprinted in Dedekind (1930–32), Vol. 3, pp. 428–438.

———— (1872). *Stetigkeit und irrationale Zahlen*; reprinted in Dedekind (1930–32), Vol. 3, pp. 315–334; English trans., "Continuity and Irrational Numbers". In Dedekind (1963), pp. 1–22.

———— (1888). *Was sind und was sollen die Zahlen?*; reprinted in Dedekind (1930–32), Vol. 3, pp. 335–391; English trans., "The Nature and Meaning of Numbers". In Dedekind (1963), pp. 29–115.

———— (1890). Letter to Keferstein, reprinted in *From Frege to Gödel*, J. v. Heijenoort, ed., Cambridge, MA: Harvard University Press, pp. 98–103.

———— (1930–32). *Gesammelte Mathematische Werke*, Vols. 1–3, R. Fricke et al., eds. Braunschweig: Vieweg.

———— (1963). *Essays on the Theory of Numbers*, W.W. Beman, trans. and ed., New York: Dover; originally published by Open Court: Chicago, 1901.

Demopoulos, W. & Clark, P. (2007). "The Logicism of Frege, Dedekind, and Russell". In *The Oxford Handbook of the Philosophy of Mathematic and Logic*, S. Shapiro, ed., Oxford: Oxford University Press, pp. 166–202.

Dummett, M. (1991). *Frege: Philosophy of Mathematics*, Cambridge, MA: Harvard University Press.

Ebert, P. & Rossberg, M., eds. (2016). *Abstractionism: Essays in the Philosophy of Mathematics*, Oxford: Oxford University Press.

Ferreirós, J. (1999). *Labyrinth of Thought: A History of Set Theory and Its Role in Modern Mathematics*, Boston: Birkhäuser; 2nd rev. ed., 2007.

———— (2007a). "The Early Development of Set Theory". In *The Stanford Encyclopedia of Philosophy*, E.N. Zalta, ed., Retrieved from https://plato.stanford.edu/entries/settheory-early/; updated version online in 2016.

———— (2007b). "The Rise of Pure Mathematics, as Arithmetic in Gauss". In *The Shaping of Arithmetic After C.F. Gauss's Disquisitiones Arithmeticae*, C. Goldstein et al., eds., Springer: Berlin, pp. 235–268.

———— (2008). "Hilbert, Logicism, and Mathematical Existence". *Synthese* 170: 33–70.

———— (forthcoming). "On Dedekind's Logicism". In *Analytic Philosophy and the Foundations of Mathematics*, A. Arana & C. Alvarez, eds., London: Palgrave.

Ferreirós, J. & Reck, E. (2020). "Dedekind's Mathematical Structuralism: From Galois Theory to Numbers, Sets, and Functions". In *The Pre-History of Mathematical Structuralism*, E. Reck & G. Schiemer, eds., Oxford: Oxford University Press, pp. 59–87.

Frege, G. (1893/1903). *Grundgesetze der Arithmetik*, Vols. 1–2; English trans., *Basic Laws of Arithmetic*, P. Ebert & M. Rossberg, eds. and trs., Oxford: Oxford University Press, 2013.

Gabriel, G. (1980). "Logizismus". In *Historisches Wörterbuch der Philosophie*, Vol. 5, J. Ritter, K. Gründer, et al., eds., Zürich: Scheidegger & Spies; https://doi.org/10.24894/HWPh.2289.

Haffner, E. (2015). "The 'Science of Numbers' in Action in Richard Dedekind's Works: Between Mathematical Explorations and Foundational Investigations". HAL archives: https://hal.archives-ouvertes.fr/tel-01144626.

———— (2017). "Strategic Use(s) of Arithmetic in Richard Dedekind and Heinrich Weber's *Theorie der algebraischen Funktionen einer Veränderlichen*". *Historia Mathematica* 44: 31–69.

Hallett, M. (2003). "Foundations of Mathematics". In *The Cambridge History of Philosophy, 1870–1945*, T. Baldwin, ed., Cambridge: Cambridge University Press, pp. 128–154.

Hellman, G. & Shapiro, S. (2019). *Mathematical Structuralism (Cambridge Elements: The Philosophy of Mathematics)*, Cambridge: Cambridge University Press.

Hilbert, D. (1922). "Neubegründung der Mathematik. Erste Mitteilung"; reprinted in *Gesammelte Abhandlungen, Dritter Band*, New York: Chelsea Publishing Company, 1965, pp. 157–177.

———— (1925). "Über das Unendliche". *Mathematische Annalen* 95: 161–190; English trans., "On the Infinite". In *Philosophy of Mathematics*, P. Benacerraf & H. Putnam, eds., Cambridge: Cambridge University Press, 1983, pp. 183–201.

Klev, A. (2011). "Dedekind and Hilbert on the Foundations of the Exact Sciences". *The Review of Symbolic Logic* 4: 645–681.

———— (2017): "Dedekind's Logicism". *Philosophia Mathematica* 25: 341–368.

———— (2018): "A Road Map to Dedekind's Theorem 66". *HOPOS: The Journal of the International Society for the History of Philosophy of Science* 8: 241–277.

Linnebo, Ø. (2018). *Thin Objects: An Abstractionist Account*, Oxford: Oxford University Press.

Linnebo, Ø. & Pettigrew, R. (2014). "Two Types of Abstraction for Structuralist". *The Philosophical Quarterly* 64: 267–283.

Mancosu, P. (2016). *Abstraction and Infinity*, Oxford: Oxford University Press.

McCarty, D. (1995). "The Mysteries of Richard Dedekind". in *Essays on the Development of the Foundations of Mathematics*, J. Hintikka, ed., Dordrecht: Kluwer, pp. 267–283.

Reck, E. (1997). "Frege's Influence on Wittgenstein: Reversing Metaphysics via the Context Principle". In *Early Analytic Philosophy*, W.W. Tait, ed., Chicago: Open Court, pp. 123–185.

—— (2003). "Dedekind's Structuralism: An Interpretation and Partial Defense". *Synthese* 137: 369–419.

—— (2005). "Frege on Truth, Judgment, and Objectivity". *Grazer Philosophische Studien* 75, 2007, Special Issue: *Essays on Frege's Conception of Truth*, D. Greimann, guest editor, pp. 149–173.

—— (2008). "Dedekind's Contributions to the Foundations fo Mathematics". In *The Stanford Encyclopedia of Philosophy*, E.N. Zalta, ed., Retrieved from https://plato.stanford.edu/entries/dedekind-foundations/; updated version online in 2016.

—— (2013a). "Frege, Dedekind, and the Origins of Logicism". *History and Philosophy of Logic* 34: 242–265.

—— (2013b). "Frege or Dedekind? Towards a Reevaluation of their Legacies". In *The Historical Turn in Analytic Philosophy*, E. Reck, ed., London: Palgrave, pp. 139–170.

—— (2017). "Dedekind as a Philosopher of Mathematics". In *In Memoriam Richard Dedekind (1831–1916)*, K. Scheel et al., eds., Münster: WTM Verlag, pp. 36–49.

—— (2018). "On Reconstructing Dedekind Abstraction Logically". In *Logic, Philosophy of Mathematics, and their History: Essays in Honor of W.W. Tait*, E. Reck, ed., London: College Publications, pp. 113–138.

—— (2019a). "The Logic in Dedekind's Logicism". In *Logic from Kant to Russell: Laying the Foundations for Analytic Philosophy*, S. Lapointe, ed., London: Routledge, pp. 171–188.

—— (2019b). "Frege's Relation to Dedekind: *Basic Laws* and Beyond". In *Essays on Frege's Basic Laws of Arithmetic*, P. Ebert & M. Rossberg, eds., Oxford: Oxford University Press, pp. 264–284.

Reck, E. & Keller, P. (forthcoming). "From Dedekind to Cassirer: Logicism and the Kantian Heritage". In *Kant's Philosophy of Mathematics, Vol. II*, C. Posy & O. Rechter, eds., Cambridge: Cambridge University Press.

Russell, B. (1904). "The Axiom of Infinity". *Hibbert Journal* 2: 809–812; reprinted in *Essays in Analysis*, D. Lackey, ed., London: Allen & Unwin, 1973, pp. 256–259.

Schlimm, D. (2017). "On Dedekind's Axiomatic Approach to the Foundations of Mathematics". In *In Memoriam Richard Dedekind (1831–1916)*, K. Scheel et al., eds., Münster: WTM Verlag, pp. 75–82.

Scholz, H. & Schweitzer, H. (1935). *Die Sogenannten Definitionen durch Abstraktion: Eine Theorie der Definitionen durch Bildung von Gleichheitsverwandtschaften*, Leipzig: Meiner Verlag.

Sieg, W. (2010). "Searching for Proofs (and Uncovering Capacities of the Mathematical Mind)". In *Proofs, Categories and Computations: Essays in Honor of Grigori Mints*, S. Feferman & W. Sieg, eds., London: College Publications, pp. 189–215.

—— (2014). "The Ways of Hilbert's Axiomatics: Structural and Formal". *Perspectives on Science* 22: 133–157.

Sieg, W. & Morris, R. (2018). "Dedekind's Structuralism: Creating Concepts and Deriving Theorems". In *Logic, Philosophy of Mathematics, and their History:*

Essays in Honor of W.W. Tait, E. Reck, ed., London: College Publications, pp. 251–301.

Sieg, W. & Schlimm, D. (2005). "Dedekind's Analysis of Number: System and Axioms". *Synthese* 147: 121–170.

———(2017). "Dedekind's Abstract Concepts: Models and Mappings". *Philosophia Mathematica* 25: 292–317.

Stein, H. (1988). "*Logos*, Logic, and *Logistiké*: Some Philosophical Remarks on Nineteenth-Century Transformations of Mathematics". In *History and Philosophy of Modern Mathematics*, W. Aspray & P. Kitcher, eds., Minneapolis: University of Minnesota Press, pp. 238–259.

———(1998). "Logicism". In *Routledge Encyclopedia of Philosophy*, Volume 5, London: Routledge, pp. 811–817.

Tait, W.W. (1996). "Frege Versus Cantor and Dedekind: On the Concept of Number". In *Frege: Importance and Legacy*, M. Schirn, ed., Berlin: DeGruyter, pp. 70–113.

Wilson, M. (1992). "Frege: The Royal Road From Geometry". *Noûs* 26: 149–180.

Yap, A. (2017). "Dedekind and Cassirer on Mathematical Concept Formation". *Philosophia Mathematica* 25: 369–389.

6 Logical Form and the Development of Russell's Logicism

Kevin C. Klement

1. Introduction

While there are no doubt more nuanced varieties worth considering, one straightforward version of logicism is the thesis that mathematical truths simply *are* logical truths. Today, a typical characterization of a logical truth is one that remains true under all (re)interpretations of its non-logical vocabulary. Roughly, this means that something can be a logical truth only if all other statements of the same form are also true. "$Fa \supset (Rab \supset Fa)$" can be a logical truth because not only it but *all* propositions of the form "$p \supset (q \supset p)$" are true. It does not matter what "F", "R", "a", and "b" mean or what specific features the objects meant have. Applying this conception of a logical truth in the context of our crude form of logicism seems to present an obstacle. "Five is prime", at least on the surface, is a simple subject-predicate assertion, and obviously not all subject-predicate assertions are true. How, then, could this be a logical truth? Similarly, "$7 > 5$" asserts a binary relation, but obviously not all binary relations hold. In what follows, I shall call this the *logical form problem* for logicism.

A proponent of logicism might respond in many ways. On the more radical side, one might reject the previous characterization of a logical truth and propose a different one. Alternatively, one might accept it as a general characterization of logical truths in a strict sense but instead argue that mathematical truths are "logical" in a different, extended, or amended sense. Perhaps mathematical truths are only *analytic* and share the universality and privileged epistemological status of logical truths without, strictly speaking, being such. However, I wish to focus my concern in what follows on what happens if one accepts the usual assumption that formal generality is at least a necessary condition for being a logical truth and also holds to a strict form of logicism that insists that mathematical truths are logical truths in this very sense. There seem to remain two options worth considering. The first, which I'll call *option A*, would admit that "five is prime" is of subject-predicate form but insist that it also has a more specific form that is universally true. After all, "$Fa \supset (Rab \supset Fa)$" is an instance not just of the form "$p \supset (q \supset p)$" but

DOI: 10.4324/9780429277894-7

also of the more generic form "$p \supset q$", not every instance of which is true. Taking this option would mean holding that such classifications as subject-predicate, binary-relational, and so on are too coarse-grained to capture fully the precise logical form of some of the truths that fall under such broad classifications. The second, *option B*, instead denies that truths such as "five is prime" and "7 > 5" really have the logical forms their surface syntax seems to suggest. While no true subject-predicate statement is logically necessary, there are mathematical truths that are worded in ordinary language in a way that seems subject-predicate, but these truths, when properly understood or analyzed, have different forms and, indeed, fully general and logically necessary ones.

It might be thought that I am here omitting a third option, which would be to consider "five", "seven", "prime", ">", and so on themselves to be logical constants. But depending on the details, I believe this suggestion collapses to one or the other of option A and option B. If "prime" is considered, syntactically, a predicate, and "five" a normal subject, then we have a version of A. By taking these words to be logical constants, one is insisting that "five is prime" is a *maximally specific* sub-form of the generic subject-predicate form. If one instead takes "five" and "prime" to be logical constants but *not* to fall under the usual syntactic categories of subject and predicate, one is in effect adopting option B, as this truth is not taken to have the logical form it seems to have.

In either approach, it should be noted that it is not enough simply to hold that statements about numbers, say, have a different logical form from statements about other things. "7 > 5" is a mathematical truth, but "4 > 5" is not, and so, for a logicist pursuing one of these options, the former must have a different (specific) logical form than the latter has, despite the apparent identity in their surface forms and despite the only difference between them being which number is involved. The logical form of a statement about a number or other mathematical entity must not just be affected by the *kind* of (abstract) entity apparently referenced but by the particularities of the specific member of the kind.

I believe the distinction between options A and B can shed light on the development of Bertrand Russell's logicism, particularly during the period of transition between his two major logicist works, *The Principles of Mathematics* from 1903 and *Principia Mathematica*, whose first volume was published in 1910. I do not mean to suggest that Russell himself explicitly considered the puzzle in such a simplistic form. If nothing else, the characterization of logical truths as those that remain true under all reinterpretations of their non-logical vocabulary is one that became standard only later,[1] perhaps in part because of the work of Russell and those

---

1 There were, of course, earlier anticipations of this approach, for example, Bolzano (1837), but there is no indication Russell was aware of them.

influenced by him, such as Wittgenstein. Nonetheless, I think Russell at least implicitly struggled with the choice between these two options and that it is useful to think of the development of his views as a migration away from option A towards option B. This migration was driven largely by his responses to the paradoxes hampering his logicist work, which forced a reorientation on how to think about "abstract objects" generally. Russell is explicit in many places that the development of his views in metaphysics, his endorsement of "logical constructions over inferred entities", and his taking apparent names of entities having "smooth logical properties" as "incomplete symbols" rather than genuine names, all of which are central to my account, were elements of his philosophy that developed while working through the problems facing his logicist philosophy of mathematics in the first decade of the 20th century (e.g., *PoM*,[2] 2nd ed., x–xi; *PLA*, 160, 234–235; *LA*, 161–169; *MPD*, 83–85). While Russell may not have explicitly formulated what I have called the logical form problem for logicism, there is evidence that his philosophy can be seen as having evolved to solve it.

I also think that framing Russell's views in this way helps to explain what is unique about his form of logicism and how it stacks up in comparison to other thinkers both within and outside of the logicist program. Along the way, I also hope to say a bit about Russell's attitude about abstraction principles and "defining" by abstraction, as these have become so important in contemporary discussions of forms of logicism and were already considered important in Russell's day.

2.   The Birth of Russell's Logicism and Peano's Logic

It would be a natural assumption to think that Russell's logicism was initially inspired by late 19th-century logicists such as Dedekind and Frege, but it would be a mistaken one. Although Russell knew Dedekind's technical work, he does not seem to have been much influenced by its underlying philosophy. While Russell was eventually influenced by Frege, he only read Frege carefully when *The Principles of Mathematics* was near completion, after Russell had already become convinced of logicism. Russell arrived at the position in a somewhat different fashion.

Russell's intellectual interests were always extraordinarily broad. His education at Cambridge was focused equally on mathematics and on "moral science" (philosophy), finishing his Tripos in the former subject in 1893 and in the latter in 1894. As early as 1895, he had a plan to produce two series of books, one on the philosophies of the various sciences and one on social questions, which he hoped would eventually "meet in a

---

2 Abbreviations are typically used for references to Russell's own works; the abbreviations are given after their titles in the References section.

synthesis" (*Auto* I: 185). The first book intended to be a part of the former series was his 1897 *Essay on the Foundations of Geometry*. Although it is often critical of Kant, this book is still recognizably situated within the Kantian tradition on geometry and is dedicated largely to making room for non-Euclidean projective geometries within that approach. The influence of the neo-Hegelian British Idealist tradition that Russell was trained in is also very much in evidence. That tradition holds that one cannot without falsification separate any intellectual subject matter from another and treat it in isolation; attempts to do so result in antinomies or contradictions. Russell's work during this period largely deals with such antinomies. On the one hand, geometrical objects such as continuous quantities and points are thought of as intrinsically identical and therefore identifiable only by their relationships to other objects of similar kinds. On the other hand, it was held by those in the neo-Hegelian tradition that all relations to other things must be grounded in the intrinsic natures of the relata—this is this so-called "doctrine of internal relations". Early Russell was prone to accepting such contradictions as unavoidable, at least prior to the unification of the logic of all sciences.

Russell's thinking on these matters changed in 1898 due to a number of factors. Discussions with G. E. Moore made him more sympathetic to a more realist philosophy rejecting the doctrine of internal relations. Whitehead, Russell's mathematical mentor, published his *Universal Algebra*, which brought to the fore aspects of mathematics that were hard to accommodate within the mostly quantity-focused conception of mathematics found in the Kantian tradition, including the algebra of logic itself. Indeed, Whitehead there defines mathematics as "the development of all types of formal, necessary, deductive reasoning" (Whitehead 1898, vi), which could itself be read as an endorsement of a form of logicism. The influence reveals itself in his changing the title of the work on arithmetic Russell was planning from "On Quantity and Allied Conceptions" to "An Analysis of Mathematical Reasoning" (*Papers* 2, 155–242). The title changed again, to "The Fundamental Ideas and Axioms of Mathematics" (*Papers* 2, 261–305), and finally to *The Principles of Mathematics*. However, even the first draft of the *Principles* (1899–1900; *Papers* 3, 9–180) did not endorse logicism. Russell had been convinced that not only mathematics generally, but even the algebra of logic in particular, made use of not-specifically-logical notions. Most significantly, Russell then understood Boolean class-logic as involving whole/part relationships. This likely changed only when Russell was introduced to Peano's work at the International Congress of Philosophy in Paris in August 1900, which he later described as one of the most important events in his intellectual development (*MMD* 12). Unlike earlier thinkers in the Boolean tradition, Peano distinguished between the logical form of claims that an individual is a member of a class and the logical form of claims that one class is a subset of another. Peano understood the latter relationship as involving an implication for all values of a variable. Russell quickly mastered Peano's logical techniques in the final months of 1900. Around the same time, he was putting the final touches on his book on Leibniz and in that process had diagnosed many of the problems as he saw them with Leibniz's philosophy as involving overly strong assumptions about the reducibility of relational propositions to subject-predicate form.

Between dropping the assumption that logical relationships are to be analyzed as part/whole relationships and concluding that relations in general need not be understood as reducible to other forms, Russell came to the view that the most important concept for mathematics is that of a variable and that formal implications, or quantified conditionals, were of crucial importance for mathematics. Indeed, by the time he finished *Principles*, he had come to the conclusion that all truths of mathematics could be seen as formal implications and that the constants used in such implications were all logical. It is at this point that Russell adopted logicism.

It is worth considering in more detail what it was about Peano's logic that made such a difference for Russell. For an example, let us consider his views on (cardinal) numbers. In the 1899 (pre-Peano) draft of *Principles*, Russell held that numbers formed a system of concepts related to each other. While they *applied* to collections, they were not the same as the collections they applied to, and arithmetical addition could not be reduced to the adding of entities to form collections. An entity "added" to itself in the latter sense (A and A) did not make up two things, whereas for arithmetical addition, it was true that $1 + 1 = 2$. Russell seems to conclude from this alone that numbers could not be defined (*Papers* 3, 15–16). It is not exactly clear how one would fill in the argument to this conclusion, but looking back, one cannot help thinking that Russell's pre-Peano logic did not allow for a coherent conception of what a *defined* concept applicable to collections would look like. It would seem one would be defining a collection of collections. Traditionally understood, the part/whole relationship, however, is transitive, so it would seem that the defining features of the collections in the collection of collections would also have to apply to the elements in the individual collections, as they, too, are "parts" of it. Exactly how this makes Russell's remarks about the incompatible notions of "addition" decisive remains somewhat obscure, but it is not difficult to see why an alternative approach was attractive to him once it became available.

Peano's distinction between a class being a member of another and being a subset of it removed most of these difficulties. Peano used an epsilon (ε) for the membership relation and a horseshoe ($\supset$)[3] for the subset relation.

---

3 Yes, $\supset$ was first used for the subset relation, not the superset relation. It was Russell himself who reversed this to the now-standard $\subset$.

Since Peano read "ε" as "is" (shorthand for the Latin/French "est" or Greek ἐστί; see Peano 1895–1908, II: 6), Russell interpreted Peano as understanding this relation as both that between a member and a class and between an object and a predicate/concept, accusing Peano as having "quite consciously" identified classes with their defining class-concepts or predicates (*PoM*, §69). This reading of Peano is probably dubious, but what is important is that since Russell himself did not go along with the identification, he thought there were two possible uses of ε: as a membership sign and also basically as a copula, or relation between an object and a concept applicable to it. Moreover, Peano introduced a symbol, $\ni$, pronounced "such that", which could be used to define new notions in terms of existing ones by binding a variable x (Peano 1895–1908, II: 7). In particular, "$x \ni \ldots x \ldots$" means the class of all xs such that $\ldots x \ldots$ is true of them, roughly akin to the modern notation $\{x \mid \ldots x \ldots\}$. Peano at least implicitly accepted the naïve class theory schema:

$$y \, \varepsilon \, (x \ni \ldots x \ldots) = \ldots y \ldots$$

That is, y is a member of the class of all xs such that $\ldots x \ldots$ iff $\ldots y \ldots$. In Russell's interpretation of Peano, however, $\ni$ could also be read as a device for defining complex predicates and the previous as asserting that y falls under the concept of being an x such that $\ldots x \ldots$ iff $\ldots y \ldots$. Either way, it provides a mechanism for forming complex mathematical definitions in terms of more basic logical vocabulary, a mechanism that did not suffer from the difficulties brought on by the simplistic logic Russell had previously been working in.

In *Principles*, Russell holds that numbers can be defined either as properties of classes or as classes of classes, settling on the latter as most consistent with mathematical practice and most convenient for a symbolic treatment of the subject. Along with this, he chose to make use of ε as the membership relation in his own symbolic work most of the time. It is fairly clear how one would go about making use of Peano's $\ni$ to give an explicit definition of the class of all empty classes (Russell's zero) or the class of all unit classes (Russell's one), and so on. Moreover, it is clear how one would define the class of all those classes which can be obtained by adding one member to some class in another class of classes, thereby allowing a definition of the "successor" relation in a way that does not conflate the addition of individuals with mathematical addition.

By giving explicit definitions of the cardinal numbers, identifying them with classes of classes alike in cardinality, Russell was deviating from Peano's own views. Peano had explicitly considered and rejected defining the number of a class as the class of classes alike in cardinality (Peano 1895–1908, III: 70), arguing that the class of all unit classes had properties specific to it considered as a class that the number one seemed to lack. Peano instead introduced cardinal numbers by an abstraction principle, roughly identical to the one now commonly called Hume's Principle, taking numbers to be objects to which all and only classes that can be put in 1–1 correspondence had a unique relation. He and his school in general defended definitions by abstraction in mathematical practice. As we now know from contemporary proponents of "abstractionism" (e.g., Hale and Wright 2001), one can get quite far using this method in capturing the objects and results of arithmetic. And, of course, this difference between Russell and Peano was not limited to cardinal numbers but also ordinals and other mathematical entities that might be similarly introduced or defined. Russell defined what he called "the principle of abstraction" thus:

> Every transitive symmetric [equivalence] relation, of which there is at least one instance, is analyzable into joint possession of a new relation to a new term, the new relation being such that no term can have this relation to more than one term, but that its converse does not have this property.
>
> (*PoM*, §220)

Russell thought that this principle did not need to be justified by a new "creative" form of mathematical definition but instead could be *proven* simply by taking the relation and new term in question to be membership and the equivalence class formed by the equivalence relation. He writes:

> Wherever Mathematics derives a common property from a reflexive, symmetric and transitive relation, all mathematical purposes of the supposed common property are completely served when it is replaced by the class of terms having the given relation to a given term.
>
> (*PoM*, §111)

Along with this, Russell rejected Peano's supposition that abstraction principles gave us access to a common predicate shared by relata of the equivalence relation, noting that it is unclear that there is a unique such predicate. In the case of cardinal numbers, for example, *being similar (equinumerous) to a* and *being similar to b* do not seem to be the same property or concept, even if *a* and *b* are similar, so neither would seem appropriate to be *the* unique number they share. No doubt Russell thought this was another instance in which Peano conflated the class concept or predicate with the class itself; only the classes themselves have extensional identity conditions.

Let us return to the logical form problem and our options A and B. There is no question that the adoption of Peano's symbolic logic was pivotal for Russell's philosophical development, and those familiar with Russell's later philosophy are no doubt aware that he took (at least his own) symbolic logic to present a better picture of the real "logical form" of the world than ordinary language does. This attitude, of course, is necessary for something like option B, in which even the language of ordinary mathematics is considered misleading about its "true" logical form. But, at least at first, Russell continued to think that surface grammar, even in ordinary language, is generally a reliable guide to logical form (as made explicit in *PoM*, §46). And in his logicist work, his solution is probably best considered a version of option A. Note that Peano's ϶ allows one to form syntactically complex terms and, under Russell's other interpretation of it, syntactically complex predicates. Taking this syntactic complexity to be indicative of more complex logical forms, this allows for forms which remain subject-predicate but are not *merely* subject-predicate. This may even make room for logically general or necessary subforms. Consider, for example:

$$a \, \varepsilon \, (x \ni x = a)$$

In one interpretation, this asserts that *a* is in the class of things identical with *a*; on the other, it claims that *a* has the quality of being identical with *a*. The only non-logical constant here is "*a*", but its interpretation seems not to matter: all instances of this specific form seem to be true, even though this form is an instance of the more generic form, $a \, \varepsilon \, b$, which does have false instances. For early Russell, of course, logical forms attached first and foremost to propositions, understood as mind- and language-independent complexes, and only derivatively to sentences. Syntactically simple expressions could be used as shorthands for complex notions. If we understand "five", "seven", ">", and "prime" as such shorthands, one could similarly argue that the form of the *proposition* expressed by "five is prime" is similarly of a more specific, but still subject-predicate, form and, indeed, one without false instances. Going into the precise details would require delving into the exact logicist definitions of "five" and "prime", which is rather involved. Hopefully it is clear enough how the basic approach might work.

3.   The Paradoxes and the Theory of Incomplete Symbols

From 1901 until the publication of *Principia Mathematica* in the early 1910s, Russell's views changed often and rapidly, deviating further and further from the rather simple approach described previously. The principal driving catalyst of these changes was the desire to solve logical and other paradoxes threatening the logical basis of his logicism. Some of these changes are evident even by the time *Principles* was published in 1903. Russell's working notes for the various solutions he tried over the years that follow have mostly been preserved and are now published in volumes 4 and 5 of his *Collected Papers*. Even a fairly rudimentary summary of the twists and turns of his thought is not possible here.[4]

---

4   Russell himself provided a summary of at least the early years of this development in a letter to Philip Jourdain, which makes for a useful comparison. See Grattan-Guinness (1977), 78–80.

Instead, I shall focus on general themes, in particular how it is that Russell was pushed more and more away from views compatible with option A responses to the logical form problem and more and more towards option B.

It is worthwhile first to note certain commitments that option A seems to require, or at least steer one towards. The approach makes use of complex terms and/or complex predicates and requires this complexity to have significance at the level of logical form. For Russell, or someone with similar commitments, this means this complexity exists at the level of the proposition, or the objective content of the truths in question. In option A, the terms and predicates are still terms or predicates: their role would seem to be to represent objects and qualities/relations, just like any other terms and predicates. How does the complexity in the term transfer to a corresponding complexity in the content?

Two possible answers suggest themselves. Perhaps there are special objects possessing an "inner" logical form so that any proposition about them, merely by virtue of being about *them*, has a different logical form than it would have if it were about an object without such an inner form or having a different inner form. The number five itself has a kind of special logical nature so that to assert that *it* is prime is to assert something of a different specific logical form than found in the false proposition that four is prime. Another possible answer would be that the more nuanced complexity in the logical form comes not from the objects meant but from a complexity in the *representation of them* in the proposition. This second kind of answer collapses to the first if the way in which the proposition represents an object must always be simply having the object itself as a constituent; in that case, the complexity in the representation would be a complexity in the represented. Russell's *early* views, however, allowed that a proposition could be about an object by virtue of having a different, represent*ing* constituent. Russell called represent*ing* constituents "denoting concepts" or, later, "denoting complexes". Others might prefer to think of Fregean senses or suchlike. In this approach, one needn't insist that the numbers four and five *themselves* have a different "inner" logical form, but one would insist rather that the propositions expressed by "four is prime" and "five is prime" nonetheless have different specific logical forms because of a difference in the constituents of the propositions having these forms that do the work of *representing* four and five. These representing entities have different forms, and this is reflected in the overall forms of the propositions. In an option A response, however, both "four is prime" and "five is prime" are still subject-predicate propositions and so share a more generic logical form. It should be noted both kinds of answers commit one to *things* having a kind of complex inner logical nature that affects the logical form of the propositions into which they enter. The only difference lies in whether these things are the meanings or the things meant.

It is now well understood that being too liberal about what abstract or logical entities one postulates to exist can lead to trouble. The simplistic option A approach sketched in the previous section is no exception. In one interpretation, "$x \ni (x \sim\varepsilon\, x)$" means the class of non-self-membered classes; in another, it means the quality of non-self-predicability. With the naïve assumptions Russell took from Peano, either leads to a contradiction when we ask whether it bears the appropriate interpretation of ε to itself. Interestingly, Russell's initial reaction to the two forms was different. Already by the time *Principles* was published, he concluded that it is a mistake to think that a complex predicate exists for every open sentence, or for what Russell called a "propositional function", by which he meant the objective content of an open sentence. While there may be propositions of the form "*u* is not *u*", there is no such predicate as being an *x* such that *x* is not an *x*; non-self-predicability is not a predicate (*PoM*, §101; cf. §84). A device for forming complex predicates allowing any arbitrary open sentence would therefore not be allowed in his logical language. Presumably Russell would not have regarded the ordinary language sentence "Blueness is non-self-predicable" as nonsense, or as not expressing any kind of proposition, but if it does express one, it could not express one of subject-predicate form; perhaps it expresses the *negation* of the subject-predicate proposition "Blueness is blue", but a negation of a subject-predicate proposition is not, or at least not always, also a subject-predicate proposition. Note that this already requires acknowledging that the grammatical form of a sentence is not always indicative of the logical form of the proposition expressed.

It took Russell longer to abandon complex terms for classes formed from arbitrary open sentences, though his notation changed from Peano's "$y \ni (\ldots y \ldots)$" to something akin to Frege's value-range notation "$\grave{\varepsilon}(\ldots \varepsilon \ldots)$" and eventually to the circumflex notation "$\hat{x}(\ldots x \ldots)$" found in *Principia Mathematica*. All other complex terms in his formal work were derivative of this notation; for example, early Russell made use of Peano's functor $\imath$, mapping a singleton to its sole member, to form definite descriptions. However, his exact understanding of these notations varied widely over the years. It is of course inconsistent to think that a class, itself understood as an individual, exists for every open sentence using a variable for individuals $\ldots x \ldots$, where the class is itself a possible value for that variable *x* and the membership of the class consists of all and only the values of *x* for which $\ldots x \ldots$ holds. Since Russell regarded mathematical objects such as numbers, and any others which might be introduced by "abstraction principles", as classes, the logical form of class-notation was central to his logicism. Are these terms to be understood as terms in anything like the usual sense, so propositions expressed by formulas using them have at least the same generic form, even if a different specific form, from other singular predications?

As Russell notes in various places, he discovered "Russell's paradox" when considering Cantor's powerclass theorem that every class has more subclasses than members (*PoM* §100; *IMP* 136; *MPD* 75). This would seem to mean that at least some, if not all, classes of individuals cannot themselves be considered individuals, or else the class of individuals would have all its subclasses as members, which is impossible if it has more sub-classes than members. At the time of finishing *Principles* (see chapter vi; appendix B), Russell distinguished classes-as-one from classes-as-many, with the latter being considered irreducibly plural. But this means that a proposition of the form "$x \ \varepsilon \ a$", where a is a class-as-many, is not a binary relation after all, as it has more than two relata. Here we again see slippage towards option B, as the apparent logical form of "$x \ \varepsilon \ a$" is not taken at face value. But, at least at first, he continued to hold that at least some membership propositions, those where the class was a class-as-one, did have their apparent logical form. Typically, as we have seen, a class was referenced by use of a class abstract of the form "$\hat{x}(\phi(x))$". At first, Russell took whether the class denoted could be considered a class-as-one, or a single thing, as determined by features of the defining function $\phi(x)$, calling those that could not be considered to define classes "quadratic forms". However, he found it difficult to isolate a specific category of forms which should be considered quadratic.

In mid-1903, Russell made his first attempt to develop a "no classes" theory in which apparent discourse about classes was to be replaced with discourse about their defining functions (*Papers* 4, 49–73; *LtF*). He re-purposed Frege's notation "$\hat{\varepsilon}(\ldots \varepsilon \ldots)$" for value-ranges of functions as a function abstract notation for the functions themselves. Yet, if these func-tions are considered discrete and individual "things" which can enter into logical forms as "units", the improvement over a realism about classes is unclear. One still needs to consider the function $\sim\phi(\phi)$ true for a given argument which is a function that is not satisfied by itself. Later, Russell thought to replace taking functions as separable entities with the notion of a substitution of one entity for another in a constant proposition, so that the work done by, for example, "x is human" could be done instead by the proposition "Socrates is human" and various substitutions for Socrates within it (*Papers* 5, 90–296). However, he found that this approach succumbed to another problem he was also already aware of at the time of publishing *Principles* (§500): Cantor's theorem also seems to pose a problem for the totality of propositions. If a different propo-sition exists for every class of things, and propositions themselves are things, Cantor's theorem suggests there must be more propositions than propositions. Russell-style paradoxes can then be generated. Consider, for example, propositions of the form "for all propositions p, if $\phi(p)$ then p", where $\phi(x)$ is a condition on propositions the proposition in question does not satisfy. Being such a proposition is itself a condition on propositions,

$\psi(x)$; now consider the proposition "for all propositions p, if $\psi(p)$ then p"; it itself satisfies $\psi(x)$ just in case it does not.[5]

A crude summary of the difficulty Russell was facing during these years is that Cantor's theorem will create a problem for any theory that tries to make a single thing "out of" many things in a generic fashion. Classes-as-one are a very simplistic way of considering many things as one, but the problem also arises for complex entities such as Russellian propositions and entities derivative from them like "propositional functions". Considered as objective complexes, propositions can be made up of any combination of entities in any form. They can have any number of constituents. So long as there is at least one such complex for any collection of individuals, there must be more such propositions than there are individuals. The same goes for defining conditions or "propositional functions", which, if taken objectively, could be of arbitrary complexity. Note, moreover, that it does not necessarily help merely to adopt a less objective account of propositions or propositional functions, that is, one that denies that they can be of arbitrary complexity or have any number of constituents. To get around the problems raised by the diagonal paradoxes Cantor's argumentation leads us to, one would have to find a principled reason to deny the existence of the "paradoxical" propositions, classes, or functions in particular. This seems to pose a definite obstacle for thinking of propositions, functions, or classes as single things, or at least as single things in the same sense as their basic constituents or urelements, and if so, then statements "about" such entities, if indeed it remains appropriate to think of them as entities, would seem to have a different form than statements about these more basic elements.

Notice, moreover, that propositions (and related entities like "propositional functions") are *precisely* the things we naturally think to *have* logical form. We noted earlier that option A responses to the logical form problem for logicism seem to have a need to posit entities that have a special logical form as part of their nature, either as special varieties of subjects and predicates or as representatives thereof. But assuming the identity conditions of these special entities are determined by their defining logical forms, and such entities exist for any arbitrary chosen proposition or propositional form, we are in a position in which a violation of Cantor's theorem or other paradoxical consequence is a definite threat. And note, moreover, that the switch from the represented entity to the representative of that entity does little good so long as the representatives themselves are entities and it is possible to disambiguate between propositions about the representative and propositions about the represented things. Consider *any* kind of complex term that makes use of

---

5 See Landini (1998) for a discussion of various forms of such "propositional" paradoxes and how they fared and developed within the context of Russell's "substitutional theory".

a propositional form as a constituent, such as Peano's "$x \ni \phi(x)$", later description notations such as "$(\imath x)\phi(x)$", or even something like "$\#x : \phi(x)$" read as "the number of x such that $\phi(x)$". Whether or not the entities *denoted* by such complex terms are so numerous as to generate a violation of Cantor's theorem, if the contribution made by such terms to the logical forms in which they appear as meanings or representatives are single things and are distinct for distinct $\phi(x)$s, there is a potential violation of Cantor's theorem and a related diagonal contradiction. Early Russell would have considered a complex term "$x \ni \phi(x)$" as contributing a denoting complex which is not itself the class but a representative that denotes the class. The complex itself would have its own individual identity determined by the defining condition $\phi(x)$, and a coextensional $\psi(x)$ would give rise to a distinct denoting complex "$x \ni \psi(x)$", even if the two denoting complexes denote the same collection of things. But if we can talk about denoting complexes themselves in addition to what they denote,[6] then we can consider properties of them, including the property W, a denoting complex of the form "$x \ni \phi(x)$" has when it itself does not satisfy $\phi(x)$. Does the denoting concept "$x \ni W(x)$" satisfy W? It does just in case it does not. This contradiction does not make use of the denotations of such representative entities. Even if classes are not single entities, if their representatives in logical forms *are* single entities, trouble brews. And the problem is just as bad for other kinds of complex terms such as those of the forms "$(\imath x)\phi(x)$" and "$\#x : \phi(x)$."[7]

In later reminiscences about his work on attempting to solve the paradoxes, Russell makes note of his suspicion that the theory of denoting was important for their solution, a suspicion he claimed turned out to be correct (*Auto* I:229; Grattan-Guinness 1977, 78). Hopefully, the connection is beginning to become clear. Prior to developing the theory of descriptions in 1905, Russell did not have a clear sense of how a complex term could be used without the logical form represented being one containing single entities of a problematic sort. The details of Russell's new theory of definite descriptions are now well known. A term of the form "$(\imath x)\phi(x)$" is not to be taken as a self-standing constituent of the logical form of the proposition represented by a sentence in which it appears but instead must

---

6 In his well-known article "On Denoting" (*Papers* 4, 414–427), in which Russell argues against Frege-style meaning/denotation or sense/reference distinctions, Russell gave what is now considered a very obscure argument against thinking that it is possible to disambiguate discourse about a denoting complex from discourse about what it means: the so-called "Grey's Elegy Argument". But it should be noted that even if this argument fails, we are simply impaled on a different horn of the dilemma, as it then becomes possible to formulate paradoxes such as this.

7 See Klement (2014) for a fuller discussion of the possibility of paradoxes such as these in a Russellian context.

be unpacked contextually. In particular, a formula of the form $\psi[(\imath x)\phi(x)]$ is taken as abbreviating:

$$(\exists x)[\phi(x) \,.\, (y)(\phi(y) \supset y = x) \,.\, \psi(x)]$$

Part and parcel of this theory is the rejection of taking "$F[(\imath x)\phi(x)]$" to have subject-predicate form, even when F is a simple predicate, despite the *apparent* similarity in form with a simple subject-predicate statement "Fa". There quite simply is no *single* constituent of the previous more complex formula corresponding to the description, no one entity with an inner connection to the form $\phi(x)$. Of course, there needs to be an object in the domain of quantification which uniquely satisfies $\phi(x)$, but there needn't be distinct such objects for every propositional form.

By the time of *Principia Mathematica* (see, e.g., PM, introd. chapter 3), Russell saw the theory of descriptions as only one component of a broader theory of "Incomplete Symbols", the other key component of which was his oxymoronically named "no classes theory of classes". According to that theory, a class abstract "$\hat{x}\phi(x)$" is similarly not taken to represent a single constituent of the resulting logical form. Instead $\psi[\hat{x}\phi(x)]$ is taken as shorthand for:

$$(\exists f)\big((x)(f!x \equiv \phi(x)) \,.\, \psi(f!\hat{x})\big)$$

To give an example, to say that a is a member of the singleton $\hat{x}(x = a)$ is to say:

$$(\exists f)((x)(f!x \equiv x = a) \,.\, f!(a))$$

There is some f true of all and only those things identical to a, and a satisfies f. Despite appearances, $a \in \hat{x}(x = a)$ has a much more complicated form than subject-predicate. There is no one single individual entity, either a meaning or a denotation, corresponding to the class abstract "$\hat{x}(x = a)$". There is some disagreement about the nature of the higher-order quantifiers and the "propositional functions" they allegedly quantify over among interpreters of Russell, but it is at any rate clear that Russell did not think of functions as additional "individuals", nor statements about functions of one type as having the same logical form as statements about those of another type. A claim about a mathematical "object" such as the number 1 is really a claim making use of a specific logical form: in the case of the number 1, this form involves uniqueness. To say that the class of ϕs is a member of 1 is really to make a complex statement involving this form, one that can be shown to be logically equivalent with the claim that there is a unique ϕ making use only of standard logical rules governing quantifiers and truth-functional connectives. Similar treatment was given to higher-order statements about classes of classes, as well as an approach to "relations in extension". Put all together, and applied to suitably contextually defined mathematical notions such as "5" and "7", one arrives at a theory in which something such as "7 > 5" is interpreted as really having a much more complicated form than that of a simple binary relation, without a unique single constituent corresponding to any of the signs "5", "7", or ">". When unpacked, this resolves into a form that makes use of nothing but variables and logical constants such as quantifiers and truth-functional propositional operators. And, moreover, this form will be without false instances.

It should be clear that, according to our earlier classification, Russell's theory of incomplete symbols is squarely an option B approach. The apparent logical form of a statement need not track the apparent surface form of a sentence we use to express it. Of course, it is possible to invent a notation, such as the logical language of *PM*, which, when written without abbreviations or convenient shorthands, does reflect the actual logical form of what is expressed with its syntax. But it would be cumbersome to make use of such a language without introducing notations that *mimic* basic logical forms such as subject-predicate, binary relational, and so on without actually being such. Most inferential patterns applicable to a true binary relation "aRb" will have an analogue for "7 > 5"; for example, generalizing to "$(\exists x)(aRx)$" from the former is akin to generalizing to "$(\exists a)(7 > a)$". This makes it seem as if we are dealing with the "same" form, but on further scrutiny, the latter has a more complicated form that this abbreviated method of representation lacks. To think about a specific number, one is thinking about a specific kind of logical form. While the rules for "unpacking" two different numerals, say, "7" and "4", at least at the first step, are analogous, so in that sense they "share" a (generic) form, what they are unpacked into has different (specific) resulting forms, explaining how it is that "7 > 5" could unpack into a universally valid logical form, whereas "4 > 5" does not so unpack, where these results presuppose nothing beyond the usual assumptions of classical logic.

Russell's theory of incomplete symbols remained a staple of his philosophy for the remainder of his career. While the 1925–1927 second edition of *Principia Mathematica* saw him experiment with a more austere propositional system and more stringent assumptions about extensionality and the existence of "predicative" functions, Russell never returned to anything like an option A approach according to which complex terms for mathematical entities were taken at anything like face value. In terms of his ontology of mathematical or abstract entities, Russell's views were mostly settled by 1910.

4. Abstract "Objects" as Fragments of Form

Any form of logicism that holds onto the view that logic has a special relationship with form must also hold that the "objects" of interest to mathematicians, numbers and suchlike, also have a special relationship to logical form. According to Russell's mature views, nearly all the apparent "things" of interest to mathematicians, including not just numbers but all classes, were considered what he called "logical fictions". The expressions that apparently stand for them are considered incomplete symbols instead: rather than referring to things, these expressions are meaningful in a different way. They contribute fragments of logical form to the overall statements they are used within, and these fragments are not themselves additional things. If there is something in common between the generic forms various categories of such fragments take—as with different (cardinal) natural numbers—and a variable of an appropriate type can be used whose values range over their differences, then it will be possible to quantify over such apparent "objects" as well. For example, *PM* *20.07 introduces a proxy for quantification over classes (including the Frege-Russell numbers) by means of higher-order quantification. Statements apparently "about" numbers can come out as true, and even quantified existence claims postulating, for example, the existence of numbers with such-and-such characteristics become possible. But at the level of ultimate, fundamental objects, there are no classes or numbers in addition to the more basic elements of more basic logical forms. We are able to invent notations that mimic what goes on at more basic forms where genuine reference to "real" objects takes place, but the actual logical forms are quite different.

It is interesting to contrast this with other popular approaches to abstract objects. At least one major strand in the "abstractionist" school[8] takes inspiration from Frege's notion of "recarving content" (Frege 1884, sec. 64). The logically complex statement to the effect that there exists a 1–1 mapping between the objects falling under concepts F and G can be "recarved" or "recast" or "reframed" as an identity statement between the number of Fs and the number of Gs. Thus, the two sides of an instance of Hume's Principle, for example, can have the same content "carved" in different ways. But carved as an identity statement, "$\#x : Fx = \#x : Gx$" really does have the same logical form as any other identity statement between objects, and, indeed, through such recarving, we can become aware of genuine objects not referred to in the other recarving. The same content can have multiple logical forms, and only on one "carving" of that content is it about objects, but somehow by virtue of this possibility, the same content seen as having a simple logical form can also be seen as having a different, logically richer logical form.

---

8 This school is obviously made up by many thinkers whose precise views differ; here I must make do with a crude summary that may oversimplify the precise attitudes of individuals within it. For some discussion, see, for example, Hale and Wright (2001), 91–116, Potter and Smiley (2004).

In Russell's view too, it is possible to take what is *apparently* a simple identity statement in a certain notation with incomplete symbols and recognize it as also having a more complicated, logically richer form as well, one such that the equivalence of the two sides of Hume's Principle, for example, can be demonstrated using only the standard logical rules for the quantifiers and propositional connectives. With Russell's general strategy of using the corresponding equivalence classes to form explicit definitions of terms introduced by "abstraction principles", the abstraction principle itself becomes a demonstrable theorem, requiring no additional assumptions beyond what is necessary for establishing that the relation in question is an equivalence-relation. What is different in the two approaches is that for Russell, the apparently simple logical form on the one side of the principle is only an apparent simplicity. The terms in the identity statement are not actual terms, and this method of reframing content to *appear* to have a different form does not make it *actually* have that form. Therefore, any claim that this reframing provides knowledge of any "new" identities is at best misleading.[9] One advantage of the Russellian approach is that abstraction principles need not be taken as in any sense basic, definitional, or logically analytic in an extended or different sense. Another advantage of the Russellian approach is that it obviates the seemingly difficult task of explaining how the same content could have different but incompatible logical forms at once.[10]

But this isn't to say that the Russellian approach does not have its disadvantages as well. Because the terms apparently standing for abstract entities are not truly referential, they cannot be used to establish new existence theorems not already demonstrable. In particular, Russell could not use any version of the Fregean bootstrapping argument to establish an infinity of individuals. As numbers are identified with classes, having extensional identity conditions, only those numbers applicable to appropriately sized collections can be proven distinct: inapplicable numbers are all "empty" and the same. This is the source of the issues surrounding the so-called "axiom of infinity" in Russell's logicism (see *IMP*, chapter xiii). In order to preserve certain truths commonly held to be mathematical ones as fully logically general, such as the Peano-Dedekind axiom that no two natural numbers share a successor, Russell had to add an assumption of an infinity of individuals as an antecedent to his own statement of these principles.[11]

---

9 Hale and Wright (2009), 190, contrast their approach instead with a "Tractarian" ontology of structured facts, but of course the metaphysics of Wittgenstein's *Tractatus* is largely inspired by Russell.

10 See Klement (2012) for a fuller comparison between these approaches.

11 For a fuller discussion of this issue and the extent to which it poses a problem for Russell's logicist project, see Klement (2019), 168–173.

Russell's attitude also was not limited to the kinds of abstract objects considered in mathematics. Mature Russell also denied the existence of semantic entities, such as Fregean senses or his own earlier "denoting concepts", although he held that something like Frege's distinction between sense and reference was preserved in his theory of descriptions, so long as it was not taken at face value or interpreted too literally (*KAKD*, 157). We have seen some of his motivation for holding this and some of the dangers of adopting too-liberal assumptions about the existence of such things. Unfortunately, other traditions have done relatively little by way of addressing such concerns, and so it is hard to form a comparison. It is often noted, for example, that Hume's Principle, unlike some other abstraction principles like Frege's Law V, is consistent, even when the terms "$\#x : Fx$" are taken as representing objects in the domain of the individual variables. Since distinct concepts often have the same number, there needn't be more numbers than concepts applicable to them, and so this by itself doesn't violate Cantor's theorem. But note that this does not address the status of the *meanings* or *senses* of such complex terms as "$\#x : Fx$"; if these are entities that enter into logical forms as *units*, it would appear that there must be many such entities, and the changes to the paradox described in the previous section about denoting complexes of the form "$x \ni \phi(x)$" to form one about those of the form "$\#x : Fx$" are trivial and easily made. That the Russellian approach sidesteps such worries is at least arguably a large point in its favor.

One aspect of Russell's views, however, that I do find worrying is his eventual insistence that propositions too be considered "logical fictions" and not as entities that enter, as units, into facts. In 1910, Russell adopted his "multiple relations" theory of judgment, whereupon such "propositional attitudes" as belief, desire, and so on are not to be thought of as a binary relation between a subject and a "proposition", as a whole, but as having a more complicated logical form. This theory spurred a lot of discussion and a lot of criticism, including by Wittgenstein, and Russell himself eventually became dissatisfied with it. For someone mainly interested in the relationship between logic and mathematics, this issue might seem not to be of central importance and may be thought better left to epistemologists or philosophers of mind. But in addition to the need for a good theory of judgment, it leaves a bit unclear exactly *what* it is that has logical form, an issue that had been clear while Russell still believed in propositions as mind-independent complexes. The nature of logical form and its relation to linguistic form is clearly an issue central to logicism. A natural answer is that logical form deals with forms of *facts*. But what are they? Is the real logical form of discourse about them just as complicated as discourse about numbers, and is positing facts as dangerous as positing propositions? The official position of the 1910 Introduction to *PM* (44–46) seems to have been that there are "complexes" or "facts" corresponding only to *elementary* truths and that no single fact or complex corresponds to a quantified truth. But notice that we are always dealing with quantifiers when unpacking incomplete symbols such as descriptions and class terms. If one claims that the "true" logical form of "$\psi(\hat{x}\phi(x))$" is not subject-predicate but a complicated form involving higher-order quantification, what is it that *has* this "true" logical form? Apparently, not a proposition, and not a fact. It cannot be something linguistic, either, because it's precisely the linguistic form that is supposed to be misleading as to the actual form. Perhaps a suitably Russellian answer can be given, but any such answer is likely to be more complicated than our discussion up until now would seem to suggest, possibly involving a fairly philosophically committal semantic theory.

In conclusion, Russell gives a compelling, if not fully complete or fully satisfying, account of the relationship between thinking about apparent mathematical objects such as numbers and thinking about logical forms, which is often taken to be the heart of logic. Personally, I find the overall approach to be at least as compelling as the often obscure and less developed suggestions to the effect that discourse about abstract objects can be taken to "recarve" content of more complicated forms. It also provides a relatively attractive solution to what I have called the problem of logical form for logicism, even if it leaves us with the impression that almost no discourse in mathematics (or possibly in other scientific endeavors) has the actual logical form we are tempted to take it as having. This impression can even lead us to doubt whether any actual discourse has the simple logical forms we often assume it to have and even to begin to lose sight of why we believe in such forms in the first place.

References

Bolzano, B. (1837). *Wissenschaftslehre*. Sulzbach: von Seidel.

Frege, G. (1884). *Die Grundlagen Der Arithmetik*. Breslau: Köbner.

Grattan-Guinness, I. (1977). *Dear Russell–Dear Jourdain*. New York: Columbia University Press.

Hale, B. and Wright, C. (2001). *Reason's Proper Study*. Oxford: Oxford University Press.

———. (2009). "The Metaontology of Abstraction." In *Metametaphysics*, edited by David Chalmers, David Manley, and Ryan Wasserman, 178–212. Oxford: Oxford University Press.

Klement, K. (2012). "Neo-Logicism and Russell's Logicism." *Russell* n.s. 32: 127–159.

———. (2014). "The Paradoxes and Russell's Theory of Incomplete Symbols." *Philosophical Studies* 169: 183–207.

———. (2019). "Russell's Logicism." In *The Bloomsbury Companion to Bertrand Russell*, edited by Russell Wahl, 151–178. London: Bloomsbury.

Landini, G. (1998). *Russell's Hidden Substitutional Theory*. Oxford: Oxford University Press.

Peano, G. (1895–1908). *Formulaire de Mathématiques*, 5 Vols. Turin: Bocca.

Potter, M. and Smiley, T. (2004). "Abstraction by Recarving." *Proceedings of the Aristotelian Society* 101: 327–338.

Russell, B. (1897). *An Essay on the Foundations of Geometry [EFG]*. Cambridge: Cambridge University Press.

———. (1903). *The Principles of Mathematics [PoM]*. Cambridge: Cambridge University Press.

———. (1919). *Introduction to Mathematical Philosophy [IMP]*. London: Allen & Unwin.

———. (1951–1967). *Autobiography*, 3 Vols. *[Auto]*. Boston: Little Brown and Company.

———. (1958). *My Philosophical Development [MPD]*. London: Allen & Unwin.

———. (1980). "Letter to Frege 24.5.1903 [LtF]." In *Gottlob Frege: Philosophical and Mathematical Correspondence*, edited by Brian McGuinness, 158–160. Chicago: The University of Chicago Press.

———. (1983). "Logical Atomism [LA]." In *The Collected Papers of Bertrand Russell, Vol. 9, Essays on Language, Mind and Matter 1919–1926*, edited by John G. Slater, 160–179. London: Routledge.

———. (1986). "The Philosophy of Logical Atomism [PLA]." In *The Collected Papers of Bertrand Russell, Vol. 8, the Philosophy of Logical Atomism and Other Essays 1914–1919*, edited by John G. Slater, 157–244. London: Routledge.

———. (1990). *The Collected Papers of Bertrand Russell, Vol. 2, Philosophical Papers 1896–99 [Papers 2]*. Edited by Nicholas Griffin and Albert C. Lewis. London: Unwin Hyman.

———. (1992). "Knowledge by Acquaintance and Knowledge by Description [KAKD]." In *The Collected Papers of Bertrand Russell, Vol. 6, Logical and Philosophical Papers 1909–1913*, edited by John G. Slater, 147–166. London: Routledge.

———. (1993). *The Collected Papers of Bertrand Russell, Vol. 3, Toward the "Principles of Mathematics" 1900–02 [Papers 3]*. Edited by G. H. Moore. London: Routledge.

———. (1994). *The Collected Papers of Bertrand Russell, Vol. 4, Foundations of Logic 1903–05 [Papers 4]*. Edited by Alasdair Urquhart. London: Routledge.

———. (1997). "My Mental Development [MMD]." In *The Collected Papers of Bertrand Russell, Vol. 11, Last Philosophical Testament 1943–68*, edited by John G. Slater and Peter Kollner, 5–18. London: Routledge.

———. (2014). *The Collected Papers of Bertrand Russell, Vol. 5, Towards "Principia Mathematica" 1905–08 [Papers 5]*. Edited by G. H. Moore. London: Routledge.

Whitehead, A. N. (1898). *A Treatise on Universal Algebra*. Cambridge: Cambridge University Press.

Whitehead, A. N., and Russell B. (1910–1913). *Principia Mathematica, 3 Vols. [PM]*. Cambridge: Cambridge University Press.

Part II
Dangerous Liaisons

7 Frege and Hilbert on Conceptual Analysis and Foundations

Some Remarks

Michael Hallett

(This chapter is dedicated to Storrs McCall, friend and colleague for 35 years.)

Introductory

Conceptual analysis was of fundamental importance in the study of the foundations of mathematics in the later 19th century. What comes immediately to mind (in no particular order) are: Gauss's analysis of complex numbers; Dirichlet's analysis of the notion of function; Cantor's analysis of counting numbers and cardinal numbers of finite and infinite sets; Frege's analysis of arithmetic, above all of natural number and 'following in a series'; and Dedekind's analyses of the continuity of the real 'line' and then of the natural number system, an analysis which shares a good deal with Frege's. On a little reflection, we can stretch this list into the 20th century and in particular into what became known as metamathematics—the analysis of deductive inference, then of a formal system, of deductive independence, and then of truth and mechanical computability, and the very notion of concept itself. We could thus add the names of Russell, Peano, Hilbert, Cassirer, Bernays, Gödel, Tarski, Turing, and Church (among others) to the list. All were trying, among other things, to grapple with the question of what these central notions are, how they can be characterised, and how they fit into developed mathematics/metamathematics, and an important part of the purpose is the attempt to clarify and demystify (or demistify) the foggy and then to present the new characterisations in proper deductive dress. The subject matter is massive, of course, and there are numberless differences, both in design, execution, and fundamental aim. However, what I intend to do in this short chapter is to open a discussion of two quite distinct examples of conceptual analysis, those of Frege and Hilbert. This is not in the least to suggest that these are the only kinds, but it's meant rather to bring out a contrast. It can be argued that many of the other sorts of conceptual analysis (those of Cantor and Dedekind come first to mind) are absorbed by the Hilbert analysis. The ultimate aim would be to consider many of

DOI: 10.4324/9780429277894-9

the other approaches mentioned in light of the two to be sketched here, though for the time being, that will have to remain just an aim.

1.   Frege: Definitional Analysis

The importance of conceptual analysis for Frege is summarised and narrowly examined in [Dummett, 1987].[1] Dummett rightly emphasises how important the notion of *definition* was for Frege in all this; even in the cases where definition was not itself the end result of analysis, the importance of giving definitions tied to that analysis (both conceptually and deductively) was paramount. For Frege, the point of definition is twofold:

1. It enables us to say quite clearly what it is we are talking about and thus to fix the reference of the fundamentally important notions on which the theory to be developed rests and thereby to fix reference throughout.
2. It also demonstrates that the things being defined are of a certain specified character and thus at the same time that certain (in the context) undesirable characteristics are avoided.[2]

Mathematics for Frege is a body of what he called true Thoughts (propositions), which must have a fixed meaning. And, as he says to Hilbert (in correspondence), the 'axioms, Basic Laws, theorems'

> ought to contain no word and no sign whose sense and reference or whose contribution to the Thought expressed does not already stand fully fixed, so that there is no doubt as to the sense of the proposition, of the Thought therein expressed. It can only be a question of whether the Thought is true, and then on what its truth rests.
>
> (Frege to Hilbert, 27.xii.1899)[3]

As I said, for Frege, axioms, Basic Laws, and theorems must all be truths, though the axioms and Basic Laws are distinguished from theorems in that they neither can be proved nor are in need of proof (see [Frege, 1884], end of §3). Conversely, whatever can be proved from more fundamental principles should *not* be taken as an axiom or Basic Law. The fundamental point is this: fixing reference is essential before we can begin to decide whether axioms or Basic Laws are *true*, because before we can do so, we have to know what they are *about*. Definitions are the most obvious way to fix reference.

---

1 This chapter might be considered a supplementary comment on this work and also on two others, namely [Blanchette, 2007] and [Hodges, 2004].
2 See n. 6.
3 Reference to the Frege/Hilbert correspondence is always to specific letters, to be found (by date) in any of the published sources. The canonical source is [Frege, 1976], and its English translation is [Frege, 1980]. The translations here and in what follows are mostly my own.

Natural numbers and arithmetic were Frege's primary concerns in the foundations of mathematics, and his work is a hymn to clever definition. Frege's idea was to fix denotation of the term 'natural number' and other essential arithmetic notions (like 'following in a series' or 'immediate successor' and the individual natural numbers) using what he took to be *logical primitives* (among them, at least later, the value-range [*Werth-verläufe*] of a function),[4] to give Basic Laws governing these things, and to show that these are sufficient (in a framework of second-order logical axioms) to prove the basic arithmetic principles concerning the numbers so defined. Frege's analysis carried out in the *Grundlagen* shows that the fundamental arithmetical principles (in effect, the second-order arithmetical axioms) all follow from what has become known as *Hume's Principle* (HP). Frege did not consider that HP itself gives a *definition* of number, for it leaves open the question of what the numbers are and remains perfectly coherent if the numbers are taken as undefined primitives. But if number is *defined* using courses-of-values, then HP becomes *provable* from the Basic Laws. In other words, Frege shows that his system 'contains' arithmetic.[5] This is supposed to assure us that no intuition, nothing informal or loose or vague or unclear, need play any part in the derivation of arithmetic.[6] Numbers are by definition logical objects and so subject only to logical (Basic) laws.

But Frege was also perfectly aware that one cannot define *every-thing*: if we construct a chain of definitions, then at some point in that chain, some terms used will not themselves have their denotations fixed by definitions; thus, there must be undefined genuine primitives. As Frege says:

> It will not always be possible to define everything properly, precisely because we must strive to get back to the logically simple, which, because it is such, is not strictly definable. I must therefore be satisfied with indicating through hints what I mean. Above all, I must endeavour to be understood.
>
> ([Frege, 1893, 3–4])

What Frege says here encapsulates what I take to be (for him) a serious problem: the Basic Laws are crucial because, without them, the definitions given will not do their work. But the Basic Laws can only be adopted if they are seen to be true, and they can't be such if the basic terms in them (the logical primitives) cannot be seen to have fixed reference, which they

4 '*Werthverläufe*' is sometimes translated as 'courses-of-values'. See [Frege, 2013, xxv].
5 For more detail, see the aforementioned [Dummett, 1987].
6 See, for example, [Frege, 1893, 1]. This exclusion is particularly important here. This was also the case with Dedekind, and for similar reasons: in Dedekind's case, in the context of continuity, excluding geometry and infinitesimals; in Frege's case, excluding intuition generally.

can't get through definition. Moreover, the possibility of the lack of referential fixity among the primitives would mean the possibility of the lack of referential fixity *everywhere*, no matter how clever the actual definitions. It should not be doubted that, for Frege, these primitives *do* have referential fixity, but for Frege's project to be viable, it has to be possible to show (perhaps by informal argument) that the Basic Laws are logical truths, and therefore truths, and it is therefore a *sine qua non* to be able to show that their basic terms refer. But how is it possible, in the absence of a definition, to indicate the right references?

Frege addresses this problem by introducing the notion of 'elucidation'. Frege writes to Hilbert:

[Elucidations] are similar to the definitions, in that they are also concerned with fixing the meaning of a sign (of a word). But in addition, they contain elements whose meaning cannot be assumed as known completely and beyond question, perhaps because they are used variously or ambiguously in the language of everyday life. In the cases where a meaning is to be given to a sign which is logically simple, then one cannot give a definition proper, but one must content oneself with fending off the unwanted meanings which crop up in the use of language, indicating the one intended. In doing this, certainly one must always count on a cooperative understanding trying to hit upon the meaning. Such statements of elucidation cannot be used in the same way that the definitions can, because they lack the necessary precision. For this reason, as I said, I confine them to the forecourt.

(Frege to Hilbert, 27.xii.1899)

So, elucidations are thus meant primarily as hints to enable an interlocutor to 'catch on', as a way of achieving mutual understanding at the fundamental level.

But this notion of elucidation is clearly inadequate for Frege's purpose. Consider this, from [Frege, 1906, 301]:

The purpose of elucidations is a practical one, and when this is achieved, one must be satisfied. In this, one must be able to bank on good will, on a cooperative understanding, on guessing; for without a figurativeness in the expression one can often not get anywhere.

Furthermore, as he says, what distinguishes elucidations from proper definitions is that the latter leave nothing to 'guesswork':

[Definitions] also serve the purpose of mutual agreement, but they achieve this in a far more complete way than elucidations do, since they leave nothing to guesswork, and do not need to reckon on cooperative understanding, or on good will.

(Op. cit., 302)[7]

It seems, then, that elucidations are engaged in some of the same work as definitions.

But one implication from the previous passages is also clear, namely that elucidations *do* leave room for 'guesswork'. Moreover, as Frege says in the letter to Hilbert, elucidations are (necessarily) given in, or imbued with, the 'language of everyday life'. But surely one of the very purposes of the extended *Begriffsschrift* project is to circumvent the ambiguities, unclarities, and misunderstandings that the 'language of everyday life' is heir to.[8] It seems as if the reliance on elucidation, at least for *fixing* reference, is tantamount to an admission that this circumvention is ultimately impossible. How can the proper definitions communicate meaning precisely if they ultimately rely on primitives, the fixity of whose meaning relies on 'cooperative understanding' and guesswork? The Basic Laws thus can't be shown to have a fixed and clear meaning, since the primitives involved in them can't be shown to have to have a fixed and clear reference. Moreover, the definitions given (e.g., that of natural number, following in a series, etc.) cannot do what they are supposed to do, for ultimately they will involve primitive elements which themselves cannot be seen to have any fixed reference (and therefore no fixed meaning), so whatever ambiguity they carry will filter down to the things properly defined. And surely it's the case that the problems presented by the permutation argument (see [Hallett, 2010, §2, especially §2.1]) simply expose this looseness.

To put it starkly, and to strengthen what was said previously, the problem for Frege is not just serious, it's decisive. By Frege's standards, we don't know what we're talking about when we mention numbers or any part of analytic (non-geometric) mathematics; as Frege himself says, if what is being talked about is never clearly fixed *exactly*, then it is simply not clear what is being asserted. For Frege,

[In] mathematics, a word without fixed meaning has no meaning at all.
([Frege, 1906, 303])

And in the same essay, Frege remarks:

When something expresses now this Thought, now that, then in truth it expresses no Thought at all.

(*Op. cit.*, 404)

---

7 Incidentally, Frege says that Hilbert's 'definitions' are not elucidations, since their ambition is to be the 'foundation stone of the science', to 'serve as premises of inferences'; see [Frege, 1906, 302].

8 See [Frege, 1879, Preface].

In short, the reliance on elucidation will not give Frege what he wants.

The point is presaged at the very beginning of the *Grundlagen* in 1884. In the opening sentence, Frege raises the question of what the number one is and supposes that we normally get as answer, 'Why, a thing'. He dismisses this answer, among other reasons, since

> it only assigns the number one to the class of things, but does not state which thing it is.

Having pointed this out, we will perhaps then 'be invited to select some thing or other that will be called one'. Frege goes on immediately:

> Yet if anyone had the right to understand by this name whatever he pleased, then the same proposition about one would mean different things for different people; there would be no common content for such propositions.

> ([Frege, 1884, Introduction, I])

In many ways, what is exposed here is familiar to us from Euclid. Recall that Euclid sets out 23 definitions at the start of Book I of *The Elements*. These are of two sorts. First, there is the kind of definition we expect in a mathematical treatise, for example, the definition of a square (Def. 22) as an equilateral, rectangular quadrilateral. To support this, Euclid then does something familiar, for in Proposition Book I, 46, he proves the *existence* of squares (more properly, that squares can be constructed), showing that the definition is not empty.

But in addition to this kind of definition, there is a kind which we find hard to construe, for example, where a point is defined as being 'that with no parts' (Def. 1) or a straight line as a line on which the points 'lie evenly with themselves' (Def. 4).[9] The attempts made in the ancient world to give explanations of the primitives (physical, visual, perceptual notions were all invoked) suggest that what Euclid was after in giving his definitions of 'point', 'line', and so on was something akin to Fregean 'elucidations', perhaps an attempt at 'fending off unwanted meanings' (as Frege says), of ruling some things out. So two ink blots cannot be taken as points, since then there will clearly be more than one straight line running between them.

There is a temptation to say that the matter of elucidation must only be of subsidiary importance for Frege, of pedagogical concern, an icing on the cake of the reduction to logic manifestly achieved in the sequence of works from the *Begriffsschrift* to the *Grundgesetze*. But this would be misleading; elucidation in a general sense informs virtually all the

---

9 For all these definitions, see [Heath, 1925, Volume 1, 153–154].

important aspects of Frege's project. There is not space here to go into this fully, but let me give just a few examples.

a. *Importance to the Structure of Frege's Project.* Frege often uses explanatory metaphors, for example, 'saturation'. One is tempted to say that for one who possesses the requisite degree of goodwill, the metaphor is designed to help understand how two things, one saturated possessing object reference and sense, the other unsaturated but possessing both a sense and a reference to a different sort of thing, combine to form a third thing which is again saturated and has sense and also an object reference (a complete Thought). There is here a reliance on an understanding of the whys and hows of chemical combination for us to see how two basically different sorts of thing can combine to form a third thing, different again. One is tempted to call the reliance on goodwill *fundamental,* for its absence might cause one to exclaim that nothing in this 'elucidation' can help to explain how the 'saturated' and the 'unsaturated' can 'combine' to do this. This is surely not a trivial matter, as anyone who has tried to explain how this works to a beginning class in the philosophy of language can attest.

b. *Consolidating the Outcome of Analysis.* There are many points in this project where Frege simply assumes that the formal results yield a conceptual analysis of the *informal* concept which one starts from. For example, isn't there an assumption that 'Heath's edition of Euclid's *Elements* and Whitehead and Russell's *Principia Mathematica* have the same number of volumes' is properly captured in the formal analysis by '$NxFx = NxGx$' and that this latter is rendered (*via* HP) by 'there is a one-one onto correspondence between the extensions of the respective concepts'? The technical insight is extremely illuminating, but it is at the same time difficult to assert that someone who acknowledges the correctness of this *informal* statement will also acknowledge that this is correctly (and fully) rendered by the *formal* statement. Doesn't one need the instruction of elucidatory insight to see this? (For a further example, see, e.g., [Blanchette, 2007, §2].)[10]

---

10 The issue exposed here is intimately tied to what is known as 'paradox of analysis', that is, whether one is analysing informal notions which are already perfectly well known or proposing new, formal ones which do not correspond exactly with the informal ones but which stand in a certain (unclear) epistemological relationship to those. In particular cases, we can ask the pressing question: are we defining something we are perfectly familiar with in a precise way, such as, say, with the concept of prime number or perhaps Euclid's notion of square, or are we doing something else? So, is something old being isolated or something new being proposed to stand in its place? So, does Frege isolate what we always knew to be the central characteristics of what a natural number is? Or is he saying 'From now on, what we mean by the term "natural number" is . . .'? The answer to this is very important, for it makes a difference to how we judge the outcome of Frege's project, the inconsistency of Basic Law V aside. Can we say that arithmetic just is part of logic, or must we say just that one can reconstruct a reasonable facsimile of arithmetic within logic? I take it that these are very different answers.

c. *Explanation of the Logical Primitives*. Following what he says to Hilbert, Frege has no real choice but to offer what he takes it will be reference-fixing elucidations of the logical primitives, the most important of which, when seen from the project to provide logical objects, are the value-ranges. There is some elucidation in [Frege, 1893, §31], followed by the less convincing remarks (which actually highlight the underlying, informal notions to be expressed in the formal) in [Frege, 1903a, §147].[11]

In any case, it seems to me that Frege is caught up in a problem. If we take the business of elucidation of primitives seriously, then setting out certain 'primitives' might be taken to be the *beginning* of a new philosophical discussion, not its end. This might be taken in some circumstances to be a good thing. But this is surely not what Frege wants, for it is the isolation of the primitives which ought to mark the philosophical *end* of his project, not a starting point.

The problem here is well captured by some things which Russell says from around this time. Russell and Frege had very different philosophical views, of course, and we should perhaps highlight here in particular Russell's early commitment to a notion of *acquaintance* which provides the basic epistemological stratum. We cannot go into this any further here.[12] Nevertheless, given what we are focusing on, Russell reveals a remarkable parallelism with Frege and his approach to analysis and primitives.

Consider this, from Russell's book on Leibniz:

> the business of philosophy is just the discovery of those simple notions, and those primitive axioms, upon which any calculus or science must be based. . . . An idea which can be defined, or a proposition which can be proved, is only of subordinate philosophical interest. The emphasis should be laid on the indefinables and indemonstrables, and here no method is available save intuition.
>
> ([Russell, 1900, §105])

This makes the search for primitives the primary philosophical activity and knowledge of these the primary philosophical knowledge. Commenting in 1899 specifically on the axioms of geometry, he writes:

> And even when precision has been attained, the meaning of the fundamental terms cannot be given, but only suggested. If the suggestion

---

11 For wider discussions of all this, see [Hallett, 2010, §2] and [Hallett, 2019, Conclusion].
12 Acquaintance, of course, later assumed a very important role in Russell's foundational view of philosophy, it being some relation between the mind and what is outside the mind, a relation which is undefined but which is perhaps 'elucidated' in some sense by the various theories Russell gave.

does not call up the right idea in the reader, there is nothing to be done.

([Russell, *1899, 412])

And later he says this in the *Principles of Mathematics* (1903):

> The discussion of indefinables—which forms the chief part of philo-
> sophical logic—is the endeavour to see clearly, and to make others
> see clearly, the entities concerned, in order that the mind may have
> that kind of acquaintance with them which it has with redness or the
> taste of a pineapple.

([Russell, 1903, Preface, xv])

And he goes on to say that the indefinables are the 'necessary residue in a process of analysis', and he compares the search for them 'with a mental telescope' to the search for the planet Neptune, when this was just, in Russell's words, an entity whose existence has been 'inferred'.

I will not take the time to draw out the analogies between this and the elements of Frege's position set out earlier. One question worth asking here is this: despite the impression given by these quotes, does Russell contemplate a notion of elucidation similar to Frege's? I return to this briefly in the following.

In sum, in his search for the referential fixity of the basic notions of arithmetic, Frege relies on some indefinables, including various logical notions required for his deduction system, among them, as he thinks, the notion of extension or value-range. For these, he lays down clear Basic Laws, which are seen as true but not capable of proof. From this it is clear that he is not only relying on unexplained logical primitives but also on some wider philosophical notions, on truth, on existence (of abstract objects), and on denotation, and perhaps, too, the semantic force of the notion of elucidation.

2.   Conceptual Analysis Without Definitions

I want to turn to a very different kind of conceptual analysis, not as well known to philosophers as Frege's or some of the others mentioned previously, but nonetheless of crucial importance in the subsequent development of mathematics and the study of the foundations of mathematics. I'm referring here to Hilbert's analysis of largely planar, rectilinear, Euclidean geometry, an analysis which led to the famous book *Grundlagen der Geometrie*, abbreviated here by '*Festschrift*'.[13] Euclid's *Elements* was

---

13 The monograph was first published in 1899 as [Hilbert, 1899]; it was reprinted as a separate monograph and then subsequently in many editions, often in revised versions. The term '*Festschrift*' is part of the title of the book it was first part of, and this term

mentioned previously, and Book I of this, like Hilbert's work, also lays out a foundation for plane rectilinear geometry. There are many obvious differences between Hilbert's and Euclid's approaches to this material. In the first place, the axiom system Hilbert sets up is a lot more extensive and explicit than Euclid's. Second, Hilbert also gives his treatment a very different direction, away from the *construction* of figures which are used as the basis of the theorems proved, based instead on (relative) *existence* assumptions (e.g., 'for every pair of points there *exists* a unique line . . .'). This is evident both in the way the propositions are formulated and in the way the respective axiom systems are shaped.[14] There is another crucial difference which we'll come to later and which ties up to the major divergence I wish to stress between Hilbert and Frege. However, to underline this, it's necessary to look briefly at a way in which Euclid's and Hilbert's approaches overlap.

2.1.   *Hilbert and Euclid's Project*

Like Euclid, Hilbert was greatly interested in the question of 'what follows from what' and indeed (in Feferman's phrase) 'what rests on what'.[15] In

is used here as a shorthand instead of '*Grundlagen*' primarily to avoid confusion with several other *Grundlagen*, notably Cantor's of 1883 and Frege's of 1884.

Hilbert's publication of the *Festschrift* in 1899 did not come out of the blue, and indeed talk of Hilbert's work on the foundations of geometry refers to much more than just the monograph. Most importantly, there were many lecture courses from 1893/94 until about 1902, and extensive notes for these are all reprinted in the volume [Hallett and Majer, 2004a], as is the original monograph itself. The most important of these, when one focuses on the *Festschrift*, is the series of lectures from 1898/1899 on Euclidean geometry, for which we have two sets of manuscript, [Hilbert, *1898/1899] and [Hilbert, *1899], the former being notes in Hilbert's own hand and the second being an *Ausarbeitung* of these notes of textbook quality made by Hilbert's Assistant Hans von Schaper. For a much fuller account of these documents, and the lecture notes generally, see [Hallett, 2008], especially the Introduction and §1.

14 For a succinct and clear account of the conceptual differences between Euclid's axiomatisation and a modern one, for example, Hilbert's, see [Mueller, 1969]. Note that the move away from construction is indirectly connected to Hilbert's rejection in his paper [1900b] of what he called there 'the genetic method'. How this was conceived is not all that clear, but Hilbert later cast it more narrowly as the 'constructive method', the paradigm being (perhaps) the recursive definition of formula in a standard first-order logic. The point here was to draw a distinction between the language in which a mathematical theory is presented and the mathematics itself. See [Hilbert, *1920, 11].

15 See [Feferman, 1993]. Feferman sets out (p. 188 of the reprint) a very general framework for discussing this question. The most basic level he outlines is when an informal theory $\mathcal{M}$ is formalised in a formal theory T, written in a precise formal language L. In this case, we can say that $\mathcal{M}$ 'rests on' T. The other important senses of 'rests on' which Feferman categorises are a formal theory T being 'justified by' a 'foundational framework' $\mathcal{F}$ (for example, a properly formalised set theory), so where either T is derivable from $\mathcal{F}$ or can be translated into a $T_{\mathcal{F}}$ in the language of $\mathcal{F}$ and $T_{\mathcal{F}}$ is derivable; and then proof-theoretic

both cases, a framework is given based around elementary principles (axioms in Hilbert's case, Postulates and Common Notions in Euclid's) which embraces the major results and imposes on them a priority through their derivations (remember the phrase 'what rests on what'). Hilbert's work is similar to Euclid's in this respect but is a *genuine* extension of Euclid.

It's important to give some detail at this juncture, since, as we will see, the fact that it *is* an extension ties directly to the kind of conceptual analysis Hilbert gives.

Euclid's system (in the carefully constructed received presentation) is officially built around a framework of 23 Definitions, five Common Notions, and five postulates, and the impression is given that this framework is adequate for the development of the elements of plane geometry, including (in Book I) the Isosceles Triangle Theorem (ITT); the congruence ('equality') of triangles; elementary results on parallels, including Playfair's Axiom (PA) and the Angle Sum Theorem (AST); the relative areas of triangles and parallelograms; and Pythagoras's Theorem (PT), the goal of Book I. So, as a sample indication of what I mean here by deductive dependence, ITT appears early in Book I (at I, 5), and its proof rests on relatively few propositions; PT, however, is proved only at I, 47 (its converse is the last proposition in Book I, at I, 48), and its proof depends on an elaborate assemblage of propositions. It has been claimed (see, for example, [Seidenberg, 1975]) that the usual presentation of Euclid's framework is misleading and that Book I, with the exception of Postulate 5 (the Euclidean Parallel Postulate, or EPP), which is only first used in the proof of Proposition 29, is built up using only the three construction postulates and nothing explicit of a propositional nature, relying instead on intuitive principles concerning movement and superposition and perhaps implicitly other unmentioned principles, such as 'Two straight lines cannot enclose a space'. Be that as it may, the fact is that Euclid *does* rely on the derivation of his propositions, and these derivations establish a logical order, from the more primitive and basic to

---

reductions of one T_1 to another T_2, the most interesting cases being where T_2 is actually a derivable part of T_1. Feferman's framework is an extremely useful and important one for analysing post-1900 projects in the foundations of mathematics, especially comparing (say) the project of the set-theoretisation of mathematics with Hilbert's programme. But it's not so useful as a framework for looking at pre-1900 analyses of mathematics. For instance, it's safe to say that the project of Euclid's *Elements* does not fit into this framework, and neither does Hilbert's *Festschrift*. Perhaps the best way to sum up Euclid's project is that he attempted to formulate a systematised $\mathcal{M}$ out of a disparate collection of results, one from which these results all follow as deduced propositions. Hilbert's situation was not the same, of course, though he, too, can be thought of as producing a systematised informal theory. Frege's project might be seen as wanting to take an informal arithmetic $\mathcal{A}$ and present a formalised version T_A inside (deducible from) a formal framework $\mathcal{F}$ (that of the *Grundgesetze*), which is arguably a form of logic. For more on Hilbert's project generally, see the opening sections of [Hallett, 2008].

the more complex. In short, the very deductive arrangement of the work 'displays', in the ordering of propositions, what the ITT (I, 5) or the PT (I, 47) rest on, fewer and simpler propositions in the case of ITT, more and more complex propositions in the case of PT.

Thus, however we might assess Euclid's official framework, there is a basis of 'initial propositions', like the first triangle congruence theorem, I, 4; the subsequent propositions are built up from them by explicit deductions; and it is possible just by looking at the deductions to see that (and how) a subsequent proposition rests on earlier ones. We might put this schematically by saying we have a basis Σ, and we can show that a proposition P 'rests on' Σ (or Σ is adequate for P) by deducing P from Σ, that is, using modern notation, $\Sigma \vdash P$. The deduction of P will reveal how much of Σ is really used.

This kind of project was of great importance in the later nineteenth and early twentieth centuries; witness Frege's work, early axiomatic set theory, and *Principia Mathematica*. Hilbert was deeply interested in this same project, too, but for less all-encompassing theories, in the case in hand, Euclidean geometry, including the modern 'Euclidean' results, such as the Saccheri-Legendre Theorems and Euclidean versions of the Desargues and Pappus/Pascal theorems. Indeed, the sort of ordering considered here is part of the purpose behind Hilbert's grouping of the axioms into five groups. We can ask questions of the form: to prove central proposition P, do we need congruence assumptions (Group III), the Archimedean Axiom (Group V), or the EPP (Group IV)? We can show we *do not* by producing a proof from the axioms which *doesn't* use the principles in question, so a proof of the form $\Sigma^- \vdash P$, where 'Σ^-' indicates that we are using less than the full set of axioms. One example would be the reconstruction of the Euclidean theory of surface measurement *without* the Archimedean Axiom (AA).[16]

I trust the similarity with Euclid comes out. But the important point here is that Hilbert extends this kind of investigation. Moreover, Hilbert also wants to deal with the *other* side of the question of 'what rests on what', namely to be able to show that 'this *can't* rest on that, or not solely'. Thus, we also have to be able to show that:

1. (Base): P (or some theoretical development, so some group of propositions T) *cannot* be derived from the collection of axioms Σ, in other words, to be able to show that

   $$\Sigma \nvdash P \text{ or, more generally, } \Sigma \nvdash \phi, \text{ for all the } \phi \text{ in } T.$$

---

16 See [Hilbert, 1899, Ch. IV, especially §21].

This is the simplest way to put the matter, though there are many variants. For example:

2. (Variant): Show that P (or some theoretical development T) *can* be deduced (carried out) using Σ but *not* using Σ^- (weakening, often slight).
3. (Variant): Show that P (or some theoretical development T) *cannot* be deduced (carried out) using Σ but *can* with Σ^+ (strengthening, often slight). One often ends up showing the following (for an axiom A_j):

$$\Sigma + A_j \vdash P$$
$$\Sigma \text{ (alone, no } A_j) \nvdash P.$$

More generally:

$$(\Sigma - \Sigma_j) + \Sigma_k \vdash P;$$

that is, we strip Σ down and then boost this stripped-down version in another way and show we can then get P.

One of the things Hilbert uses his analyses for is to establish the following: having shown that $\Sigma^* \nvdash P$, we can ask: how do we modify Σ^* to find a $\Sigma^\dagger$ such that $\Sigma^\dagger$ *can* give a proof of P? What emerges from this is that there are often very *different* ways of getting P, by building different $\Sigma^\dagger$ from conceptually very different groups of axioms. One example of this is the Planar Desargues Theorem (Planar DT). This theorem does not follow from the basic planar axioms (so from the axioms governing the things that the theorem seems to be about), but it does follow if these are extended by adding either some *spatial* incidence axioms or some *planar* congruence axioms. The conceptual choice involved here is significantly different. In general, there might be definite mathematical or physical or philosophical reasons for adopting one alternative over another, but choosing one route in one context doesn't preclude choosing another in a *different* context. What is mathematically important is noting the choice. Showing 'what rests on what' need not be a straightforward matter.

4. (Variant): Lastly, Hilbert also investigates situations like the following:

$$\Sigma^- + P \vdash A_j,$$

where P is a central theorem of the system Σ; A_j is a central axiom, essential in the usual proof of P; and Σ^- is a version of Σ with A_j removed. This represents the beginnings of what is now known as the reverse mathematics programme. More generally, we might ask: (a) if we weaken P, can we still get a proof of A_j? (b) Can we find a 'universal' Σ^- which will work for many of the central propositions P?

All this gives a taste of the kinds of investigation of logical, deductive ordering which Hilbert conducts and the variety of results he achieves. A hint of this is given by Hilbert in a letter to Frege in December 1899:

> One more preliminary remark: If we wish to understand one another, then we must not forget the quite different nature of the intentions which guide us. I was forced to set up my system of axioms by necessity. I wanted to make it possible to understand those geometrical theorems which I regard as the most important results of geometrical research, that the Parallel Axiom is not a consequence of the other axioms, likewise not the Archimedean Axiom, etc. I wanted to answer the question whether the theorem that in two equal rectangles with the same base the sides are also equal can be proved, or whether it has to be a new postulate, as it is in Euclid. I wanted to create the possibility of understanding and answering such questions as why the angle sum in triangles is 2 right angles and how this fact is related to the Parallel Axiom. I believe that my *Festschrift* shows that my system of axioms was shaped to answer such questions in a quite definite way, . . . and that in many cases these questions have very surprising and quite unexpected answers.

Note that in this passage, Hilbert refers to 'theorems', that is, propositions which are *proved*. Moreover, at the centre of the kind of investigation of logical ordering we have outlined, and also behind most of the results and investigations raised in the letter to Frege, is the *independence* question, that is, showing at some point or other that $\Gamma \nvdash \phi$ for some selection of axioms Γ and some dedicated proposition ϕ. So when Hilbert says to Frege 'I wanted to create the possibility of understanding and answering such questions', what he means is that he endeavours to set up a framework which allows for the *proof* of the kind of theorem in question and above all of the independence results at their heart.

These theorems are very different from the theorems which Euclid proves. So the genuinely novel feature of Hilbert's work which I want to stress here, and which Hilbert stresses in his letter, is the provision of an extended framework for Euclidean geometry which will enable us to embrace both Euclidean-style proofs and also theorems of this new sort as well.

Hilbert, of course, is *not new* in tackling independence questions, the most significant result here being the independence of EPP in its various forms, mentioned explicitly in the letter to Frege quoted previously. But the crucial thing is that Hilbert's work *generalises* this radically, as is clear from the schematic representations recently given. The treasury of results is massively expanded by widespread use of the independence procedures. But it is important to underline the methodological shift which makes this possible and which will bring us to the central difference between Hilbert's conceptual analysis and that of Frege.

2.2. A Framework for Independence

The basis of Hilbert's method seems quite familiar to us now, and we would perhaps describe it as using *models* of a (formal) theory in which all members of a group of sentences Σ are *true*, and then a certain target sentence *P* is *false*, models which can be chosen arbitrarily. As pursued by Hilbert, the mathematical heart of this is to find novel interpretations of the sentences concerned which bring this about, and the consummate mathematical invention is manifest in the wide-ranging and sophisticated nature of the models that Hilbert creates. All of this work leads to many interesting *geometries* other than Euclidean and non-Euclidean, for example, non-Archimedean, non-Desarguesian, non-Pascalian, non-Pythagorean. It is not quite the modelling of a formal theory which we would recognise now, but it goes a good way towards this. I say 'as pursued by Hilbert', because he generalised in a radical way the method used by others (e.g., Beltrami and Poincaré) in considering the EPP, a way deeply associated with the novelty of his version of conceptual analysis when compared to Frege's.

Before turning to this generalisation, let me review very briefly the history of the independence of the EPP, at least as I see it.

In the first part of the nineteenth century, Bolyai and Lobachevsky (independently) had developed a different kind of geometry, which Saccheri (unknowingly) and Lambert and Gauss (knowingly) had latched onto before Lobachevsky and Bolyai, in which, to highlight the most striking things, there are infinitely many parallels to a given line through a given point, there is a unique limiting parallel to a given line, and the angle-sum of a triangle is proportional to its area. For want of a term, we call this geometry 'BL geometry', after Bolyai and Lobachevsky. A question raised for those soaked in traditional geometry is whether such a geometry is coherent or internally inconsistent, as Saccheri and Legendre thought it must be. In 1868, Beltrami noticed that the geometric behaviour of certain curvilinear triangles on a particular two-dimensional analytic surface in ordinary (analytic) Euclidean three-space behave just as triangles are described to behave in the BL geometry. If, as Beltrami says, we *reinterpret* the BL statements to be about such triangles, then we have something of a guarantee that the BL geometry is coherent and not flawed, though it's a limited guarantee, for Beltrami should have added the caveat 'at least as far as BL had so far been developed'. The important point is, though, that the strange behaviour of the BL geometry which Saccheri and others had noted is all *modelled* in parts of these Euclidean structures. Beltrami's observation was then confirmed by further investigations of Beltrami himself, Klein, and Poincaré, all of which led to the development of simpler 'models' of BL geometry.

What this shows is what we would now naturally think of as the *satisfiability* of this geometry *within* Euclidean geometry (given the previous caveats), and not its *deductive* consistency, and this assumes implicitly, of course, the satisfiability of Euclidean geometry itself.

The question of deductive consistency was addressed *explicitly* by Poincaré in a paper published in 1891. He pointed out the following.

Suppose we take the language of BL (call it $\mathcal{L}_{BL}$) and translate it into the language of Euclidean geometry (call it call it $\mathcal{L}_{EG}$) using one of Poincaré's simple models (say the Poincaré disc $\mathcal{D}$, with straight lines taken as arcs of circles running inside the disc orthogonal to the circumference of $\mathcal{D}$). A condition of Poincaré's translation is that axioms of BL are taken to theorems of Euclidean geometry, and it must preserve logical form (a matter overlooked by Poincaré). Such a translation (preserving logical form and relativising quantifiers in the right way) can be used to show that any proof of a contradiction in BL will be translated into a proof of a contradiction in Euclidean geometry. Since this latter ('ordinary geometry', as Poincaré calls it) is 'clearly consistent' (so Poincaré), then BL is also consistent. The same procedure can be adapted to show the unprovability of EPP from the other Euclidean assumptions.

What underlies Poincaré's explanation, I think, is the view that BL is really just describing a small part of the wider geometric world of Euclidean geometry. So, if we ask the question 'What is Euclidean geometry about?', the answer would be 'points, straight lines, segments, triangles, . . .'; if we then ask the question 'What is BL about?', we would get the answer 'Insofar as it's coherent, it's about things like $\mathcal{D}$ and certain arcs within it, or about certain lines on the Beltrami surface, or . . .'. The point is, though, that here we have Euclidean geometry all the way through.[17] Poincaré's position is already a generalisation of the Beltrami considerations, since satisfiability is used explicitly to show *deductive consistency*.

2.3. *Hilbert and Variable Interpretation*

But Hilbert's position is not Poincaré's. What characterises the latter is the dominance of Euclidean geometry, and its points, straight lines, circles, and so on and the non-Euclidean notions are understood through this Euclidean lens. This *could not have been* the Hilbert view, since for him (and he is clear about this), there is no interpretation to begin with.

There are various pieces of evidential support for this.

First, as early as 1893/94, Hilbert states that a (geometric) theory is really only what he calls a '*Fachwerk*' (something like a framework) of concepts, which is left open for interpretation. According to a story related by Blumenthal, he was already saying in 1891 that *any* primitives

---

17 This might be the origin of Poincaré's famous 'geometric conventionalism', but this is not the place to discuss the matter.

would do, and to emphasise this, he gives the outlandish example of allowing tables, chairs, beer mugs as points, straight lines, and planes.[18]

Second, Hilbert repeats this some years later in his correspondence with Frege. There he calls a theory a 'schema of concepts' which can be filled with 'material [*Stoff*]' in whatever way we choose (and the more ways, the better!), and he uses the same kind of provocative example of underlying 'material' as we find in the Blumenthal anecdote, this time using, love, law, chimney-sweep. (See letter to Frege 29.xii.1899.)[19]

Third, Hilbert avoids giving any *definitions*, even allusive explanations ('elucidations') of the primitives, and in fact indicates to his students that the hardest task they will have will be that of setting aside what they think they know about points, lines, and planes (or beer mugs), in other words, that they must *operate with no fixed interpretation*. (For this, see [Hilbert, *1898/1899, 6].) This is summed up in the following passage from a letter to Frege:

> You [Frege] say further: 'The explanations in §1 [of the *Festschrift*] are apparently quite different, for here the meanings of the words "point", "line", . . . are not specified, but assumed as known.' Here lies the cardinal point of the misunderstanding. *I wish to assume nothing as known in advance*; I regard my explanation in §1 as the definition of the concepts point, line, plane, if one adds to this all the axioms of groups I–V as characteristic marks. If one is look-ing for other definitions of 'point', perhaps through paraphrases like extensionless, etc., then I must oppose such approaches in the most decisive way; one is looking for something one can never find because nothing is there; and everything gets lost and becomes confused and vague, and degenerates into a game of hide-and-seek.
> (Hilbert to Frege, 29.xii.1899)[20]

Thus, it is the *axioms*, if anything, which do the defining. Nothing is assumed known about the primitives 'in advance' of giving the axioms, something which is also stressed in Hilbert's 1899 lectures; there (as in the *Festschrift*), he names the primitives ('points', 'straight lines', 'planes') in the usual way but says that it's only with the axioms that we begin to assign properties to these things.

---

18 See [Blumenthal, 1935, 402–403].

19 The view of axiomatic theories as '*Fachwerke*' can be found in many subsequent places in Hilbert's writing, for example, in [Hilbert, 1918], and in lectures from 1921–22, [Hilbert, *1921/1922].

20 The italics are mine. The reference to 'characteristic marks' is a reply to Frege's com-ment that Hilbert's axioms cannot give definitions of the primitives, counter to Hilbert's insistence, since they lay down no 'characteristic marks' of the things being defined.

This is clearly consonant with the outlandish example that Blumen-thal cites and indeed makes sense of it, for surely we are *not*, as we might think at first, taking beer mugs and all we normally associ-ate with them to be planes. What properties those particular objects (tables, chair legs, mugs) will have when taken as the basis of a *geome-try* is regulated solely by the axioms, and it is very much to be doubted whether their physical constitution (wood, pewter, or glass) will play any part.

Last, let us also be clear that these axioms are relational in nature, in the sense that the axioms standardly state ways in which primitives in different classes (typically, the points, straight lines, and planes) *behave with respect to each other*. For example, the first incidence axiom states that given any two points, there is a unique straight line running between them. Axiom I, 5 says that if two points lie in the same plane, then all the points on the unique line joining them also lie in that plane. Then there is an axiom which says that if two planes have a point in common, then they have a straight line in common. So, if we have no information other than the axioms, as Hilbert insists we do not, then this means that we only start to lay down the properties characteristic of points by saying some things about straight lines and planes as well and how they are all related. As Hilbert says to Frege:

> My view is just this, that a concept can only be logically fixed through its relations to other concepts. These relations, formulated in definite statements, I call axioms, and thus I arrive at the view that these axioms (perhaps with the addition of names for the concepts) are the definitions of these concepts.
>
> (Letter of 29.xii.1899)

Moreover, if there is something important in a traditional definition which, on consideration, we want to include, then this can be built in as an axiom.[21] An example is found in Hilbert's plane axiom I, 5, mentioned previously, which corresponds fairly exactly to one traditional definition of a plane surface, namely that the unique straight line joining any two points in the plane lies wholly in it, too.[22]

And there is one other important thing: it is not the separate axioms which count so much as the axioms taken together. Note Hilbert's insistence in the passage cited previously that the 'characteristic marks' are given by the axioms in Groups I–V, that is, *all* the groups. For

---

21 A nice example might be one from the history of non-Euclidean geometry, for example, Legendre's Axiom (LA). Legendre was appalled by the notion, which BL geometry per-mits, of there being a point in the interior of an angle such that there can be a straight line through it which cuts neither leg of the angle. LA rules this out. For a statement of LA, and the relevant remarks of Legendre, see [Hartshorne, 2000, 322–323].

22 Gauss was seriously interested in this 'definition'. See [Hallett and Majer, 2004a, 397].

example, the Playfair version of the Parallel Axiom, which Hilbert adopts, tells us that for *any point* outside *a straight line, there's a unique co-planar straight line* parallel to the given line through that point.

Once we observe this, since the axioms are all we have, it's clear that the more axioms we lay down, the more we refine or change the primitive notions, then adding *different* axioms (or axiom groups) will give us *different* conceptions of line, plane, and linear congruence. So, in short, we treat 'point', 'line', 'plane', 'betweenness', and 'congruence' *together*, as a system, and we can take smaller or larger sets of axioms. If the final aim (that is, with all the axioms together) is to frame a system which matches in theorems and achievements Euclid's geometry, this way of presenting it also considers many different systems along the way: and each time an axiom is added or one altered or subtracted, we have a different system with a different conception of point (or line or . . .). For instance, adding the linear congruence axioms tells us that we can compare the linear distances between points; adding the PA tells us something very important about the relationship between lines and points outside those lines; adding the Archimedean Axiom tells us that points are all finitely accessible from each other whatever fixed length we take as unit. Each addition to, or subtraction from, the axiom system changes our concept of point (or line or . . .). Similarly, if we introduce a non-Euclidean Parallel Axiom or the *denial* of the Archimedean Axiom, then we have *different* concepts of point. As Hilbert says to Frege (letter, 29.xii.1899), in each of these systems, 'point' is something different. And it should be added here that the investigation of logical dependency briefly described previously is of a piece with this view. For this opens up many more interesting candidates for axioms.

For Frege, none of this can happen: the reference of 'point' must be fixed *before* we give axioms at all.[23] For Hilbert, though, consideration of meaning can only begin *once* we have axioms. We are, for him, clearly *defining* in this way, though it's wrong to say (as Hilbert sometimes does) that in doing this we are 'defining' points, straight lines, planes, and so on: what we're defining, rather, is a geometrical *system*, in fact, many systems.[24]

We could say much more, but this is sufficient to bring out the difference with Frege's conception. One way to summarise this is to say

---

23 Recall this declaration from Frege's *Grundlagen* quoted earlier:

> Yet if anyone had the right to understand by this name whatever he pleased, then the same proposition about one would mean different things for different people; there would be no common content for such propositions. ([Frege, 1884, Introduction, I])

24 In fact, we are defining higher-order complex concepts, and we have to show that they are instantiated. See the discussion in [Hallett, 2012].

that Frege's approach is atomic and parallels (or perhaps is of a piece with) his molecular notion of sense, which we might sum up as saying that the individual linguistic items are taken to be meaningful in themselves, and then go to make up the meanings of whole sentences, axioms being some of these.[25] In other words, we start from meaningful 'atoms' and build up to meaningful (and true) axioms. Hilbert's view is totally opposed to this. Indeed, Frege says clearly that if the references of the primitives are not fixed, as they are not in Hilbert's view, then the 'axioms' will be in effect pseudo-axioms. (See [Frege, 1903b] and [Frege, 1906].) For Hilbert, sense is not built up from the primitives to the complex propositions but, if anything (insofar as the meaning of individual statements taken in this atomistic way is concerned), the other way around. We start from a group of axioms and then focus down on what for Frege are the atoms. Whether we say the purpose is to isolate theories or structures, and meaning in general, the approach is certainly not atomistic.

Hilbert's 'axiomatic method' had a significant impact. The context of what is presented here is geometry, but Hilbert saw this way of presenting mathematical concepts and mathematical theories as quite general, applying across mathematics as a whole. Take the example of the field axioms: there are some primitives, and the axioms specify the way these are all tied to each other by the given operations, and we can add further axioms, and so on. We might end up with axioms for complete ordered fields, but many fields will be considered on the way to this. The point to be emphasised here, though, is that a great many different things could be taken as primitives, including line segments and certain sets of rationals. To put it dramatically, in this view, there are no uniquely fixed real numbers.

Another rather striking example is given by elementary logic, as axiomatised in both [Hilbert, *1917/1918] and [Bernays, *1918] (see also [Bernays, 1926]). For one thing, the axioms bring out the interdependence of the connectives. And investigations of independence are carried out by reinterpreting the connectives (which are taken in logic to have fixed interpretations) as operators in simple numerical algebras.

3. Concluding Elucidations

I want to conclude by looking at some of the consequences of the Hilbert view, consequences which connect to Frege's view.

---

25 These remarks are designed largely for rhetorical effect to bring out a distinction, and they brush over many of the complexities of Frege's theory of meaning, not least the role of the Context Principle.

One way to state the difference between Hilbert and Frege is to say that Hilbert was interested in using axiom systems to single out the underlying *structures* and not in the constitution of the *elements* of the structures, whereas Frege shows no serious interest in this. This was also a very different approach to geometry from that of Euclid, as Mueller points out (see [Mueller, 1969, 299]).

Nevertheless, one might wonder whether there is such a huge difference in the matter of definitions between Hilbert's position and that of Euclid. After all, Euclid's definitions of the geometrical primitives (such as 'point') play no role in the theoretical development of the *Elements*. To take just one example, Euclid never uses the 'lies evenly' criterion in proving that there is a straight line connecting points A and B. Hence, one might ask the following: if Euclid's definitions of the primitives *play no role* in the development of the theory itself, is there really any difference between Hilbert's geometry *without* definitions of the primitives and Euclid's, which *has* definitions of primitives but which play no significant role?

But I think this is to misunderstand. Perhaps the best way to see this is to see Euclid's definitions not so much as telling you what points and so on *are*, thus telling you what geometry is really about, but rather as an attempt to *rule things out*.[26] So, an ink blot cannot be a point because it has parts. But Hilbert's position is that in principle, *nothing* is to be ruled out as a point, particularly not things which have parts, as tables do! We might *want* to take ink blots as points, or nail heads, or distant stars, 'pinpoints' of light, or stretched threads as straight lines, or light rays, even though we know that these will never 'lie evenly with the points on them', because they will always bend a little under the force of gravity. Of course, one does this with a measure of salt and also in the knowledge that if one wants to cut down the amount of salt used, we can take the corresponding centres of mass as the points, or some such thing. But the point is clear: *we should in principle rule nothing out.*

Physical interpretations aside, what Hilbert, the mathematician, exploits is an element of mathematical practice which had grown in importance during the nineteenth century. One example, mentioned previously, is taking bundles of lines as points as a way of 'adding' points in projective geometry. Bundles *clearly* have parts and thus would have been excluded by a definition like Euclid's. Hilbert relies on this very widely in the subsequent way in which he constructs models for geometries; if we are freed from the convention that points should have no 'parts', then we are at liberty to take certain infinite series

---

26 Recall Frege's claim that elucidations are intended 'to fend off' the unwanted meanings.

in a function space as points and use this to violate the Archimedean Axiom, and so on.

In short, Hilbert's way of 'theoretical development independent of definitions' is *very different* from Euclid's. This is the core of Hilbert's difference with Frege, for it is precisely here that we find the new kinds of theorems of geometry that Hilbert describes to Frege. We seem to have the very converse of the Frege view about elucidation: the more we exclude things, as Frege/Euclid wanted to do, the more we narrow down the class of interpretations permissible, and so the more we cut ourselves off from mathematically interesting knowledge. And this is the point where Hilbert the consummate mathematician is to be seen, representing the study of 'elementary mathematics from an advanced standpoint'.

It might be argued that theorems of this new kind are not geometrical, rather meta-results, and the knowledge so achieved is not really geometrical knowledge but rather logical or meta-geometrical knowledge. But I think this is wrong. When we learn that EPP cannot be proved from the other axioms of geometry, we have surely learned something straightforwardly geometrical, just as we have when we learn (following Euclid) that the AST can be deduced using that axiom. Similarly, we have learned something geometrical when we learn that one can only prove the parallel axiom from the AST (thus, the converse) in the presence of the Archimedean Axiom, as Hilbert's student Dehn showed. (See [Dehn, 1900].) In his correspondence with Frege, Hilbert calls these results 'geometrical theorems', and he was correct to do so.

One of the things revealed by results like these is the surprising connectedness of geometrical propositions and hence geometrical concepts, and this is at the root of the difference between the analyses of Hilbert and Frege. All this, I submit, is part of the 'conceptual analysis' of Euclid's geometry, just as the taking apart of a steam engine, trying to see what is responsible for what, what depends on what, what could be dispensed with, and what is essential, whether the same goals could be achieved with different materials or in a different fashion, is part of the (conceptual) analysis of that engine.

Last, I want to return to elucidation. Earlier in the chapter, we quoted Russell:

> And even when precision has been attained, the meaning of the fundamental terms cannot be given, but only suggested. If the suggestion does not call up the right idea in the reader, there is nothing to be done.
>
> (Russell's unpublished English version, [Russell, *1899, 412])

This comes in the middle of the following:

> He [Poincaré] suggests, in various passages, other phrases, chiefly drawn from the theory of groups, which he rightly considers to be more precise. I was aware of these phrases, and in a mathematical work I should have employed them. But in a philosophical work they would have been wholly out of place, and would not have helped to solve the problems in hand. This is one of the difficulties of mathematical philosophy, that the whole language of mathematics has to be abandoned, owing to the difference of the point of view. This difference may, I think, be described roughly as follows. In Mathematics, what is important is the relation of terms. When two sets of terms have the same mutual relations, they are equivalent for mathematical purposes. What the terms are in themselves, is irrelevant; only their relations are important. But in philosophy, it is essentially the terms themselves that are important. We must ask, what do our terms *mean*? not, how are they related to other terms? Whenever a term is analyzable, philosophy should undertake the analysis. This is a task for which mathematical language is in general very ill-suited, and in which precision can only be attained with great difficulty. *And even when precision has been attained, the meaning of the fundamental terms cannot be given, but only suggested. If the suggestion does not call up the right idea in the reader, there is nothing to be done.* Thus philosophical precision is a very different matter from mathematical precision. It is more difficult to attain, and far more difficult to communicate. And this is why the language of groups cannot help us to a philosophical account of the foundations of geometry.
>
> (Russell's unpublished English version,
> [Russell, *1899, 411–412])[27]

What Russell suggests here is that what philosophers look for is the meaning of the terms *individually*, which is what the philosophical analysis is designed to help with, and this applies even to the *fundamentals* (conceived individually) where there cannot be analysis any more. As Russell says, 'in philosophy, it is essentially the terms themselves that are important. We must ask, what do our terms *mean*? not, how are they related to other terms?' In Russell's view, we will have to resort in the end to an understanding of these primitives through

---

27 I have set the short passage originally quoted in italics. The published French version is [Russell, 1899, 702–703].

some sort of direct acquaintance, akin, to use Russell's example, to our acquaintance with the taste of a pineapple. We *could* here give further elucidation, for instance, through something like 'Well, the taste is slightly more acidic than the taste of a peach, but not as acidic as that of an orange, sweeter than that of a tomato, but juicier than that of a banana, . . .'. But this can't be a replacement for the kind of thing Russell has in view. For one thing, note that this further elucidation is relational; the tastes of peaches, oranges, tomatoes, and so on are brought into comparison as well, and pursuing relational accounts is not what the philosopher should be doing, although it is what the mathematician does.

Thus, Russell appears to be saying not that further explanations are *not* possible but rather that the *philosopher* must cut these off, no doubt because they will just invoke *more* theory, relating them to other notions themselves not explained, and so on. In the case of supposedly fundamental mathematical terms, the supplementary explanations involve further mathematical theories, which suggests that we could (often) get further explanations of the primitives (further elucidation) if we're prepared to allow further mathematical theory.

So what we see in this longer passage is a glimpse (by way of rejection) of the alternative which Hilbert decisively *adopts*. Indeed, the use of the further mathematical theory in the interpretation of the primitives plays a crucial part in the building of models, in particular with the notion of relative interpretability. I said earlier that Hilbert *eschews* Fregean elucidation as a way of fixing reference in some kind of *absolute* way. However, freedom of interpretation allows Hilbert many *different* 'elucidations' of the primitives, using a wide (indeed unlimited) range of auxiliary mathematical theories to carry this out. Although I can't go into this here, I think this is deeply connected to the later emphasis on the 'deepening of the foundations' which is strikingly asserted in Hilbert's paper on the axiomatic way of thinking [1918].

References

Bernays, P. (*1918). Beiträge zur axiomatischen Behandlung des Logik-Kalküls. Habilitationschrift, presented to the Georg-August Universitat Göttingen. First published in [Ewald et al., 2013, Chapter 1, Appendix, 231–270].

Bernays, P. (1926). Axiomatische Untersuchung des Aussagen-Kalküls der "Principia Mathematica". *Mathematische Zeitschrift*, 305–320.

Blanchette, P. (2007). Frege on consistency and conceptual analysisi. *Philosophia Mathematica*, 15: 321–346.

Blumenthal, O. (1935). Lebensgeschichte. Published in [Hilbert, 1935, 388–429].

Czermak, J., editor (1993). *Philosophy of Mathematics. Proceedings of the Fifteenth International Wittegenstein-Symposium, Part 1*. Verlag Hölder-PichlerTempsky, Vienna.

Dehn, M. (1900). Die Legendre'sche Sätze über die Winkelsumme im Dreieck. *Mathematische Annalen*, 53: 404–439.

Dummett, M. (1987). Frege and the paradox of analysis. Published in [Dummett, 1991, 17–52].

Dummett, M. (1991). *Frege and Other Philosophers*. Clarendon Press, Oxford.

Ebert, P. and Rossberg, M., editors (2019). *Essays on Frege's Basic Laws of Arithmetic*. Oxford University Press, Oxford.

Ewald, W., editor (1996). *From Kant to Hilbert. Two Volumes*. Oxford University Press, Oxford.

Ewald, W., Sieg, W., and Hallett, M. (2013). *David Hilbert's Lectures on the Foundations of Logic and Arithmetic, 1917–1933*. Springer Verlag, Heidelberg, Berlin, New York. David Hilbert's Lectures on the Foundations of Mathematics and Natural Science, Volume 3.

Feferman, S. (1993). What rests on what? The proof-theoretic analysis of mathematics. Published in [Czermak, 1993, 147–171]. Reprinted (with corrections and additions) in [Feferman, 1998, 187–208].

Feferman, S. (1998). *In the Light of Logic*. Oxford University Press, New York and Oxford.

Frappier, M., Brown, D., and DiSalle, R., editors (2012). *Analysis and Interpretation in the Exact Sciences: Essays in Honour of William Demopoulos*. Springer, Dordrecht, Berlin.

Frege, G. (1879). *Begriffsschrift, eine der arithmetischen nachgebildete Formelsprache des reinen Denkens*. Verlag von Louis Nebert, Halle-an-die-Saale. Reprinted in [Frege, 1964]; English translation in [Heijenoort, 1967, 1–82].

Frege, G. (1884). *Die Grundlagen der Arithmetik*. Wilhelm Koebner. Reprinted by Felix Meiner Verlag, Hamburg, 1986, 1988. English translation by J. L. Austin as *The Foundations of Arithmetic*, Basil Blackwell, Oxford. Second Edition, 1953. The Austin edition is a bilingual one, and the page numbers are the same for both the German and English texts.

Frege, G. (1893). *Grundgesetze der Arithmetik, Band 1*. Hermann Pohle, Jena. Reprinted together with [Frege, 1903a] in one volume by Georg Olms, 1966. English translation [Frege, 2013].

Frege, G. (1903a). *Grundgesetze der Arithmetik, Band II*. Hermann Pohle, Jena. Reprinted together with [Frege, 1893] in one volume by Georg Olms, 1966. English translation [Frege, 2013].

Frege, G. (1903b). Über die Grundlagen der Geometrie. *Jahresbericht der deutschen Mathematiker-Vereinigung*, 12: 319–324, 368–375. Reprinted in [Frege, 1967, 262–272]. English translation in [Frege, 1984], where the original page numbers are also given.

Frege, G. (1906). Über die Grundlagen der Geometrie. *Jahresbericht der deutschen Mathematiker-Vereinigung*, 15: 293–309, 377–403, 423–430. Reprinted in [Frege, 1967, 262–272]. English translation in [Frege, 1984], where the original page numbers are also given.

Frege, G. (1964). *Begriffsschrift und andere Aufsätze. Zweite Auflage. Mit E. Husserls und H. Scholz' Anmerkungen herausgegeben von Ignacio Angelelli*. Wissenschaftliche Buchgesellschaft, Darmstadt.

Frege, G. (1967). *Kleine Schriften. Edited by Ignacio Angelelli*. Georg Olms Verlag, Hildesheim. Reprinted in 1990 with additional comments and corrections by the editor.

Frege, G. (1976). *Wissenschaftlicher Briefwechsel. Edited by G. Gabriel, H. Hermes, F. Kambartel, F. Kaulbach, C. Thiel and A. Veraart.* Felix Meiner, Hamburg.

Frege, G. (1980). *Philosophical and Mathematical Correspondence.* Basil Blackwell, Oxford. Abridged from [Frege, 1976] by Brian McGuinness, and translated by Hans Kaal.

Frege, G. (1984). *Collected Papers on Mathematics, Logic, and Philosophy.* Basil Blackwell, Oxford. Edited by Brian McGuinness.

Frege, G. (2013). *Basic Laws of Arithmetic, Derived Using Concept-Script, Volumes I and II. Translated and Edited by Philip Ebert and Marcus Rossberg, with Crispin Wright.* Oxford University Press.

Hallett, M. (2008). The 'purity of method' in Hilbert's *Grundlagen der Geometrie.* Published in [Mancosu, 2008, 198–255].

Hallett, M. (2010). Frege and Hilbert. Published in [Potter and Ricketts, 2010, 413–464].

Hallett, M. (2012). More on Frege and Hilbert. Published in [Frappier et al., 2012, 135–162].

Hallett, M. (2019). Frege and creation. Published in [Ebert and Rossberg, 2019, 285–324].

Hallett, M. and Majer, U., editors (2004a). *David Hilbert's Lectures on the Foundations of Geometry, 1891–1902.* Springer Verlag, Heidelberg, Berlin, New York. Hilbert's Lectures on the Foundations of Mathematics and Physics, Volume 1.

Hartshorne, R. (2000). *Geometry: Euclid and Beyond.* Springer, New York, Berlin, Heidelberg.

Heath, T. L. (1925). *The Thirteen Books of Euclid's Elements. Three Volumes. Second Edition.* Cambridge University Press, Cambridge.

Heijenoort, J. v., editor (1967). *From Frege to Gödel: A Source Book in Mathematical Logic.* Harvard University Press, Cambridge, MA.

Hilbert, D. (*1898/1899). *Grundlagen der Euklidischen Geometrie.* Lecture notes for a course held in the Wintersemester of 1898/1899 at the Georg-August Universität, Göttingen. Niedersächsische Staats- und Universitätsbibliothek, Göttingen. First published in [Hallett and Majer, 2004a, 221–301].

Hilbert, D. (*1899). *Elemente der Euklidischen Geometrie. Ausarbeitung* by Hans von Schaper of the lecture notes Hilbert *1898/1899. Niedersächsische Staats- und Universitätsbibliothek, Göttingen, and the Mathematisches Institut of the Georg-August Universität, Göttingen. First published in [Hallett and Majer, 2004a, 302–406].

Hilbert, D. (1899). Grundlagen der Geometrie. In *Festschrift zur Feier der Enthüllung des Gauss-Weber-Denkmals in Göttingen.* B. G. Teubner, Leipzig. Republished as Chapter 5 in [Hallett and Majer 2004a].

Hilbert, D. (1900b). Über den Zahlbegriff. *Jahresbericht der deutschen Mathematiker-Vereinigung,* 8: 180–185. Reprinted (with small modifications) in the second to seventh editions of [Hilbert, 1899]. English translation [Ewald, 1996, Volume 2, 1092–1195].

Hilbert, D. (*1917/1918). Prinzipien der Mathematik. *Ausarbeitung* by Paul Bernays of notes for lectures in the Wintersemester 1917/1918, Mathematisches Institut of the Georg-August Universität, Göttingen, vii + 246 pages, typewritten. First published in [Ewald et al., 2013, Chapter 1].

Hilbert, D. (1918). Axiomatisches Denken. *Mathematische Annalen*, 78: 405–415. Reprinted in [Hilbert, 1935, 146–156]; English translation in [Ewald, 1996, Volume 2, 1105–1115].

Hilbert, D. (*1920). Probleme der mathematischen Logik. *Ausarbeitung* by N. [sic] Schonfinkel and Paul Bernays of notes for lectures in the Sommersemester 1920, Mathematisches Institut of the Georg-August Universität, Göttingen, i + 46 pages, typewritten. First published in [Ewald et al., 2013, in Chapter 2, 342–375].

Hilbert, D. (*1921/1922). Grundlagen der Mathematik. *Ausarbeitung* by Paul Bernays of lecture notes for a course held in the Wintersemester of 1921/1922 at the GeorgAugust Universität, Göttingen. Library of the Mathematisches Institut, 100 + 9 + 38 pages, typewritten. First published in [Ewald et al., 2013, Chapter 3, 431–527].

Hilbert, D. (1935). *Gesammelte Abhandlungen, Band 3*. Julius Springer. Reprinted by Chelsea Publishing Co., New York, 1965.

Hodges, W. (2004). The importance and neglect of conceptual analysis: Hilbert-Ackermann iii-3. In Hendriks, editor, *First-Order Logic Revisited*, pages 129–153. Logos-Verlag, Berlin.

Mancosu, P., editor (2008). *The Philosophy of Mathematical Practice*. Oxford University Press, Oxford.

Mueller, I. (1969). Euclid's *Elements* and the axiomatic method. *British Journal for the Philosophy of Science*, 289–309.

Poincaré, H. (1891). Les géométries non euclidiennes. *Revue général des sciences pures et appliqués*, 2: 769–774. Revised and reproduced in part in [Poincaré, 1908].

Poincaré, H. (1902). *La science et l'hypothèse*. Ernst Flammarion, Paris. English translation as [Poincaré, 1905], and retranslated in [Poincaré, 1913].

Poincaré, H. (1905a). *La valeur de la science*. Ernst Flammarion, Paris. English translation in [Poincaré, 1913].

Poincaré, H. (1905b). *Science and Hypothesis*. Walter Scott Publishing Company. English translation by W. J. G. of [Poincaré, 1902]. Reprinted by Dover Publications, New York, 1952.

Poincaré, H. (1908). *Science et méthode*. Ernst Flammarion, Paris. English translation in [Poincaré, 1913], and retranslated by Francis Maitland as *Science and Method*, Dover Publications, New York.

Poincaré, H. (1913). *The Foundations of Science*. The Science Press, New York. English translation by G. B. Halsted of [Poincaré, 1902], [Poincaré, 1905] and [Poincaré, 1908], with a Preface by Poincaré, and an Introduction by Josiah Royce.

Potter, M. and Ricketts, T., editors (2010). *The Cambridge Companion to Frege*. Cambridge University Press, Cambridge.

Russell, B. (1899). Sur les axiomes de la géométrie. *Revue de métaphysique et de morale*, 7: 684–707. Reprinted in [Russell, 1990, 434–451]. An English version also appears in [Russell, 1990, 394–415], drawn mainly from Russell's English manuscript, supplemented by some pages retranslated from the French.

Russell, B. (*1899). The axioms of geometry. First published in [Russell, 1990, 390–415].

Russell, B. (1900). *A Critical Exposition of the Philosophy of Leibniz*. Cambridge University Press, Cambridge.

Russell, B. (1903). *The Principles of Mathematics, Volume 1*. Cambridge University Press.

Russell, B. (1990). *The Collected Papers of Bertrand Russell, Volume 2: Philosophical Papers, 1896–99*. Routledge, London and New York.

Seidenberg, A. (1975). Did Euclid's *Elements*, Book I, develop geometry axiomatically? *Archive for History of Exact Sciences*, 14: 263–295.

8 The Chimera of Logicism
Husserl's Criticism of Frege[1]

Mirja Hartimo

Introduction

In his *Philosophy of Arithmetic* (1891, henceforth PA), Husserl discusses Frege's view of the notion of number in detail. He summarizes Frege's goal as follows:

> A founding of arithmetic on a sequence of formal definitions, out of which all the theorems of that science could be deduced purely syllogistically, is Frege's ideal. [*Eine Fundierung der Arithmetik auf eine Folge formaler Definitionen, aus welchen die sämtlichen Lehrsätze dieser Wissenschaft rein syllogistisch gefolgert werden könnten, ist das Ideal Freges.*]
>
> (PA 118/124)

Husserl thus captures Frege as a logicist in the contemporary sense of the term (the term was not used at the time). Equally clearly, Husserl is not himself a logicist. Immediately after having stated Frege's ideal, Husserl continues:

> Surely no discussion is necessary to show why I cannot share this view, especially since all of the investigations which I have carried out to this point present nothing but arguments in refutation of it.
>
> (PA 118–119/124)

I will discuss the investigations that Husserl here refers to in the first section of this chapter. In them, Husserl argues against different kinds of attempts to define equivalence. Among these is what is today called 'Hume's Principle', central to neologicism; thus, it will become clear that Husserl's arguments also represent a challenge to the neologicist program. In the second section, I will present Husserl's more precise argument

1 Thanks are due to John Corcoran, Øystein Linnebo, and an anonymous referee.

DOI: 10.4324/9780429277894-10

against Frege as presented in *Philosophy of Arithmetic*. It culminates in a view that, for Husserl, any attempt to logically define abstract concepts, such as numbers, is "chimerical" because it fails to capture what is meant by number words in ordinary life and sciences. After this, I will shortly examine Husserl's own proposal for how numbers should be analyzed.

In his famous review of Husserl's *Philosophy of Arithmetic*, Frege found Husserl's approach psychological. In the remainder of this chapter, I will touch upon the debate about the psychologism of logic, casting it as a debate about logicism. My aim is to show how it could be seen to lead Husserl to his later transcendental phenomenological concept of correlation and, with it, to a more developed criticism of logicism. I will discuss Frege's review of Husserl in Section 4. In the review, Frege criticized Husserl for failing to provide a clear-cut distinction between the subjective and the objective. In Section 5, I will review Husserl's own understanding of his development. He thinks that he arrived at his position expressed in *Logical Investigations* having wrestled with problems in two different directions: on one hand, his logical investigations pushed him beyond the domain of quantity, towards a universal theory of formal deductive systems, and, on the other hand, he was troubled with the epistemological problems of connecting logic with psychological foundations. I will argue that the former led Husserl to his idea of pure logic and the latter to reconstructing the relationship between the objective and the subjective by means of his late notion of "correlation" (that he later claims he had discovered in 1898 when working on *Logical Investigations*). Whereas the idea of pure logic shows the mathematicians' abstract structures to be independent of psychology, the latter shows how they should be related to the subject of knowledge without falling prey to psychologism. I will argue that, considered together, these two claims imply a refined criticism of Frege. This is twofold: First, Frege's procedure that emphasizes the need to define the notion of number does not agree with the mathematicians' more axiomatic view of what mathematics is about (and take into account the development of mathematics in the late 19th century), and second, the logicist methodology cannot provide *philosophical insight* into the essence of mathematics.

1.   Over-Zealousness for Rigor

Husserl's argument against Frege grows out from his more general complaint about the attempts to define equality [*Zahlengleichheit*] and difference [*Zahlenverschiedenheit*]. While Husserl thinks that the search for rigor is undoubtedly useful, in this case, he finds it exaggerated:

> In over-zealousness for a presumed rigor, attempts were made to define concepts that, because of their elemental character, are neither capable of definition nor in need of it. Of this sort are the so-called "definitions" of equality and difference with respect to number whose

refutation will now engage us. And they have indeed a special claim on our interest precisely because they have led to a class of definitions of the number concepts themselves. These definitions, baseless and scientifically useless, have nevertheless, in virtue of a certain formal character, found favor among mathematicians and among the philosophers influenced by them.

(PA 96/101)

Husserl's view is that the definition of the concept of equality [*Gleichheitsbegriff*] is, if not impossible, then simply useless. He next goes through a series of attempts to define it, arguing against them one by one. Husserl's first target is Hermann Grassmann's definition "Two things are said to be equal if the one can replace the other in any assertion," which Husserl holds to be in essentials the same as Leibniz's principle "Things are the same as each other, of which one can be substituted for the other without loss of truth" [*Eadem sunt quorum unum potest substitui alteri salva veritate*], used by Frege as the basis for the definition of number concept in *Grundlagen* (PA 96–97/101–102). Husserl's criticism is twofold: First, instead of a definition of equality, it is a definition of identity. If there is a difference between the two things, then they cannot be substituted for each other. Second, it leads to an infinite regress because it relies on the notion of substitutability, which presupposes equality.

Without further ado about Leibniz's principle, Husserl moves on to discuss the definitions of equality. He ascribes what is now known as Hume's Principle to an Austrian mathematician, Otto Stolz (1842–1905): "Two multiplicities are said to be equal to one another [or, more correctly: are said to be equally many or equal in number], provided that each single thing in the first can be correlated with one in the latter, and none of the latter remains uncorrelated" (PA 98/103). Husserl's primary criticism is that, on the one hand, equality needs no further explanation, and, on the other hand, the definition renders what is obvious and intrinsically familiar obscure by what is distant and strange (PA 99/104).

Husserl then concedes that sometimes linguistic usage of the term "equal" vacillates so that it should be fixed. To do this, Husserl suggests specifying the property that is "the central focus of our interest" and regarding which the two contents are said to be equal (in other words, he is talking about something like sortal identity). His examples of the "focuses of our interest" are quantity of the structures (length, surface, volume) in metric geometry or the position in topology. In his view, these kinds of definitions of equality are useful and necessary in science (PA 100–101/105–106).

The equality of arbitrary multiplicities is a special case of this kind of equality. In it, "[t]wo collective wholes are compared when one tries to bring them element-by-element into reciprocal *coincidence*", and Husserl continues that

> [i]n fact, the procedure described is analogous to that the geometer uses to prove the congruence or incongruence of spatial formations. Indeed it is, one recognizes, the same in all cases where an analyzed whole is compared with another one of the same type
>
> (PA 102/107)

This way, one may confirm the numerical equality (but not the number itself) of two groups (PA 104/109). Indeed, the simplest way of comparing two multiplicities is by counting the units of each group. This, however, is a mechanical process that we follow without thinking about numbers themselves. The enumeration of the elements will convince us about the equality of the numbers but also these numbers themselves (PA 104–105/109–110). But the problem is that the "possibility of the reciprocal one-to-one correlation of two multiplicities is *not* the same thing as their number-equality, but only *guarantees* it" (PA 105/110):

> The knowledge that the numbers are equal absolutely does not require the knowledge of the possibility of their correlation, much less, then, that the two would be identical. The definition contested is, therefore, far removed from providing a nominal definition that fixes the signification of the expression "equality of two multiplicities with respect to number." All that we can grant is that it formulates a *necessary and sufficient criterion in the logical sense,* valid for all cases, for the obtaining of equality.
>
> (translation modified PA 105/110)

To show this, Husserl analyzes equality by thinking of how exactly we make judgments about equality of two multiplicities in ordinary life. In his view, equality of two multiplicities is something that can be directly grasped. He analyzes it with the notion of "collective combination," a notion that anticipates his later "categorial intuition." Collective combination, according to Husserl, is the "sort of combination which is characteristic of the totality [*Inbegriff*]" (PA 20/21). A totality, in turn,

> *originates in that a unitary interest—and, simultaneously with and in it, a unitary noticing—distinctly picks out and encompasses various contents.* Hence, the collective combination also can only be grasped by means of reflexion upon the psychical act through which the totality comes about.
>
> (PA 74/77)

It is notable that Husserl assumes that there is an act of grasping a totality, which is analyzed as a collective combination. Instead of "building" or "constructing" collections with the collective combination, Husserl takes for granted our ability to grasp abstract concepts and analyzes this ability by means of collective combination. In this regard, he already anticipates his later view of transcendental phenomenological account of constitution.

Grasping the combination of contents includes in it distinct acts of grasping of individual contents. The combination of the contents is thus a higher-order act (just like categorial intuition introduced in *Logical Investigations*). If these totalities are combined with other totalities, then an act of third order is needed, and so forth (PA74/77). Collective combination is thus grasping totalities, like seeing Gestalts, of different orders.

Grasping the equality of two multiplicities is a collective combination of pairs in which one element comes from one multiplicity and the other from the other multiplicity. It is a higher-order collective combination in which the elements are paired (PA 106/111) and then seen as an intuitive unity of the two-fold group that is comprehensible in one glance (PA 108/113). Essential to it is that the corresponding elements are linked in our thought, "that they are colligated [*kolligiert*]" (PA 109/114); the precise way of doing it can vary depending on the circumstances and the type of linkage. Even though the comparison between two multiplicities by way of one-to-one correspondence is thus possible, the simplest and preferable way of doing it is by counting the objects in each multiplicity. This has a virtue of bringing us a conviction of the numbers in themselves in question, not only of equality of the two multiplicities (PA 104–105/109–110). For Husserl, this consideration shows that the knowledge that the two numbers are equal does not require the knowledge of the possibility of their correlation.

> To introduce a relation as a technique only intending a one-to-one correlation means to forget and to miss its very purpose. For this would result in the directing of attention away from what is at issue here: namely, away from comparison as to number.
>
> (PA 109/114)

Furthermore, the definition of number equality in terms of reciprocal, one-to-one correlation entails an unfortunate consequence of "a total misconstrual of the concept of number itself" (PA 111/117). Elaborating on this, Husserl moves on to discuss the views in which numbers are construed as equivalence classes of groups. Not surprisingly, his criticism is directed against the notion of equivalence on which they are based: "But it is not true that 'equivalence' and 'equal in number' are concepts with the same content. Only this much is true, that their *extensions* are the same" (PA 115/121). Logicists' extensionalism thus misses out something essential in these notions; namely, it fails to capture "the same number in the true and authentic sense of the word" (PA 116/122):

Before launching into criticizing Frege, Husserl first sets out to explain a view based on equivalence theories in "the form of a maximally coherent theory" (111/117), a view he ascribes to Stolz. He explains how to form equivalent classes of objects, how to order them, and then how to obtain the number concept from them:

> Each class encompasses the entirety of conceivable concepts. Each class encompasses the entirety of conceivable groups of the same cardinal number. To different classes correspond different numbers. That we assign one and the same number to all groups of one class can only result from there being a characteristic which is common to all groups in that class. But what they all have in common, and what distinguishes them from all remaining conceivable groups, is surely nothing other than the circumstance that they belong precisely to the same class, i.e., that they stand in the relationship of mutual equivalence.
>
> (113/119)

Husserl then holds that "in order to attain uniform designations suited to the scientific use of language," the classes have to be represented concretely with groups of strikes 11, 111, 1111, . . . through repetition of mark "1." These groups are then named in sequence by 2, 3, 4, . . ., so that the groups formed out of marks are the natural numbers. "Thus there results a group of marks equivalent to the group, and this is the natural number. The numbers form an ordered sequence corresponding to the sequence of classes" (113/119). In line with his earlier critique, Husserl holds that this definition does not capture the number concept "in the true and authentic sense of the word," since in "language in life and science" numerical assertions are directed upon the concretely present group and not to its relationship to other groups. We can apply number concepts without any recourse to them as based on one-to-one correlations. He explains:

> Do we call a group of nuts lying before us "four" because of the fact that it belongs to a certain class of infinitely many groups that can be mutually put into a univocal, one-to-one correspondence?
>
> Very likely no one has ever thought of such a thing in this context, and we would be hard put to find any practical occasion whatever that would make it of interest. What does in truth interest us is the fact that here is one nut and one nut and one nut and one nut. We immediately give this awkward and complicated representation (and it all the more deserves this characterization once we come to larger groups!) a form more convenient to thinking and speaking by thinking it under mediation of the general group form one-and-one-and-one-and-one, which has the name "four."
>
> (PA 116/122)

Husserl's problem thus is that the approaches based on equivalence theories do not capture the way we use number words in everyday life and sciences. Instead, the equivalence theories postulate extraneous, artificial rigor to the intuitive number concept.

2.   Husserl's Criticism of Frege

Frege's "noteworthy effort" in his "often cited and ingenious book [*Foundations of Arithmetic*]" is closely related to the equivalence theory explained previously, and thus much of the previous criticism holds also of it. In addition, Husserl's further disagreement with Frege's view is based on the disagreement over the primitive, undefinable concepts. There are ultimate, elemental concepts and relations that cannot be defined. Husserl lists these:

> Concepts such as quality, intensity, place, time, and the like, no one can define. And the same is true of elemental relations and the concepts grounded on them. Equality, similarity, gradation, whole and part, multiplicity and unity, and so on, are concepts that are totally incapable of a formal logical definition. What one can do in such cases consists only in pointing to the concrete phenomena from or through which the concepts are abstracted and laying clear the nature of the abstraction process involved.
>
> (PA 119/124–125)

This holds about the number concepts as well:

> We have precisely such a case before us in the number concepts, and therefore we can find absolutely nothing inherently blameworthy when mathematicians, at the apex of their system, "describe the route by which one arrives at the number concepts" instead of giving a logical definition of those concepts. It is only required that those descriptions be correct, and ones which also achieve their goal.
>
> (PA 119/125)

In Husserl's view, the number concept is so close to the indefinable concepts of multiplicity and of unity that "in its case one can scarcely speak of any 'defining'" (PA 119/125). Number is thus a primitive term that should not be given a logical definition but whose psychological origin should be clarified. Husserl thus concludes that "[t]he goal Frege sets for himself must therefore be termed chimerical" (ibid.).

Husserl then discusses Frege's view in more detail, first pointing out that, for Frege, the statement of number involves an assertion about a concept, not a group of objects. He then explains how Frege defines the number-equality by means of what is to be considered equal and how parallelism and geometrical similarity supply the concept of equality, which, when considered as for concepts, is used by Frege to define the notion of number. Husserl's criticism is that

> what this method allows us to define are not the contents of the concepts *direction*, *shape* and *number*, but rather their *extensions*. . . . We note, however, that all the definitions become correct statements if the concepts to be defined are replaced by their extensions. Correct, but certainly entirely obvious and useless statements as well.
>
> (PA 122/128)

Thus, if one adopts the logicist aim, in Husserl's view, there is nothing wrong in Frege's definition. But he thinks the project as such is misguided and hence useless. To conclude, in addition to his general criticisms of the theories based on equivalence, Husserl's specific criticism of Frege consists of the disagreement over what the primitive concepts are and then, in Husserl's view, that that these primitive concepts should be clarified by description of their "origin"—what Frege and, after him, Wittgenstein would hold that can only be elucidated. To be sure, for Husserl, on the pain of circularity, the method of clarification of the primitive terms of the formal disciplines cannot use the methods of these disciplines, which is the original motivation to develop phenomenological methodology.

In the rest of the chapter, Husserl discusses Kerry's approach. After it, Husserl concludes that the concept of equivalence has nothing to do with the definition or analysis of the concept of number and in a footnote adds that after having written the chapter, a number of new views that follow the tendency characterized in it had appeared: Husserl mentions G. Heymans, *Die Gesetze und Elemente des wissenschaftliche Denkens*, 1890, and Dedekind's *Was sind und was sollen die Zahlen?* (1888). About the latter, he writes that "As much as I admire the inner formal coherence of the developments in the theory of this remarkable mathematician, it still seems to me to deviate far from the truth in its bizarre artificiality [*absonderlichen Künstlichkeit*]" (PA 125/131–132).

3.   Constitution of Number by "Collective Combination" and "Something"

To have a clearer grasp of what Husserl has in mind, let us further examine Husserl's own proposal for how to account for numbers. The basis of his approach is the acknowledgement that we think and talk about abstract entities in everyday life and science, and therefore, he seems to think, they exist:

> life and science certainly are not lacking in occasions to consider the abstract as such, and accordingly speech also has made provision for the clearly marked designation of it, in which the general term is used

and is simply modified by means of determining expressions, by the attachment of syncategorematic signs, etc.

(PA, 137/144)

The problem at hand is to describe the way in which the numbers assumed to exist are constituted in everyday life. This does not involve comparing groups and their sizes:

> According to our perspective, the representation of the number of a determinate group does not arise through our comparing the objects of that group with one another and subsuming them under the generic concept emerging from that comparison (horse, apple, tone, pencil), but rather through our bringing them—whatever it is we may be counting—always under the same concept, that of the "something," and simultaneously grasping those objects collectively together, objects thought under mediation of that concept and designated as same in terms of it. Thus there originates the general form of multiplicity—one-and-one-and . . . one—under which the multiplicity concretely before us falls, i.e., under the number belonging to it.

(PA, 143/150)

In Husserl's view, abstraction has nothing to do with comparison with other groups, nor the relation of sameness—these are not "essential psychological factors in the representation of the number," and "in many cases they do not come to our attention at all"(PA, 143/150–151). Instead the abstract concepts originate in perceiving or imagining a group of objects, so that all other properties of the presented objects except their individuality are abstracted away. Thus numbers are constituted by means of "collective combination" and abstraction from the contents [*Inhalte*] so that only *that* they are contents is retained. The equality is then a consequence of abstraction, not its basis and presupposition. (PA, 144/152)

4. Frege's Rejoinder

Frege wrote a famous review of Husserl's *Philosophy of Arithmetic*, published in *Zeitschrift für Philosophie und Philosophische Kritik* in 1894. After summarizing Husserl's view of the concept of number, obtained by abstracting from the particular constitution of the individual contents that are collected together in the multiplicity, while retaining each one as if it is a something, Frege first claims that any conception is naïve "if according to it a number-statement is not an assertion about a concept or the extension of a concept" (1984, 323). Frege writes:

> The most naive opinion is that according to which a number is something like a heap, a swarm in which the things are contained lock,

stock and barrel. Next comes the conception of a number as a property of a heap, aggregate, or whatever else one might call it. Thereby one feels the need for cleansing the objects of their particularities. The present attempt belongs to those which undertake this cleansing in the psychological wash-tub. This offers the advantage that in it, things acquire a most peculiar suppleness, no longer have as hard a spatial impact on each other and lose many bothersome particularities and differences. The mixture of psychology and logic that is now so popular provides good suds for this purpose.

(Frege 1984, 323)

Frege then continues to complain that on Husserl's view everything becomes a presentation, even references of words. "But might not the moon, for example, be somewhat hard to digest for a state of consciousness?" (ibid., 324). He thus thinks that Husserl's "intensionalism" leads to idealism. This view, says Frege, is connected to an arbitrary view of abstraction: "We pay less attention to a property and it disappears," so that "the something derived from the one object nevertheless does differ from that derived from the other object, although it is not easy to say how" (ibid., 324). In Frege's view, since Husserl's view of grasping totalities shows them to be constituted by collective combination and abstraction, these totalities are psychologically constructed presentations. This indeed is the way Husserl at times describes the acts through which the totality in question comes about. Frege seems to think that the view is also applicable to physical objects so that even they, reliant on psychological operations, become presentations. Frege seems to understand Husserl's presentations (*Vorstellungen*) to be like his own view of ideas (*Vorstellungen*) that are in the mind. This explains why, in Frege's view, Husserl's view results in viewing the moon to be in the mind. In any case, in such a view,

the difference between presentation and concept, between presenting and thinking, is blurred. Everything is shunted off into the subjective. But it is precisely because the boundary between the subjective and the objective is blurred, that conversely the subjective also acquires the appearance of the object.

(ibid., 324–325)

Frege then focuses particularly on three obstacles in the view, namely that Husserl cannot explain how the sameness of units can be reconciled with their distinguishability, that Husserl cannot account for the numbers zero and one, and, third, that the approach cannot explain the nature of the large numbers (ibid., 330–334). The last problem takes Frege to discuss Husserl's account of symbolic presentations in the later chapters of *Philosophy of Arithmetic*. There Husserl explains that since they cannot be intuited directly, the large numbers are presented indirectly, by means of symbols. Frege points out that this still restricts Husserl to symbolic constructions and hence to finite numbers (ibid., 334); he complains that Husserl still confuses the subjective and the objective (ibid., 335), even though Husserl admittedly occasionally refers to "real numbers" or numbers "themselves" (ibid., 336).

Husserl's primary criticism against Frege is directed at the sense of his logicist project and the futility of the attempt at defining (extensionally) what does not need to be defined. Frege in turn does not defend his logicist project against Husserl's criticisms but attacks Husserl's account as vague, psychological and idealistic and, in particular, his view of abstraction as an arbitrary "wash-tub" of properties. While for Husserl the debate was about logicisim, Frege shifts the debate to be about psychologism, and here the commentators have followed Frege's lead (for a more detailed account of the debate between Frege and Husserl, see Hill 1994). In what follows, I will look at the debate from the point of view of logicism to show how Husserl's position in *Logical Investigations* answers Frege's criticism and implies further criticism of logicism.

5. *Husserl's Subsequent Development and Frege's Role*

In the Foreword to *Prolegomena to Pure Logic* (Husserl 1975, henceforth *Prolegomena*) Husserl admits that the psychological foundations of the concept of number never came to satisfy him, and he wrote that "I became more and more disquieted by doubts of principle, as to how to reconcile the objectivity of mathematics, and of all science in general, with a psychological foundation for logic" (Prolegomena 6/2). This critical remark about his own earlier approach has given rise to a debate about the extent to which Frege's review influenced Husserl. The general understanding, including mine, is that Husserl turned away from his psychological conception of logic before Frege's review was published, so Frege's review could not exercise major influence on Husserl (see Hill & Rosado 2000, xiii; Miller 1982, 19–23; Mohanty 1984; Peucker 2002, 155–157; Willard 1984). But since Husserl did relate to Boyce Gibson that "Frege's criticism was the only one he was really grateful for. It hit the nail on the head" (1971, 66), it would be unreasonable to maintain that Frege's review had *no* impact on Husserl. My suggestion here is that while Frege probably did not impact Husserl on the issue of (logical) psychologism, he may have influenced Husserl in his development towards transcendental phenomenology and at the same time towards a more refined criticism of logicism.

The changes in Husserl's views after *Philosophy of Arithmetic* would merit a detailed examination. But, for the sake of brevity, I will cut the long story short by looking at it as he himself describes it in the foreword to *Logical Investigations*. In it, Husserl explains how his view was influenced by two different developments: First, he explains that his formal research had taken him towards "a universal theory of formal deductive systems," and this made him realize that

> [t]here were evidently possibilities of generalizing (transforming) formal arithmetic so that, . . . This made me see that quantity did not at all belong to the most universal essence of the mathematical or the "formal", or to the method of calculation which has its roots in this essence.
>
> (1975, 5/1)

Instead of quantities, Husserl started to view mathematics to be about abstract structures. The change in Husserl's view is from what Hilbert called "genetic" to the "axiomatic" approach to the foundations. Instead of building arithmetic from the notion of number, he now conceives mathematics to be about abstract structures (for more detail, see, e.g., Centrone 2010, 154–159; Hartimo 2019b). In *Prolegomena*, Husserl's argument against (logical) psychologism culminates in establishing the existence of independent and objective abstract notions and domains, as I will explain in more detail in the next section.

The second set of problems that moved Husserl was the epistemological problem: "I felt myself more and more pushed towards general critical reflections on the essence of logic, and on the relationship, in particular, between the subjectivity of knowing and the objectivity of the content known" (1975, 6/2). This set of problems led Husserl to a renewed account for how the objective structures of pure logic are related to the subject of knowledge. I will discuss this in the section after the following.

6.   *Anti-Psychologism and the Idea of Pure Logic*

Husserl's argument against psychologism in *Prolegomena* is, in a nutshell, that unless logic as an art or technique is founded on a priori, theoretical idea of logic, it leads to relativism and ultimately to skepticism. The argument thus hinges on the establishment of the a priori theoretical idea of logic that is independent of psychology. Husserl derives this idea from considerations of what an essence of a scientific theory is and especially what it is for mathematicians in the late 19th century. This takes Husserl to include mathematics, the formal objects and structures, into his idea of logic. Hence, the idea of pure logic has two sides, the side of meaning and the side of objects. The two-sided idea of pure logic has then three tasks: fixing the pure categories of meaning and objects and their law-governed combinations; establishing the formal theories that are based on these categories, and, third, building the theory of these theories. While the first task specifies the primitive notions, the next task defines the axiomatic theories on both sides, and the last one establishes an idea of a general framework in which to formally examine these theories and structures and their relationships to each other. Husserl assumes that the last task is what mathematicians are generally striving for, but at his time, it was only partially realized in various theories such as Riemann's theory of manifolds, Cantor's set theory, Lie's theory of transformation groups, Grassmann's theory of extensions, and Hamilton's theory of quaternions (*Prolegomena*, §70). Husserl's idea of logic thus comprises the emerging modern structural mathematics. Its primary aim is to describe structures rather than capturing the notion of deduction (cf. Hartimo 2018). The domains of the axiomatic theories are objective and independent from the mathematicians' activities. I have argued elsewhere (ibid.) that Husserl understands the theories at best to be categorical (i.e., all their possible realizations are isomorphic with each other), which makes their domains pure structures that are independent of even individual theories and hence clearly independent from anything subjective. Husserl's argument against psychologism thus uses theories developed in mathematics and presupposes that mathematicians aim at describing objective pure structures. The mathematicians' goal embodies the paradigm of objectivity for Husserl and saves us from psychologism. Thus whereas in *Philosophy of Arithmetic*, Husserl's starting point was the usage of number words in everyday life and sciences, in *Logical Investigations*, his point of departure is in the modern structural mathematics and its metaphysical commitments that are described by means of pure theory of grammar. Instead of the use of language in the everyday world, the emphasis is now on the mathematicians' view of the abstract notions.

7. Correlation

The idea of pure logic serves as the bedrock for Husserl's argument against psychologism. But Husserl's next problem was the epistemological one: He was not able to connect these abstract structures to the subject of knowledge:

> Where one was concerned with questions as to the origin of mathematical presentations, or with the elaboration of those practical methods which are indeed psychologically determined, psychological analyses seemed to me to promote clearness and instruction. But once one had passed from the psychological connections of thinking, to the logical unity of the thought-content (the unity of theory), no true continuity and clarity could be established.
> (*Prolegomena* 6/2, translation modified)

Husserl thus failed to connect the axiomatic theories and the abstract "unities" defined by them to the subjectivity. In Husserl's more mature phenomenology, this receives a central role known as the "correlation between subjective and objective." The objective refers to the world (of various kinds of attitudes), the subjective is the study of constitution of them, and I want to suggest here that Husserl's answer to Frege is to adopt a transcendental phenomenological view of correlation. Again, I follow Husserl's own view of his development, for in *Crisis*, he claims to have found the notion when working on *Logical Investigations*:

> The first breakthrough of this universal a priori of correlation between experienced object and manners of givenness (which occurred during work on my *Logical Investigations* around 1898) affected me so deeply that my whole subsequent life-work has been dominated by the task of systematically elaborating on this a priori of correlation.
>
> (*Crisis* 166n)

Thus, if Husserl's own words are to be trusted, the notion of correlation was already in place in *Logical Investigations*. Since the correlation is Husserl's mature approach to how the objective is related to the subjective, this could be seen as a response to Frege, who criticized Husserl namely for failing to distinguish between the subjective and the objective. It is perhaps not particularly far fetched, then, to claim that this could be the very issue on which "Frege hit the nail on" Husserl's head, as Husserl relayed to Boyce Gibson.

In *Prolegomena*, Husserl does not use the term correlation, but it is essentially what he discusses in the end of the work and which leads to the establishment of the division of labor between mathematicians and philosophers. According to the division of labor, mathematicians should do what they do best, that is, construct theories and prove theorems, while philosophers should examine the essence and sense of these theories. Husserl's explanation merits quoting it in full:

> The construction of theories, the strict methical solution of all formal problems, will always remain the home domain of the mathematician. . . . No one can debar mathematicians from staking claims to all that can be treated in terms of mathematical form and method.
>
> If the development of all true theories falls in the mathematician's field, what is left over for philosophers? Here we must note that the mathematician is not really the pure theoretician, but only the ingenious technician, the constructor, as it were, who, looking merely to formal interconnections, builds up his theory like a technical work of art. As the practical mechanic constructs machines without needing to have ultimate insight into the essence of nature and its laws, so the mathematician constructs theories of numbers, quantities, syllogisms, manifolds, without ultimate insight into the essence of theory in general, and that of the concepts and laws which are its conditions. . . . Philosophical investigation has quite other ends, and

therefore presupposes quite other methods and capacities. It does not seek to meddle in the work of the specialist, but to achieve insight in regard to the sense and essence of his achievements as regards method and manner. The philosopher is not content with the fact that we find our way about in the world, that we have law-governed formulae which enable us to predict the future course of things, or to reconstruct its past course: he wants to clarify the essence of a thing, an event, a cause, and effect, of space, of time etc., as well as that wonderful affinity which this essence has with the essence of thought, with the essence of knowledge, which makes it knowable, with meanings which make it capable of being meant etc. And if science constructs theories in the systematic dispatch of its problems, the philosopher enquires into the essence of theory and what makes theory as such possible etc. Philosophical research so supplements the scientific achievements of the natural scientist and of the mathematician, as for the first time to perfect pure, genuine, theoretical knowledge. The *ars inventiva* of the special investigator and the philosopher's critique of knowledge, are mutually complementary scientific activities, through which complete theoretical insight, comprehending all relations of essence, first comes into being.

(*Prolegomena* §71)

I apologize for the long quote, but it is important. According to this division of labor, mathematicians should do the formal work, while philosophers should not meddle with their work but to attempt to understand the essence of what the mathematicians are doing. Right before this passage, Husserl discussed the idea of logic that in his view culminated in the theory of theories but was only partially realized in various existing theories such as Riemann's theory of manifolds, Cantor's set theory, Lie's theory of transformation groups, Grassmann's theory of extensions, and Hamilton's theory of quaternions (*Prolegomena*, §70). The idea of logic is the goal of the work of these "special investigators," or "ingenious technicians." Husserl's subsequent individual logical investigations, in turn, supplement the mathematical research by clarifying the essence of the mathematicians' theories and what makes these theories possible.[2] The mathematicians thus work out the objective theories, while the philosophers examine their givenness to the subject, that is, their constitution, to use Husserl's later terminology.

---

2 Referring to his *Logical Investigations*, Husserl writes in the end of *Prolegomena* that, "The following individual investigations preparatory to the philosophical side of our discipline, will further elucidate what the mathematician will not, and cannot do, but what must nevertheless be done" (§71).

The division of labor solves Husserl's problem of how to reconcile objectivity of mathematics with the psychological foundations without falling prey to logical psychologism. It does this first by stating that objectivity of mathematics and its relationship to the subjectivity are problems that belong to different disciplines and thus demand different kinds of methodologies. For Husserl, the "logical unity of the thought content" is conceived as something that mathematicians work toward. It is mathematicians' goal to produce complete and coherent axiomatizations. He thus assumes a naturalist, "mathematics-first" account of mathematics that takes mathematics on its own without seeking revisions on a priori philosophical grounds. The philosophers' task in turn is to examine the essence and givenness of mathematicians' theories. This means describing the givenness of these mathematical notions, the task that Husserl then (in a preparatory manner, as he puts it) goes on to carry out in the six logical investigations. The relationship between the objective and the subjective is such that the objective theoretical unities are pregiven (in mathematics), and, presupposing that, their givenness to the subjectivity is philosophically described. Mathematicians aim to construct objective theories independently of the relationship these theories have to the subjectivity. The formal theories are not constructed on the top of the psychological descriptions, but the objectivity of mathematics is assumed, and then its givenness to the subjectivity is examined. Thus the phenomenological foundations do not explain how mathematical theories should be constructed but essentially how they are constituted (for the difference between the two, see Hartimo 2019a).

Thus one could claim that in *Prolegomena*, Husserl responds to Frege's criticism by offering a clear distinction between what is mathematical and what is philosophical. While this led Husserl to further develop phenomenological methodology, for the present chapter, it is of interest how it clarifies Husserl's attitude toward Fregean logicism.

It does it in two ways. First, Husserl explicitly points out that by the time of writing *Logical Investigations*, mathematics is not about numbers and quantities anymore:

> Only if one is ignorant of the modern science of mathematics, particularly of formal mathematics, and measures it by the standards of Euclid or Adam Riese, can one remain stuck in the common prejudice that the essence of mathematics lies in number and quantity.
>
> (*Prolegomena* §71)

Accordingly, while in *Philosophy of Arithmetic*, Husserl discussed grasping totalities, Husserl now seems to think that the essence of mathematics is captured by the axiomatic (Husserl uses the term "nomological") theories. Hence, his interest is in notions such as the unity of a theory (*Prolegomena* §§62–64) or a pure manifold (*Prolegomena* §70) and then on the notions needed in the philosophical analysis of them, namely meanings, objects, parts and wholes, essences, and truth. He continues to think that the axiomatic theories should be connected to numbers and to numerals, but this is for the sake of application, not to capture the essence of the theory (for a detailed explanation, see Hartimo 2018).

Second, the division of labor clearly demarcates between philosophical and mathematical method. Frege's logical construction uses a formal method, which is not the proper method to evaluate the philosophical essence of mathematics. At best, it offers an ingenious technical formulation to the definable notions of arithmetic, which Husserl would probably still find useless and chimerical. The indefinable, primitive notions Frege could only elucidate, whereas Husserl proposes a new method with which to analyze their constitution.

Husserl's view of Fregean logicism in this later formulation amounts thus to two claims: First, Frege focuses on the notion of number in a way that does not do justice to the abstractness of modern structural mathematics. Husserl's view of mathematics is closer to naturalists' view; it is "mathematics-first" and assumes that mathematicians' practice displays the objectivity of mathematics (which is not thus built upon a psychological foundation or a wash-tub of abstraction, as in Frege's critique, nor on logical foundation, as Frege would have it). Second, logicism has a misguided view of the philosophical method. Using mathematical method to give foundations for mathematics is circular, and it is not able to provide insight into the essence of mathematics. Husserl's subsequent formulations of the phenomenological reduction can then be understood as a further refinement of the strict demarcation between objective (given naturalistically) and how it is given to the subjectivity (examined by phenomenology).

Conclusion

Husserl's attempt to capture the use of number-words in everyday life and the sciences anticipates his "mathematics first" view spelled out in *Prolegomena*. In it, a properly objective view of mathematics comes from the mathematicians' practices. For Husserl, philosophers should not meddle with the mathematicians, but their task is to give insight into the essence of mathematics. While Husserl might hold that logicist definitions are useful in raising the standards of rigor in foundations of mathematics and in advancing the formal development of logic, for philosophical purposes, he finds them mainly useless, uninformative, and ultimately circular. They fall short in giving a *philosophical* analysis of the essence of mathematics.

References

Centrone, S. (2010). *Logic and Philosophy of Mathematics in the Early Husserl*. Synthese Library 345. Dordrecht: Springer.

Frege, G. (1894). "Rezension von E. Husserl: Philosophie der Arithmetik." *Zeitschrift für Philosophie und philosophische Kritik*, 103: 313–332. English translation: "Review of Dr. E. Husserl's Philosophy of Arithmetic." *Mind*, New Series, 81 (1972): 321–337. Translated by E. W. Kluge. References in the text are to the English translation.

Gibson, B. W.R. (1971). "From Husserl to Heidegger: Excerpts from a 1928 Diary by W. R. Boyce Gibson." Edited by Herbert Spiegelberg, *Journal of the British Society for Phenomenology* 2: 58–81.

Hartimo, M. (2018). "Husserl on Completeness, Definitely." *Synthese* 195: 1509–1527. https://doi.org/10.1007/s11229-016-1278-7.

Hartimo, M. (2019a). "Constitution and Construction." In Christina Weiss (ed.), *Constructive Semantics—Meaning in Between Phenomenology and Constructivism*. Logic, Epistemology and the Unity of Science Series. Dordrecht: Springer.

Hartimo, M. (2019b). "Husserl on Kant, and the Critical View of Logic." *Inquiry. An Interdisciplinary Journal of Philosophy*, https://doi.org/10.1080/00201 74X.2019.1651089.

Hill, C. O. (1994). "Frege's Attack on Husserl and Cantor." *The Monist*, 77(3): 345–357.

Hill, C. O. and Haddock R. (2000). *Husserl or Frege. Meaning, Objectivity, and Mathematics*. Chicago and La Salle, IL: Open Court.

Husserl, E.(1973). *Philosophie der Arithmetik*. Husserliana Band XII. Ed. L. Eley. The Hague: Martinus Nijhoff. English translation in *Philosophy of Arithmetic, Psychological and Logical Investigations with Supplementary Texts from 1887–1901*, translated by Dallas Willard. Dordrecht, Boston, London: Kluwer, 2003.

Husserl, E. (1975). *Logische Untersuchungen. Erster Band. Prolegomena zur reinen Logik*. Husserliana XVIII. Ed. E. Holenstein. The Hague: Martinus Nijhoff. English translation in *Logical Investigations*, translated by J. N. Findlay. Volume One. New York: Humanities Press, 1970.

Husserl, E. (1976). *Die Krisis der europäischen Wissenschaften und die transzendentale Phänomenologie: Eine Einleitung in die phänomenologische Philosophie* (1936). Husserliana VI. Ed. W. Biemel. The Hague: Martinus Nijhoff. English translation: *The Crisis of European Sciences and Transcendental Phenomenology: An Introduction to Phenomenological Philosophy*, translated by David Carr. Evanston: Northwestern University Press, 1970.

Lohmar, D. (2000). *Edmund Husserls "Formale und transzendentale Logik" (Werkinterpretationen)*. Darmstadt: Wissen Bildung Gemeinschaft.

Miller, P. J. (1982). *Numbers in Presence and Absence: A Study of Husserl's Philosophy of Mathematics*. The Hague, Boston, London: Martinus Nijhoff Publishers.

Mohanty, J. (1984). "Husserl, Frege and the Overcoming of Psychology." In Kah Kyung Cho (ed.), *Philosophy and Science in Phenomenological Perspective*. Phaenomenologica 95. Dordrecht: Springer, 143–152.

Peucker, H. (2002). *Von der Psychologie zur Phänomenologie, Husserls Weg in die Phänomenologie der Logischen Untersuchungen*. Hamburg: Feliz Meiner Verlag.

Willard, D. (1984). *Logic and the Objectivity of Knowledge: A Study of Husserl's Early Philosophy*. Athens, OH: Ohio University Press.

9 Peano's Philosophical Views between Structuralism and Logicism

Paola Cantù

1. Introduction[1]

The main objective of this chapter is to analyze Peano's philosophical views as they emerge from several mathematical practices that are usually associated with logicism: for lack of a better name, I will call this view *structural algebraism* and will try to show some differences with respect to Frege's own approach and to a foundationalist notion of logicism.

The latter is based on some general claim about a relation of inclusion or intersection between logic and mathematics and could come in different versions, depending on the relata that one considers. For example, the relation could subsist between: a) the symbolic notation adopted in logic and in mathematics, b) the primitive propositions occurring in them (which may be identical or reducible one to the other or translatable one into the other), c) the respective inference rules, d) the axiomatic presentation of the theories, e) the philosophical claims that ground them, or f) their metatheoretic properties. The usual definition of logicism based on some claim about the relation between mathematics and logic has often been criticized as a product of a philosophical question that is extrinsic to mathematical and logical practice (Reck 2013b; Wilson 1992; Tappenden 1997). No matter how interesting the question about foundationalist logicism could be, this chapter will follow a different path, taking logicism to be a mathematical practice that emerges from a logical investigation of definitions, from a certain notion of interpretation of the symbols of a formal language, from the distinction between relations and functions, and from the difference between primitive and derived terms or propositions in axiomatics. On the one hand, Peano was primarily a mathematician, and this explains why it is reasonable to reformulate the question about Peano's logicism as a question about which mathematical practices he preferred and why. On the other hand, several of the mentioned

1 I would like to thank the editors, Marco Panza and Jan Von Plato for precious insights and comments on a preliminary version of this chapter.

DOI: 10.4324/9780429277894-11

mathematical practices were actually highly debated by Peano, as he discussed them with members of the School[2] involved in the joint enterprise of the *Formulario* or with contemporaries, especially Frege. The latter is a perfect candidate in this respect, because Frege and Peano corresponded with each other and compared their respective results (Frege 1980). Besides, they have often been associated in the literature, as if they shared a similar understanding of logic (van Heijenoort 1967).[3]

Of these two ways to investigate an author's stance towards logicism, the foundationalist is better suited to evaluating an explicitly foundational philosophy, the latter to evaluating a philosophical view that is embedded in a logico-mathematical practice. The former aims to compare different historical realizations of a unique philosophical and foundational notion of logicism that is preliminarily assumed to be correct, whereas the latter believes that logicism is but a schematic representation of the richness and variety of mathematical practices that accompanied the transformation of axiomatics from the end of the 19th century onwards.[4] Following the first line of investigation, I hope not only to give further support to the thesis that "philosophical reflections can grow out of mathematical practice" (Reck 2013a, 257) but also to claim that Peano, as well as other scholars of his group, had explicit philosophical views, even if not easily translatable in the foundationalist jargon.

A third way to study Peano's relation to logicism could be an historical investigation of his relation to Leibniz and to Schröder. Although this chapter cannot provide a detailed examination of this question, it will be clarified that Leibniz's influence did not urge Peano to endorse a form of logicism based on a conceptualist form of semantics.[5]

---

2   It has been questioned whether one could speak of a School in the absence of an academic filiation and of an explicit communication strategy (Lolli 2007) and given the difficulty in determining who should be counted as a member: Peano's students at the University of Turin, the collaborators to the *Formulario*, or the more than forty people who "worked closely" with Peano (Luciano and Roero 2010). For the purpose of this chapter, I will restrict the focus to Padoa, Pieri, Burali-Forti, Vacca, and Vailati.

3   Recent literature has questioned van Hejenhoort's point of view and underlined the differences between Peano and Frege's understanding of formalism. See, for example, von Plato (2021) and Bertran-San Millan (2021).

4   Anyway, it would be anachronistic to talk of logicism before 1929, when the term "Logizismus" began to refer to a philosophical view on the relation between mathematics and logic (Carnap 1929; Fraenkel 1928), whereas the term "Logistique" had been introduced by Couturat as a generic term for symbolic logic (Grattan-Guinness 2000, 479, 501).

5   Logicism is often considered a perspective that originated in Leibniz's conceptual analysis, which aimed to determine the simple ideas from which other ideas are composed. Some scholars even go so far as suggesting that Leibniz contra Kant was at the origin of a new non-representational understanding of semantics, where terms mean concepts and only by means of concepts refer to objects. Frege's or Russell's logicism would result exactly

Peano's philosophical view will be described as a *structural algebraism*, that is, a methodology that is based on: a) a metatheoretical use of definitions; b) a particular theory of meaning (formal symbols mean words and sentences of the mathematical language); c) a distinction between a mathematical and a pre-scientific (logical) notion of function; d) an epistemological and dynamic understanding of axiomatics; and e) a joint investigation of logic, language, and mathematics, considered both as theoretical and didactic practices.

2.   Peano's Logicism Reconsidered

There are quite different views in the literature on Peano's philosophy of mathematics and logic: it has been claimed that Peano was a logicist (Grattan-Guinness) and an anti-logicist (Kennedy). Some scholars attribute to Peano an explicit philosophical approach (Vailati, Kennedy); others claim that he had no philosophical approach at all (Lolli), or at most an implicit and never openly declared view on philosophy (Geymonat, Grattan-Guinness).

Basing his argument on a passage by Peano stating that natural numbers cannot be defined from mere logical concepts (Peano 1891b, 91), Kennedy claimed that Peano had an explicit philosophical view but only about the philosophical views he did not share (Kennedy 2002, 9). Quoting the 1895 Review of Frege's *Grundgesetze* asserting that thanks to the *Formulario*, "mathematics is now in possession of an appropriate instrument to represent all its propositions, and to analyze the various forms of reasoning" (Peano 1895, 190), Grattan-Guinness remarked that Peano looked like a logicist, or rather like an "opportunist logician", because he introduced important logical notions and made considerable contributions to the foundations of mathematics but lacked a general program that could induce him to pursue all the logical consequences of his own insights (Grattan-Guinness 2000, 247).[6] Lolli claimed that the question about Peano's philosophy of mathematics is pointless, because Peano was neither interested nor competent in philosophy, as he himself declared in the review to Schröder's *Vorlesungen* (Peano 1891a, 164). That the *Formulario* was not a contribution to the foundations of logic and mathematics was already Frege's view, and actually Lolli (2007, 3) recalled one of his famous expressions as he described the *Formulario* as

---

from the combination of a Leibnizian conceptual analysis and conceptualist semantics (Coffa 1982).

6   "Peano presented arithmetic in a symbolic language which contained logical techniques rather than grounded it in an ideal language which expressed such features" (Grattan-Guinness 2000, 247).

a project of information storage and retrieval.[7] Geymonat attested that Peano did not want to write about philosophical issues and remained silent when asked about his own epistemological conception (Geymonat 1955).

The situation is even more complicated if one considers philosophical remarks by other collaborators of Peano, for example, Pieri, Padoa, and Burali-Forti. Each author seems to have a particular philosophical view, which agrees on certain points with a common view of the school but also differs quite radically on other aspects. One of the earlier interpreters of Peano's philosophy was Giovanni Vailati (1906b), who was unanimously recognized as the philosopher of the group. He claimed that Peano's ideas were similar to the views held by pragmatists, and even if (as he was a pragmatist himself) he could have had the tendency to attribute to Peano his own views, there are some effective similarities with the pragmatist epistemology.

Mario Pieri and Alessandro Padoa explicitly discussed the question of the relations between mathematics and logic from a foundational perspective. Pieri offered a restrictive interpretation of logicism, which Grattan-Guinness (2000, 371) considered the expression of the general position of the school: mathematics consisted of the application of some logical principles to special relations—like Russell's applied mathematics. Alessandro Padoa considered logic and mathematics distinct but seemed to consider the criteria for demarcating between the two partly conventional or at least historically determined.

Giving up on the idea that it may be possible to offer a simple yes-no answer to the question as to whether Peano was a logicist on the basis of paying selective attention to some of his more "philosophical" remarks, I will instead try to show that he had a coherent epistemology that can be investigated by an analysis of several mathematical practices. As Frege puts it, the question is whether Peano's *Formulario* was simply "the work that has to be done in order to write out a sentence as simply as possible in symbols" or a full effort to continue the reduction into simpler components right down to the simplest elements, developing a logical investigation on proofs, and explaining "how, from one of those formulae, or from two of them, a new one is obtained" (Frege 1897, 366, English translation 238).[8] But what mathematical practices, if any, can be chosen as significant objects of investigation in this respect?

---

7 As Frege puts it in his 1897 paper *On Mr. Peano's Conceptual Notation and My Own*, Peano's project "seems orientated towards the storage of knowledge rather than towards proof, towards brevity and international intelligibility rather than towards logical perfection" (Frege 1897, 365–366, repr. in *Kleinen Schriften*, 223; English Translation in Frege 1984, 237).

8 On the notion of simplicity in Peano, Padoa, Frege, and Russell, see Bellucci et al. (2018).

Apart from the practice of proving, which has already been studied in some detail,[9] other interesting topics could also be analyzed. In Frege's logic, functions are primary, and relations are defined as a special case of functions, that is, as the mapping from the Cartesian product of two sets to the set formed by the truth values True and False, whereas in modern predicate calculi (following Russell), relations are primary (usually defined as subsets of the Cartesian product of two sets), and functions are introduced as special cases of relations that satisfy functionality (Zalta 2015, §2.1). Another issue concerns the presence or absence of metatheoretical investigations, which van Heijenhoort considered one among many reasons to distinguish between different conceptions of logic as a language or as a calculus (van Heijenoort 1967). A comparison with Frege and Russell, who are presented by van Heijenhoort as representatives of the conception of logic as language, will show the need for some clarifications and further distinctions. The main point will concern a detailed analysis of Peano's theory of meaning and of his notion of interpretation, which radically differs from Frege's notion, even if both assign a central role to the distinction between function and argument,[10] as well as from the contemporary model-theoretic notion.[11]

Drawing attention to these topics, several further philosophical issues come to the forefront: the influence of algebraic investigations, the scope and limits of formal notation, and the distinction between interpreted and non-interpreted symbols. This is in turn correlated with the understanding of the notions of equality, meaning, truth, axiom, and the meta-theoretical role of definitions—topics that will ground my characterization of Peano's philosophical view as a structural and algebraist approach to logic, without any strong demarcation between logic, mathematics, and linguistics, and conceiving axiomatics as a dynamical investigation and comparison of scientific theories (Cantù and Luciano 2021). This epistemological understanding of axiomatics and the syntactic notion of interpretation that grounds it will show some relevant differences from Frege's conception of what logic is (or should be).

---

9 See von Plato (2014; 2017; 2018) but also Bertran-San Millan (2020), who claims that Peano lacked, at least in his early writings, Frege's interest in a logical calculus and the separation of inference rules from axioms and suggests (based on Badesa 2004) that this was because Peano used as a basis for his early mathematical logic Schröder's logic, which was not indented as a deductive system. Another reason might be that Peano's notion of interpretation, as reconstructed in the following (§4) did not favor a separate treatment of the calculus of classes and of the calculus of propositions, which should both be obtained by substitution of natural language words and sentences to the formal symbols of the logical section of the *Formulario*.

10 On Frege's distinction between function and argument in *Begriffsschrift*, see Badesa and Bertran-San Millán (2017).

11 I would like to thank Georg Schiemer for bringing this point to my attention. See also Giovannini and Schiemer (2019).

3.   Function, Operation, and Relation

Peano first introduces the symbol of function and then introduces the symbol of relation. The notion of function is considered synonymous with that of operation and of mapping or transformation.[12]

> Given a symbol *h* of operation, or function or mapping, which are all synonymous, one has to consider the class of individuals on which one can operate with the symbol *h* and the class of individuals that one obtains as a result. . . . We will thus write *bfa* for "symbol of function that being written before an *a* produces a *b*" or "operation that transforms the *a* in *b*", or "correspondence between the *a* and the *b*" or "*b* function defined in *a*" or "*b* function of the *a*".
>
> (Peano 1894, §23, 27–28)

Peano considers four properties of this functional symbol. The first says that if *h* maps a set *A* into a set *B*, then an element *x* of *A* is mapped by *h* into an element *hx* of *B*. The second property is functionality: if $x=y$, then $hx=hy$, that is, the feature that is nowadays used as a defining condition for functions. Two further properties are mentioned: let's call them co-domain extendibility (if *B* is contained in a set *C*, then *h* maps *A* into *C*, i.e., it can be defined on a superset of its original co-domain) and domain contractibility (if *C* is contained in *A*, then *hx* maps *C* into *B*, i.e., can be defined on a subset of its original domain) (Peano 1894, §23, 27 ff.). The insistence on the latter two properties already reveals Peano's interest in a dynamical use of functional symbols and particularly in questions concerning generally valid ways to modify the definition of a function by transformation of its domain and co-domain.

Relations are defined by means of functions (§3): if *xαy* expresses a relation between *x* and *y*, then it can be considered a function of *y* that produces the *x* that are in relation with *y*.[13] The first examples that Peano mentions are the algebraic relations of equality, greater than, and less than.[14]

---

12 I will not discuss here similarities and differences between Peano's and Dedekind's approaches. For a recent discussion, see von Plato (2019) and Kahle (2021).

13 For example, the relation $y \leq x$, with $x,y \in N$, can be expressed as a function *f* from *N* to $P(N)$ such that to each $x \in N$ is associated a unique $f(x)=\{y: y \leq x\}$.

14 Russell and Dedekind always took the order relation as a primitive. Peano also considers the possibility of defining it by means of an operation; that is, he envisages both an additive and an order approach (it depends on the theory to be axiomatized). Rodolfo Bettazzi, who was not, strictly speaking, a scholar of Peano but had many contacts with him and his school, preferred the additive approach in his axiomatization of quantities (Bettazzi 1890). In the *additive approach*, the primitives are an equality or inequality relation and an operation of addition; the order relation is defined by means of the former. In the *order approach*, the primitives are an order relation and an operation (of

Padoa and Pieri considered the notion of function as an ordinary logical notion, and at some point this became the dominant view in the Peano School (Pieri 1906, 204–206). Yet Peano long hesitated to consider the notion of function as a logical concept.[15] Is this a sign of anti-logicism? As a mathematician, Peano distinguished two quite distinct practices concerning the use of function symbols: sometimes one considers a general notion of function without specifying the domain of variability of the variable, and sometimes one operates on functions whose domain is fixed.

> In mathematics there is no single definition for example of "multiplication", nor does there exist in the *Formulario* an equality of the form × = (expression composed by other signs). But there exists a definition of multiplication between two natural numbers, then between two relative numbers, then between two rational numbers, and so on. In the *Formulario* one can easily find more than 30 definitions of $x \times y$, with different hypotheses. Thus to the sign of function it is not connected a domain onto which the function is determined, which is also called variability domain of the function. As a matter of fact, it is not possible to talk about equality of two functions, because two functions may produce identical results in one domain and different results in another domain. But two arbitrary functions u and v always have a common domain that can be expressed by $x \ni (ux = vx)$. We cannot talk about the number of functions that satisfy a given condition: no function is invertible, and so on. When mathematicians talk about equality, number, inverse of a function, the term "function" denotes the system $(u; a)$, where u is the function considered in § 1 and a is the variability domain. We call it "definite function".
>
> (Peano 1906, §4, 80, my translation)

The latter is the usual notion of function that occurs in mathematical practices: Peano calls it "definite function", but what is the former? Is it the logical notion of a propositional function? Or rather a pre-formal notion of function that logic tries to capture?

A comparison with Hermann Grassmann's *General Theory of Forms* might suggest a different reading: the general notion of function is

---

successor, division, passage to the limit . . .). The additive approach had its origins in the Greek theory of proportions and was interestingly applied to physical measurement by Helmholtz: to measure a class of physical magnitudes, one should identify a physical operation that can be executed on such quantities and then investigate its algebraic properties (commutativity, associativity, etc.), so as to choose the appropriate system of numbers to measure them (Cantù 2018).

15 Vailati explained that properties of functions and correspondences have been included in the list of logical formulas, "because, *from a certain point of view*, they belong to logic rather than mathematics" (my emphasis) (Vailati 1893).

an abstraction from the notions of functions occurring in different mathematical domains: it retains their common properties and remains undetermined with respect to further mathematical features. Grassmann distinguishes between formal and real operations. Formal operations of addition and multiplication are operations characterized by a structure of commutative group (in the case of addition) and of a ring (in the case of addition and multiplication taken together). Formal operations are denoted by the usual arithmetical symbols + and · but are not defined with respect to a specific mathematical domain. These general features of operations are studied in the *General Theory of Forms* (Cantù 2020).

Then there are real operations of addition and multiplication that occur when the formal notion is applied to a specific domain (e.g., natural numbers, vectors). Real operations may have different properties depending on the domain they are applied to: for example, an operation of multiplication between natural numbers is commutative, whereas the multiplication between vectors is not. According to Grassmann, the *General Theory of Forms* is the study of some fundamental relations and operations that occur in all branches of mathematics but is not itself part of mathematics, because it contains formal operations that are underdetermined and that could receive full determination only when they are applied to a specific domain and become real operations in mathematical specific theories.[16] Is the *General Theory of Forms* a logical theory? As with proportion theory in Euclid's *Elements*, it does not imply the creation of a new genus of objects (Cantù 2008, §3) but merely assembles a list of propositions that "relate to all branches of mathematics in the same way" (Grassmann 1844, 33; 1995, 33). Peano seems to consider logic a list of all propositions that are later used and applied in the *Formulario* and that can be put together in an introductory volume to mathematical theories.

> Mathematical logic is the only instrument that could express and treat propositions of ordinary mathematics. It is not an end in itself.
>
> (Peano 1913)

Both the calculus of the classes and the calculus of propositions appear as applications of the general abstract algebra to the domain of logic (as usually treated in the Boolean tradition), as Table 9.1 suggests.

---

16 A different but similar way to distinguish between functions as expressions of algebraic symbols and functions acquiring mathematical meaning can be found in Lagrange: "algebraic analysis is pure and general, because the arguments are taken as being previously indeterminate" (Panza 2015).

Table 9.1 Peano's Logical Symbols and Interpretation

Symbols	∪	∩	⊃	Λ
Reading for classes	union	intersection	is contained in	nothing
Reading for propositions	disjunction	conjunction	implies	absurd

Some remarks from Gödel's *Philosophical Notebooks* confirm that he considered Peano's general notion of function as an alternative to Frege's and Russell's view, because "Peano seems to assume that all functions have a meaning (*Sinn*) for any argument" (Crocco et al. 2017, 51, my translation). Gödel interpreted this peculiar notion of function as the most general operation that can be applied to any two things, having usual mathematical and logical relations as special cases.

> I. What is the most general operation that can be applied to two arbitrary things *a, b* and that contains as special cases "*ε*" and "application"—and that corresponds to the operation of moving side by side [*Nebeneinanderschiebens*]? That there is such an operation follows from the fact that each concept has an extension in almost all directions and that the application is the extension of ∈. This relation [is] maybe definable by: 1) the simplest kind of combination, 2) that which owes its existence immediately to the existence of this pair, 3) that which is perceived (that towards which one turns its gaze) when one directs its vision to *a* and *b*, 4) the relation that subsists between *a* and *b* (or the connection that subsists between them), e.g. the vector *a − b*. In particular it is to be expected that for two things emerges a pair, for two propositions a product, for something that needs completion what is obtained by the completion (as for example by an operation), for classes and numbers maybe the sum, for concepts the product. Notably, depending on the type, *ab* would denote *a ∈ b*, *a⌐b*, *b ∈ a*, *a | b* (for relations and functions).
>
> (Crocco et al. 2017, 53–54, my translation)

A relevant difference with respect to Frege's and Russell's theories (especially in the typed versions) is that there is not, in principle, a distinction between a domain of entities that occur only as arguments of a function and entities that have a functional nature, as is the case in the Fregean analysis of predication, where there is a clear distinction between predicates conceived as unsaturated functions and subjects that saturate them, or in Russell's tradition, where at each type level, there are objects that can occur only as arguments of a function. The lack of this asymmetry, which is presumably related to Peano's algebraic concern for a representation of logical truths by means of equalities and for geometric duality, could correspond to what Gödel had in mind as he wrote:

> II. If one explains the concept of function in Peano's way on the basis of this operation of combination, then each thing is both a function and a transformation, because each thing can be combined with other things (the concept application, ∈, etc., in this operation of combination is "limited" to certain classes of things). Yet, on the other hand there are functions in a strict sense, i.e. objects whose essence consists in the possibility of being combined, whereas for the others the fact of being combined is so to say something "external".
>
> (Crocco et al. 2017, 54, my translation)

Peano seems to distinguish the pre-formal non-mathematical notion of indefinite function from the formal logical notions of union and intersection of classes or the notions of conjunction and disjunction of propositions. To further clarify the relations between them, and to investigate whether the pre-formal notion should be considered logical, we will have to analyze in some more detail Peano's linguistic notion of meaning (see §4). Yet the distance from Frege's approach is already evident. Frege requires all functions to be defined and requires, in particular, the existence of a value in output for any value taken as argument. As remarked by Gödel, Peano used the symbol of (indefinite) function for the maximum of a function, because he introduced it without ascertaining the arguments on which the function could be defined and whether it could have a value for each of these arguments (Cantù 2016). Peano, unlike Frege, wanted to preserve as much as possible the ordinary way in which mathematicians use functional symbols,[17] introducing them before having fully determined their specific properties in a given domain. One could say that Peano's functional symbols, like the symbols occurring in Grassmann's *General Theory of Forms*, are endowed with a sort of minimal, incomplete meaning described by the four properties mentioned previously (that of being a mapping, of being functional, of being extendible with respect to the codomain, and of being contractible with respect to the domain). This algebraic interpretation of Peano's approach, which explains why he had no problems in accepting conditional and case-by-case definitions, is supported by another passage of Gödel's *Max Phil*:

> In the case of an incomplete definition of "function" (Peano), there is also the possibility that it might be defined for certain arguments,

---

17 For a further analysis of Peano's notational practices and symbolism, see Schlimm (2021).

undefined for certain arguments, defined by cases for certain arguments.

(Crocco et al. 2017, 55–56, my translation)

The distance from Frege's perspective is even greater if one considers that Peano did not share Frege's association between the concept/object distinction and the function/argument distinction.[18] But this dissimilarity is explained by a radically different understanding of meaning: instead of concepts and objects, formal symbols acquire their full meaning when they are interpreted ("read as") ordinary mathematical language terms and sentences of a specific mathematical theory.

4. Meaning and Interpretation

Criticizing Peano in *Begrüundung meiner strengeren Grundsätze des Definierens* (1897/98), Frege explicitly mentions that he is using the term "meanings" [*Bedeutungen*] in the sense of Peano's "significations" (1969, 166, fn.; English translation in Frege 1979, 153, fn.) Yet what "significations" should mean is not fully clear neither in Frege's commentary nor in Peano's own writings. The term "to mean" that Peano writes in Italian, French, and Latino sine flexione, respectively [*significare, signifier, significat*], is first used in *I principii di geometria* (Peano 1889b) and in the *Arithmetices Principia* (Peano 1889a), even if other expressions like "to indicate" [*indica*] or "can be read as" [*si legge come*] are used as if they were synonymous. Apart from some inconsistencies, two main uses can be distinguished.

First, Peano uses the term when a symbol stands for a longer symbolic expression, for example, when "ab.cd significat (ab)(cd)" (Peano 1889a, vii). Here he introduces a typographical abbreviation, suggesting that a symbol means another symbol when the latter can be replaced *salva veritate* by the former in the appropriate context. Second, Peano uses the term when a symbol means different things, depending on the context in which it is interpreted (or applied). For example, in *I principii di geometria* and in the *Arithmetices Principia*, he mentions that the symbol Λ means "nothing" [*nulla, nihil*] in the context of classes and "absurd" [*assurdo, absurdum*] in the context of propositions. Another example is offered by the symbol ⊃, which means "it is deduced" [*si deduce, deducitur*] in the context of propositions and "is contained" [*è contenuto, continetur*] in the context of classes. A formal mathematical symbol (non-interpreted and to which no fully determinate concept could correspond) means another

---

18 Yet some scholars have recently pointed out that Frege did not make this association either in his early writings but only from the *Grundgesetze* onwards (Badesa and Bertran-San Millán 2017).

symbol (a word of the natural language) that is always used in a specific mathematical or logical context and is thus endowed with the meaning it usually has in that context. Recalling the distinction introduced in §3, this relation of meaning is the one that allows the passage from formal to real operations both in the case of logical symbols (as here) and in the case of mathematical symbols, where the same symbol could be read as the multiplication between natural numbers N or between integer numbers n.

In other words, a symbol of this formal general language means a word of the language of a specific mathematical theory: meaning is again a relation that concerns only the linguistic level. Symbols can be interpreted as varying on linguistic terms. Paolo Mancosu's remarks on Padoa's notion of interpretation support the idea that Peano had a linguistic notion of interpretation.

> In contemporary model theory we think of an interpretation as specifying a domain of individuals with relations on them satisfying the propositions of the system, by means of an appropriate function sending individual constants to objects and relation symbols to subsets of the domain. It is important to remark that in Padoa's notion of interpretation something else is going on. An interpretation of a generic system is given by a concrete set of propositions with meaning. In this sense the abstract theory captures all of the individual theories, just as the expression x + y = y + x captures all the particular expressions of the form 2 + 3 = 3 + 2, 5 + 7 = 7 + 5, etc.
>
> (Mancosu et al. 2009, 7)

The example should not induce one to believe that this *sui generis* notion of interpretation holds only for individual constants; it holds also for relation symbols. An important difference with respect to the model-theoretic interpretation should be spelled out in some more detail. The model-theoretic notion requires two steps: the individuation of the right domain of objects and the choice of the appropriate function. Different interpretations could then arise from the choice of different functions on the same domain of objects. Besides, the interpretation maps symbols into individual objects and sets of individual objects.

In Padoa's as well as in Peano's notion of interpretation, there is just a one-step move, because the symbols are interpreted by substitution with a set of propositions from ordinary mathematical language, and these propositions have a fixed domain and a unique function that maps the proper names into individual objects and the predicates into sets of such objects. Padoa's remarks about the difference between general and specialized theories seem to confirm this idea:

> The system of undefined symbols can then be regarded as the abstraction obtained from all these interpretations, and the generic theory

can then be regarded as the abstraction obtained from the specialized theories that result when in the generic theory the system of undefined symbols is successively replaced by each of the interpretations of this theory. Thus, by means of just one argument that proves a proposition of the generic theory we prove implicitly a proposition in each of the specialized theories.

(Padoa 1900a, 121)

This move from generic to specialized theories, from uninterpreted formal symbols to a concrete set of meaningful propositions of the ordinary mathematical language, from the symbols for formal operation to those for real operations, requires the passage from a list of formal sentences that summarize some common traits of different specific theories to specific theories themselves. And this holds for Peano in the case of mathematical and logical theories. This explains Peano's development of piecemeal definitions of numeric systems, as well as the distance from Frege's need to have a unique definition of each symbol. What guides the admissible substitutions and avoids confusion are for Peano the actual mathematical practices, embedded in their ordinary language formulation. Even if the distinction between function and argument is clearly identified, the meanings of symbols are ordinary mathematics words and propositions and not concepts or objects.

What, then, can be said about Peano's logicism? Is it a kind of heretic logicism that has nothing to share with the idea of reducing arithmetic to logic? There are at least two senses in which mathematics still has to be "preceded" by logic in Peano's work, even if logic emerges as the result of a bottom-up process of successive abstraction from specialized theories (as their invariant). Contrary to what is usually said, this view was not first developed by Padoa but had already been developed by Peano in his *Calcolo geometrico*[19] and commonly shared by the members of the school already at the beginning of the 1880s.[20]

---

19 Yet in the *Calcolo Geometrico*, Peano seemed to consider logic part of mathematics, probably because he adopted the algebra of logic perspective and language, from which he later departed: "Deductive logic, which forms part of the science of mathematics, has not previously advanced very far, although it was a subject of study by Leibniz, Hamilton, Cayley, Boole, H. and R. Grassmann, Schroder, etc. The few questions treated in this introduction already constitute an organic whole, which may serve in much research. Many of the notations introduced are adopted in the geometric calculus" (Peano 1888, vii. English Translation in Peano 2013, x).

20 Vailati's 1892 paper *On the Fundamental Principles of Straight Line Geometry* clearly expressed the geometrical origin of this notion of interpretation. The point is to deduce the principles of geometry from a minimal set of conventions as postulates, which do not refer to any privileged interpretation of the considered signs of relations and operations but only to combinational properties of relations and operations themselves: for example, the relation of two points *a,b* on a straight line, the relation of succession

First, logical signs occur in mathematical language, and they are usually given one or the other of the two previously mentioned interpretations (see e.g., Table 9.1 in §3): it is therefore useful to let them precede the treatment of mathematical theories in a treatise so that repetitions could be avoided and proofs economized. Just as in proportion theory, we can prove theorems that hold both in geometry and in arithmetic: "by means of just one argument that proves a proposition of the generic theory we prove implicitly a proposition in each of the specialized theories" (Padoa 1900a, 121).

Second, logic precedes mathematics in the sense that it is used to actually do mathematics, because: a) logical principles of inferences are used in proving theorems;[21] b) certain general properties of uninterpreted symbols (for example, the codomain expandability or the domain contractility of the symbol of function) are used in the transformation of functions into new functions or used to define them case by case; and c) because the kind of abstraction that one has at the level of formal, uninterpreted, or rather underdetermined symbols derives from a wider metamathematical understanding of logic as an algebraic investigation of structural similarities between different operations and functions.

Before analyzing Peano's metatheory (§5), I would just like to add some brief remarks on abstraction, a question that is quite relevant for logicism, especially for neo-logicism.[22] It has been claimed that the Peano School first recognized definitions by abstraction as a specific form of abstraction and that Frege was the first to use them "as abstraction principles in the service of a philosophical program such as logicism" (Mancosu 2016, 110). Besides, it is well known that Russell's definition of number as the class of equinumerous classes did not raise Peano's enthusiasm, even if it was readily accepted by Padoa and Pieri. Should this mean that Peano was in some sense an anti-logicist? The question is complex.

First, some of the reasons why Peano disliked definitions by abstraction do not depend on philosophical prejudices but on some defects of the corresponding mathematical practice. Any adequate definition should satisfy a criterion of homogeneity (Peano's condition is more liberal than conservative, as adopted in the traditional account of definitions), but if, for example, one defines rational numbers as the class of equinumerous classes, then one has to introduce a non-homogeneous definition of the quotient between numbers.

---

between two instants, or the relation of "greater than" between two quantities (Vailati 1892, 71–72).

21 von Plato (2018; 2017; 2014) has recently shown how rich Peano's *Formulario* is from this point of view.

22 For reasons of space, I cannot develop here a detailed discussion of Peano's genuine or non-genuine abstraction and its relations to his theory of definitions. For the analysis of Peano's definitions by abstraction, see Mancosu (2016) and Mancosu (2018).

Second, it is questionable whether for Peano the value of the abstraction function was really the equivalence class itself rather than one of its members taken as a representative of the whole class (Mancosu 2018). Even if Peano said that a definition by abstraction introduces a new entity, it is disputable whether this entity should be considered the class of equivalence of the abstraction function. On the one hand, the required relation could be weaker than an equivalence relation (Burali-Forti 1894).[23] On the other hand, Padoa and Vailati often mentioned that whenever a new word or a new expression is introduced in mathematics, the latter need not necessarily mean a new concept or object: for example, in a case-by-case definition of a function, the value of the abstraction function could be the class of equivalence for certain cases and a term for other cases.

If one agrees with Ignacio Angelelli that there were genuine abstractions in Peano, and no principles of abstraction that "dispense with abstraction and allow us to do well without worrying about abstraction" (Angelelli 2004, 18), then the value of an abstraction function could not have generally coincided with the equivalence class (even if this is certainly the case in Padoa 1907).

5. Metatheory and Axiomatics

Whether the Peano School contributed to the development of metatheory is a controversial issue in the literature. Marco Borga claimed that there was no metalogic in the Peano School, because not even the notion of logical law was deeply investigated, and there was no real development of any modern meta-theoretical inquiries (Borga 1985). Casari, on the contrary, claimed that Peano contributed both to metalogic and to the metatheoretical investigation of arithmetic:

> it is he who, making the concept of numerical system depend on the concepts of formal language and of the satisfaction of formal conditions, created, albeit quite unaware, the premises for the possible variation of the very concept of numerical system; this, in fact, is transformed when the language adopted and the accepted relationship of satisfaction, vary.
>
> (2011, 145)

If Peano definitely had a metalogical interest in the investigations of independence (von Plato 2019; Bertran-San-Millan 2021), one reason that could explain why the issue was controversial is the fact that when

---

23 For a modern revival of this idea, see Joinet (2017), who claims that definitions by abstraction could be introduced in the presence of relations that are much weaker than equivalence relations.

one investigates metatheory in Peano, one should not look for a meta-linguistic study of general properties of formal deductive systems, but, as in the case of Frege, one should check whether there is some "theoretical scrutiny of a particular theory using notions of 'language', 'reference', 'interpretation', etc." (Tappenden 1997, 213). A further reason that could explain the divergence among scholars is the fact that Peano, Padoa, and other members of the school disagreed not only on the notion and use of definitions by abstraction (see §4) but also on other topics that are quite relevant for a metatheoretical inquiry. The following issues, though only sketched out briefly, will shed some light on the difference between the meta-logical approach developed by Padoa and Peano's meta-mathematical perspective.

Concerning definitions, there were strong dissimilarities about axioms as implicit definitions. Peano claimed that they determine rather than define what is common to an infinity of systems that satisfy the axioms (1891b, 93–94; 1898, 1–2); Pieri remarked that they define the primitive ideas denoted by the symbols occurring in the axioms (e.g., number, unity, addition in arithmetical axioms) (1901, 378, 387); Vailati said that geometric axioms define a class of objects that are indirectly defined by means of other definitions (e.g., Hilbert's notions of points, straight lines, planes) or a single primitive relation (e.g., Hilbert's betweenness) (Arrighi et al. 2010, 151); Vacca claimed that they define the meaning of primitive terms (e.g., the symbols 0, N_0, + in the arithmetical axioms) (1896–99, 186); Burali-Forti asserted that they define an entity in itself, that is, belonging to a class whose individuals verify the primitive propositions (e.g., number in the arithmetical axioms) (1894); Padoa specified that what determines the meaning directly of the primitive ideas and indirectly of the derivative ideas are not the axioms themselves but the interpretations of the symbols occurring in them that verify the axioms (e.g., the interpretations given to the symbols N and + in the arithmetical axioms) (1900a; 1900b).

Concerning equality and logical identity, Peano strongly disagreed with Frege and refuted the idea that logic should have a unique domain with a fixed equality on it, but Padoa seemed rather sympathetic to this view and, following Hilbert, suggested reflexivity and substitutivity as a definition of identity (Cantù 2007; 2010; Hilbert 1905).

Peano was mainly interested in inference rules that were actually used in mathematics, whereas Padoa was generally interested in syllogistic reasoning (Padoa 1912). Even when there was some explicit metatheoretical investigation of the properties of axiomatic systems, different members were especially interested in different features: Pieri in consistency and Peano and Padoa in independence, whereas Burali-Forti, at least in the last edition of his *Logic*, was very dismissive.

Even if the various members of the school shared similar views on the classification of definitions (by operators, by abstraction, by induction, and so on), they diverged on the kind of definitions that should be preferred in a mathematical treatise: Burali-Forti was initially quite critical of abstract definitions and claimed that only nominal definitions express the concept, whereas the other kinds of definitions merely express an intuition; later on, he accepted definitions by abstraction, convinced by Russell's claim that one could always transform them into nominal definitions; Peano admitted that each definition should be evaluated with respect to the specific aims of the theory in which it occurs; Padoa defended definitions by abstraction and investigated the nature of equivalence relations that ground them.

Finally, slightly different formulations of the notions of formal language and logical consequence occur in the works of different authors (especially Peano, Padoa, and Pieri).

We have already mentioned Peano's tendency to view logic as a tool to prove the propositions of ordinary mathematics (Peano 1913, 48) but also as an investigation of deduction, even if, according to Frege, he did not go far enough in this direction (Frege 1984, 237). Two examples of Peano's use of definitions in logical and geometric writings, respectively, explain why the role of logic in mathematics was not only that of deducing theorems from the axioms of a specific mathematical theory but also that of directing the axiomatic presentation of mathematics in the *Formulario*. My claim will be that Peano, like 19th-century mathematicians,[24] understood logic in a broad sense as an epistemological enterprise, interested both in the investigation of the logical components of mathematics (an interest that Peano shared with Frege) but also in the architecture of mathematical axiomatic theories (an interest that survived in successive encyclopedic works, as in Bourbaki's *Elements of Mathematics*) (Segre 1955, 35).

The first example shows that Peano was deeply interested in what he called "possible" definitions, that is, alternative definitions of the same symbol leading to a different architecture of the axiomatic presentation of the logic of classes. Λ is a symbol that can be defined in two different ways in the same logical theory (of classes): as the class that contains the objects that are common to any class or as the only element of a set obtained by intersection of a property and its contradictory, that is, as the result of the application of an operation and its inverse.

The second example shows the epistemological significance of shifting from one to another interpretation of the same string of symbols in the geometrical calculus.

---

24 See, for example, Bolzano, who considered logic a theory of science, that is, "the collection of all rules which we must follow, if we want to do a competent piece of work, when we divide the total domain of truths into individual sciences, and present them in their respective treatises" (*Wissenschaftslehre*, I, §1, 9; Bolzano 1973, 38).

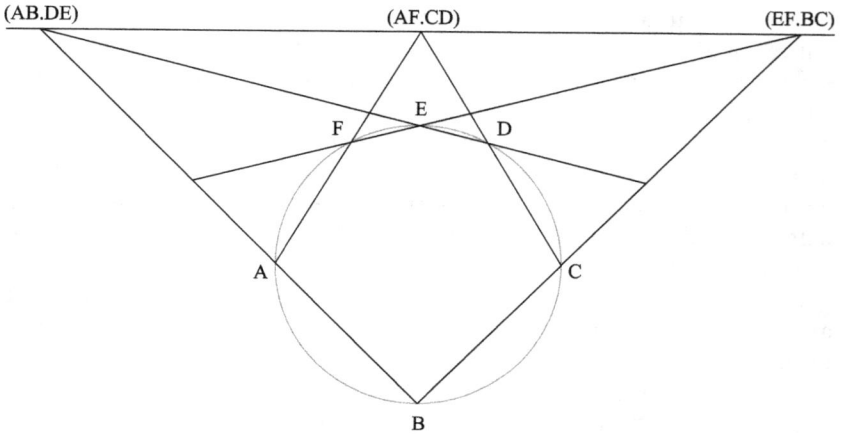

The equation (AB.DE)(BC.EF)(CD.AF)=0 expresses a relation between six elements of first species ABCDEF; it is of second degree in each of these elements, and is satisfied if two of the elements coincide. Suppose five of the elements are given. This is an equation of second degree in the sixth, representing a locus of second order that contains the five elements given. The preceding relation is nothing but the relation between six points of a conic expressed by the hexagram of Pascal.

(Peano 2000, 78–79)

The previously mentioned equation can be interpreted (in Peano's sense) as saying something about the six points A, B, C, D, E, F, that is, that they are inscribed in a conic (they occur in a quadratic equation), or as saying something about the three points of intersection of the segments joining the conic points in two by two (AB.DE), (BC.EF) and (CD. AF), that is, that they are collinear (the equation says that the product is zero).[25] Shifting from one interpretation to the other is a way to grasp the theorem of Pappus-Pascal, which affirms that if six arbitrary points are chosen on a conic and joined by line segments to form a hexagon, then the three pairs of opposite sides of the hexagon (extended if necessary) meet at three points which lie on a straight line.

---

25 The different interpretation is related to the fact that in the second case, the primitives are the segments AB, DE, BC, EF, CD, AF, which allows us to apply the rule that the product of two segments is the point of intersection of their lines. When one is saying that the product of three points is zero if they are aligned, one is actually saying that the product of the intersection of the segments AB and DE, BC and EF, and CD and AF is zero. So we can read the equation (AB.DE)(BC.EF)(CD.AF)=0 as expressing a property of the points A, B, C, D, E, F or a property of the points AB.DE, BC.EF, CD.AF.

The two examples explain both the origin of Peano's notion of meaning (different "readings" of the same symbol by substitution of sentences of the ordinary mathematical language correspond to different interpretations) and the epistemological fruitfulness of the move from one interpretation to another in the dynamical, axiomatic construction of a theory or in the epistemological understanding of a result expressed by means of mathematical equations. A mathematical theory (in some informal sense) is not fully contained in or exhausted by its axiomatic presentation, which should work as a tool to grasp similarities and differences with respect to other theories or other parts of the same theory. In this sense, I claim that Peano's interest in logic was truly meta-mathematical, because he investigated different possible definitions of the same symbol and considered different possible interpretations of formal symbols through substitution of ordinary-language sentences.

6. Peano and Leibniz

It is well known that Peano often mentions Leibniz in the *Formulario* and in his writings on the universal language. Many references have been detected and commented on in the literature: scholars oscillate between considering Peano's readings of Leibniz as guided by a true interest for his logico-linguistic enterprise (Roero 2011, 89) or by the search for an authoritative forerunner of his own work (Luciano 2012, 41). This was not uncommon at the time, and Peano was strongly influenced by Grassmann's geometric calculus, which was developed independently from Leibniz's *analysis situs*.[26] But Peano not only related his geometrical calculus to Leibniz's *analysis situs*: he also explicitly described Leibniz's project of a *Speciosa Generalis* as a sort of universal language or writing system, where the symbols guide reasoning; he made appeal to Leibniz's *calculemus* as a means to put useless controversies to an end; and—using Schröder's words—claimed he had solved the Leibnizian problem in the *Arithmetices Principia*.[27] Not only does Peano mention Leibniz very often in the *Formulario*, but as he became more and more interested in the history of logic and mathematics, he suggested that Vacca, and afterwards Couturat, conduct research on Leibniz's unpublished manuscripts at the Hannover Library (Couturat 1901, Roero 2011). But is Leibniz's undeniable influence on Peano a proof of his logicism?

First, I claim that Leibniz's heritage, combined with the influence of the algebra of logic tradition, generates a linguistic understanding of

---

26 For a further discussion of this point, see De Risi (2007), and Cantù (2003, 319–320).

27 Incidentally, Schröder disagreed on that point, claiming that the members of the Peano School were still using "sailing boats" by the time "steamboats" had been invented (Schröder 1898).

semantics, which is neither representational nor conceptualist but rather is based on the idea that the terms of a symbolic formal language have several possible interpretations, each one determined by substitution of words and propositions of the natural language for the symbols of the formal language.

Second, Leibniz was interested not only in conceptual analyses as ways to determine the relation between simple and complex ideas but also in logical calculi as ways to determine the relation between simple and complex terms that are used to express those ideas. It is exactly this understanding of a logical calculus that can be found in Peano, who aimed to express propositions as equations and deduction between propositions as a resolution method for systems of equations (Peano 1889a; Lolli 2011). Peano, like Leibniz and Grassmann, oscillates between the aim of building as many characteristics as there are domains of investigation and the aim to build a characteristic of all characteristics: "what holds for the characteristic language holds for the characteristic calculus too: there is a tension between specific calculi, like the geometric calculus, and the idea of a general calculus ratiocinator that should operate on real characters" (Cantù 2014, Grassmann 1847).[28] Here, I further claim that what holds for the characteristic and the calculus also holds for semantics: a conceptualist type of semantics at the level of specific theories is combined with what I called a linguistic type of semantics at the formal level (see §3).

There are two main differences with respect to what is usually described as a conceptualist (or as a three-level: words/concepts/objects) semantics. First, a term is primitive with respect to a given set of symbols: being a primitive is a relative and not an absolute property of symbols (there is no need for a primitive term to express some fundamental idea at the level of the *ordo essendi*). Second, Peano's symbols do not express exactly the same concept in all contexts: the symbol of equality is used to express a relation of equivalence between individuals in one section and a mutual implication between propositions in another section. This is usually taken to be a mistake or a limit in Peano's move toward a conceptualist form of semantics. On the contrary, we have taken it to be a sign of the fact that Peano developed a linguistic approach to semantics. Symbols do not mean concepts. Like entries in a dictionary that get their meaning only when they are inserted in a given linguistic context, so the meaning of symbols can be determined only through a preliminary substitution with linguistic

---

28 Whereas in Leibniz the tension involved the right order of concepts (should it be first established as a metaphysical *ordo essendi* in the general characteristic?), Peano believes that a specific calculus can be developed without preliminarily establishing a general characteristic: the foundational and didactical advantage of each part of the *Formulario* is already evident before a complete dictionary of all uses of linguistic terms can be achieved [Peano 1896, 4].

sentences, and in each substitution, they express the concepts expressed by the corresponding words in ordinary mathematical language.

Peano adopts a linguistic semantics for the symbols of the formal language, as he declares that symbols "mean" words of the ordinary mathematical language that have a fixed meaning, which corresponds to their use in the mathematical practice. Even if Peano shared this tendency with the algebra of logic tradition, he was further influenced by Leibniz's characteristic with respect to the idea that the signs should "naturally" mirror what they stand for. Yet this should not be interpreted as being grounded in a representational or conceptualist type of semantics: symbols do not stand for concepts nor for things but for natural language words: for example, ε (being a member of) stands for the Greek word "εστι", ι (definite description) stands for the word "illo", and the symbol for the inverse operation should be an inverted iota, mirroring the inversion of the function.[29] It is in this sense that Leibniz's idea of a characteristic containing "real" characters is not completely abandoned in Peano's perspective. This emerges with even more force in Peano's investigations into universal languages. As the root of a word in ancient languages (e.g., Arabic or Sanskrit) already expresses a part of the meaning of the words that are obtained by adding prefixes and suffixes to it, so formal symbols already express a part of the meaning of the natural-language words that can be substituted for them in specific mathematical theories. At least in this sense, Peano can be considered an heir of Leibniz's logicism: the idea that signs "naturally" mirror what they stand for is compatible with Peano's linguistic semantics.

7. Conclusion

Peano's approach could thus be characterized as a form of structural algebraism, influenced by Leibniz (§6), based on the distinction between a logical and a mathematical notion of function (§3), on a peculiar notion of interpretation and meaning (symbols mean linguistic entities), (§4) on a genuine use of abstraction (§4), and on a strong metatheoretical interest for alternative definitions that is related to his epistemological and dynamic understanding of axiomatics (§5).

Why I call this approach *algebraism* should be clear from what was said in §3 about Grassmann's influence,[30] and more could be said about its relation to Schröder's algebra of logic. Its structural features were

---

29 Quine considers Peano's exploitation of inverses to reflect "his sensitivity to the inner logic of natural language" (1987).

30 See, for example, Grattan-Guinness's (2000, 126) distinction between the algebraic logic of Schröder and the mathematical logic of Frege, but note also that he includes Peano in the tradition of mathematical rather than algebraic logic.

illustrated in §5, where I suggested that Peano's main interest in axiomatics was to identify the logical as the common "structuring element" of different theories. But the primacy of the notion of function over that of relation and the influence of geometric duality on the development of the notion of interpretation also suggests a structural approach, intended not as a philosophical theory based on the idea that all there is to mathematical objects is their structure but as a methodological structuralism that considers that objects could carry many structures and that their relations should be investigated.

This structural algebraism cannot be understood without taking into account that Peano's enterprise was an investigation that concerned at the same time logic, mathematics, and linguistics, as well as theoretical and didactic practices.[31] But can it then be considered as a variety of logicism? If logicism is considered related in its various forms to a mathematical practice that is centered on the analysis of the relation between logical and mathematical notions, on the study of the conditions and limits of a formal treatment of mathematical objects based on a genuine abstraction, and on the effort to highlight the logical inferences drawn between primitive and derived propositions of an axiomatic system, then structural algebraism could still be a variety of logicism, especially if one also takes into account Leibniz's interest in looking for a general characteristic and for a "natural" correspondence between symbols and what they stand for. If logicism is taken to be a reductionist enterprise aiming at introducing a unique symbolic calculus that can be interpreted on a unique domain of objects and at giving a unique and definitive presentation of arithmetic by means of logical concepts, then algebraist structuralism is definitely not a variety of logicism.

To conclude, Peano's approach to logic does not coincide with a mathematical treatment of logic, nor with a general and systematic "investigation of the logical components of mathematics" (Zermelo 1908, 1). In other terms—using a distinction by Jourdain (1914, viii) that was later popularized by van Heijenhoort (van Heijenoort 1967)—Peano did not develop a *calculus ratiocinator* nor a *lingua characteristica*, or at least he had an objective that was broader than both. van Heijenhoort's characterization neglects the epistemological and metamathematical aspect of Peano's enterprise that was highlighted through the analysis of what he says about definitions. The choice between different kinds of definitions cannot be done in purely abstract terms without taking into account

---

31 For an analysis of the strict relation between Peano's logics and linguistics, as well as of the deep interaction between Peano's philosophical views and didactics, see the contributions published in Roero (2010) and Lolli (2001). As an example, I will just quote Peano's remarks on definitions: "the best definition is the one that the teacher prefers" (1915, 112).

specific mathematical practices and specific theoretical systems, without considering syntactic and semantic virtues that a mathematical definition should satisfy, as well as the consequences of a given definition on the order and virtues of successive definitions in the presentation of a theory. These epistemological interests cannot be separated from Peano's didactic motivation.

Peano's "mathematical logic" differs both from the algebra of logic tradition that used mathematical symbols to express logical calculi and from logical investigations centered on the effort to understand the functional nature of predication. Logic is rather, in a Grassmannian sense, the invariant that emerges by algebraic syntactic substitutions, the root of mathematics intended in a linguistic rather than reductionist sense or, in Bourbaki's words, the structuring element of mathematics. The content of the logical inquiry includes the analysis of mathematical notions and theories as well as the mathematical practices in which they are formed and the procedures that characterize them. As Vailati himself remarked, a distinctive feature of the *Formulario* was a joint interest in history, linguistics, mathematics, and logic that can be evinced from the historical annotations, from the application of linguistic categories to the analysis of function (pre-functions and post-functions are introduced by analogy with prefixes and suffixes in natural language), and from the effort to study the dynamics and the evolution of theories as if they were living organisms or, we could add, as if they were living languages.[32]

References

Angelelli, I. (2004). "Adventures of abstraction". *Poznan Studies in the Philosophy of the Sciences and the Humanities*, 82: 11–36.

Arrighi, C., Cantù, P., de Zan, M., and Suppes, P. (eds.). (2010). *Logic and Pragmatism: Selected Essays by Giovanni Vailati*. Stanford: Center for the Study of Language and Information.

---

32 "In the *Formulario* of Peano the importance given to historical data has steadily increased, especially under the inspiration of one of the principal collaborators, Vacca (among other things an enthusiastic investigator of the development of mathematics in the Far East); and the importance attributed to articles of this kind now constitutes one of the most noteworthy among the distinctive characteristics of the method of treatment of the various branches of mathematics that the said *Formulario* presents. Theories are therein expounded, not as in the ordinary treatment, under their 'static' aspect—as one could express it—their aspect of repose; but under that of movement and development—not in the conventional attitudes of stuffed animals, with glass eyes; but as organisms, which live, eat, struggle, reproduce: or at least like figures in a cinematograph, with some naturalness of progression and development. To this tendency to recognize the identity of theories, beyond or under differences of expression, symbolism, language, representative conventions and the rest, is to be attributed also the constant interest of the mathematical logicians in linguistic questions" (Vailati 1906a).

Badesa, C. and Bertran-San Millán, J. (2017). "Function and argument in *Begriffsschrift*". *History and Philosophy of Logic*, 38(4): 316–341.

Bellucci, F., Moktefi, A., and Pietarinen, A.-V. (2018). "Simplex sigillum veri: Peano, Frege, and Peirce on the primitives of logic". *History & Philosophy of Logic*, 39(1): 80–95.

Bertran-San Millán, J. (2021). "Frege, Peano and the interplay between logic and mathematics". *Philosophia Scientiae*, Special Issue *Giuseppe Peano and his School: Logic, epistemology and didactics*, ed. by P. Cantù and E. Luciano, 25(1).

———. (Forthcoming). "Frege, Peano and the construction of a deductive calculus". *Logique et Analyse*.

Bettazzi, R. (1890). *Teoria delle grandezze*. Pisa: Spoerri.

Borga, M. (1985). "La logica, il metodo assiomatico e la problematica metateorica". In *I contributi fondazionali della scuola di Peano*, edited by D. Freguglia, P. Borga, and M. Palladino. Milano: Franco Angeli.

Burali-Forti, C. (1894). *Logica Matematica*. Milano: Hoepli. (2nd edition 1919).

Cantù, P. (2003). *La matematica da scienza delle grandezze a teoria delle forme. L'Ausdehnungslehre di H. Grassmann*. PhD diss., Università degli Studi di Genova.

———. (2007). "Il carteggio Padoa-Vailati. Un'introduzione alle lettere inviate da Chioggia". *Chioggia. Rivista di Studi e ricerche*, 30: 45–70.

———. (2008). "Aristotle's prohibition rule on kind-crossing and the definition of mathematics as a science of quantities". *Synthese*, 174(2): 225–235.

———. (2010). "Sul concetto di eguaglianza: Peano e la sua scuola". In *Peano e la sua scuola fra matematica, logica e interlingua. Atti del Congresso Internazionale di Studi (Torino, 6–7 Ottobre 2008)*, edited by C. S. Roero, 545–561. Torino: Università degli Studi di Torino.

———. (2014). "The right order of concepts: Grassmann, Peano, Gödel and the inheritance of Leibniz's universal characteristic". *Philosophia Scientiae—Studies in History and Philosophy of Science*, 18(1): 157–182.

———. (2016). "Peano and Gödel". In *Kurt Gödel: Philosopher-Scientist*, edited by G. Crocco and E.-M. Engelen, 107–126. Aix en Provence: Presses Universitaires de Provence.

———. (2018). "The epistemological question of the applicability of mathematics". *Journal for the History of Analytic Philosophy*, 6(3): 96–114. Special Issue ed. by Scott Edgar and Lydia Patton: Method, Science, and Mathematics: Neo-Kantianism and Analytic Philosophy.

———. (2020). "Grassmann's concept structuralism". In *The Prehistory of Mathematical Structuralism*, edited by E. Reck and G. Schiemer. Oxford: Oxford University Press.

Cantù P. and Luciano E. (eds.) (2021). *The Peano School: Logic, Epistemology and Didactics*. Special Issue of *Philosophia Scientiae*, 25/1.

Carnap, R. (1929). *Abriss Der Logistik, Mit Besondere Berücksichtigung Der Relationstheorie Und Ihre Anwendungen*. Vienna: Springer.

Casari, E. (2011). "At the origins of metalogic". In *Giuseppe Peano between Mathematics and Logic*, edited by F. Skof, 143–156. Milano: Springer.

Coffa, J. A. (1982). "Kant, Bolzano, and the emergence of logicism". *The Journal of Philosophy*, 79(11): 679–689.

Couturat, L. (1901). *La logique de Leibniz: d'après des documents inédits*. Repr. Hildesheim: Olms, 1985.

Crocco, G., Van Atten, M., Cantù, P., and Engelen, E.-M. (2017). *Kurt Gödel Maxims and Philosophical Remarks Volume X*. https://hal.archives-ouvertes.fr/hal-01459188.

De Risi, V. (2007). *Geometry and Monadology: Leibniz's Analysis Situs and Philosophy of Space*. Basel: Birkhäuser.

Fraenkel, A. (1928). *Einleitung in Der Mengenlehre*, 3rd ed. Berlin-Heidelberg: Springer.

Frege, G. (1897). "Über die Begriffsschrift des Herrn Peano". *Berichte über die Verhandlungen der Königlich Sächsischen Gesellschaft der Wissenschaften zu Leipzig* 48: 361–378.

———. (1969). *Nachgelassene Schriften und Wissenschaftlicher Briefwechsel*. Hamburg: Meiner, vol. 1. Engl. Transl. *Posthumous Writings*, Blackwell, 1979.

———. (1980). *Philosophical and Mathematical Correspondence*. Edited by Gottfried Gabriel, Hans Hermes, Friedrich Kambartel, Christian Thiel, and Albert Veraart. Chicago: University of Chicago Press.

———. (1984). *Collected Papers on Mathematics, Logic, and Philosophy*. Edited by Brian McGuinness. Oxford: Blackwell Publishing Ltd.

Geymonat, L. (1955). "I fondamenti dell'aritmetica secondo Peano e le obiezioni "filosofiche" di B. Russell". In *In Memoria Di Giuseppe Peano*, edited by A Terracini, 51–63. Cuneo: Liceo Scientifico di Cuneo.

Giovannini, E.N. and Schiemer, G. (2019). "What are implicit definitions?". *Erkenntnis*, 1–31.

Grassmann, H.G. (1844). *Die Wissenschaft der extensiven Grösse oder die Ausdehnungslehre, eine neue mathematische Disciplin dargestellt und durch Anwendungen erläutert*. Wigand. Repr. in *Gesammelte mathematische und physikalische Werke*, edited by F. Engel. Leipzig: Teubner. Vol. I.1,1894.

Grassmann, H.G. (1847). *Geometrische Analyse geknüpft an die von Leibniz erfundene geometrische Charakteristik*. Gekrönte Preisschrift von H. Graßmann. Leipzig: Weidmannsche Buchhandlung.

Grassmann, H.G. (1955). *A New Branch of Mathematics: The "Ausdehnungslehre" of 1844 and Other Works*, edited by L.C. Kannenberg. Open Court, 1995.

Grattan-Guinness, I. (2000). *The Search for Mathematical Roots, 1870–1940: Logics, Set Theories and the Foundations of Mathematics from Cantor Through Russell to Gödel*. Princeton, NJ: Princeton University Press.

Heijenoort, J.v. (1967). "Logic as calculus and logic as language". *Synthese*, 17(3): 324–330.

———. (1905). "Über die Grundlagen der Logik und der Arithmetik". In *Verhandlungen Des Dritten Internationalen Mathematiker-Kongresses in Heidelberg Vom 8. Bis 13. August 1904*, edited by A. Krazer, 174–185. Leipzig: Teubner.

Joinet, J.-B. (2017). *Collusions and Agonal Quotients: Generalizing Equivalence Relations and Definitions by Abstraction*. https://hal.archives-ouvertes.fr/hal-02369662.

Jourdain, P. (1914). "Preface". In *Louis Couturat, The Algebra of Logic*, iii–xiii. Chicago and London: Open Court.

Kahle, R. (2021). "Dedekinds Sätze und Peanos Axiomata". *Philosophia Scientiae*, Special Issue *Giuseppe Peano and his School: logic, epistemology and didactics*, ed. by P. Cantù and E. Luciano, 25(1).

Kennedy, H. C. (2002). *Twelve Articles on Giuseppe Peano*. San Francisco: Peremptory Publications.

Lolli, G. (2007). "La Logica tra le due Culture". Paper presented at the Conference: *Le Culture in Italia: Una, Due, Nessuna*, Pisa, March 6–8.

———. (2011). "Peano and the foundations of arithmetic". In *Giuseppe Peano between Mathematics and Logic*, edited by F. Skof, 47–68. Milano: Springer.

Luciano, E., and Roero C. S. (2010). "La Scuola di Giuseppe Peano". In *Peano e la sua scuola fra matematica, logica e interlingua. Atti del Congresso Internazionale di Studi (Torino, 6–7 Ottobre 2008)*, edited by C.S. Roero, xi–xviii. Torino: Università degli Studi di Torino.

Luciano, E. (2012). "Peano and his school between Leibniz and Couturat: The influence in mathematics and in international language". In *New Essays on Leibniz Reception* edited by R. Krömer and Y. Chin-Drian, 41–64. Basel: Springer.

Mancosu, P. (2016). *Abstraction and Infinity*. Oxford: Oxford University Press.

———. (2018). "Definitions by abstraction in the Peano School". In *From Arithmetic to Metaphysics: A Path Through Philosophical Logic*, edited by A. Giordani and C. de Florio, 261–288. Berlin and Boston: De Gruyter.

Mancosu, P., Zach R., and Badesa C. (2009). "The development of mathematical logic from Russell to Tarski, 1900–1935". In *The Development of Modern Logic*, edited by L. Haaparanta, 318–470. Oxford: Oxford University Press.

Padoa, A. (1900a). "Essai d'un théorie algébrique des nombres entiers précédé d'une introduction logique à une théorie déductive quelconque". In *Bibliothèque Du Congrès International de Philosophie*, Vol. 3, 309–365. Paris: Colin.

———. (1900b). *Riassunto delle conferenze sull'algebra e la geometria quali teorie deduttive*. Roma: Università di Roma.

———. (1907). "Dell'astrazione matematica". In *Questioni filosofiche a cura della Società Filosofica Italiana*, 91–104. Bologna: Chiantore-Formiggini.

———. (1912). "Analisi della sillogistica". *Rivista di Filosofia Neo-Scolastica*, 3–4: 337–345.

Panza, M. (2015). "From Lagrange to Frege: Functions and expressions". In *Functions and Generality of Logic: Reflections on Dedekind's and Frege's Logicisms*, edited by H. Benis-Sinaceur, M. Panza, and G. Sandu, 59–95. Cham: Springer International Publishing.

Peano, G. (1888). *Calcolo geometrico secondo l'Ausdehnungslehre di Grassmann*. Torino: Bocca.

———. (1889a). *Arithmetices Principia. Nova Methodo Exposita*. Torino: Bocca.

———. (1889b). *I Principii Di Geometria Logicamente Esposti*. Torino: Bocca.

———. (1891a). "Recens.: *Dr. Ernst Schröder, Vorlesungen über die Algebra der Logik (Exacte Logik)*, I Bd. Leipzig, Teubner, 1890. II Bd. Erste Abtheilung, 1891". *Rivista di Matematica*, 1: 164–170.

———. (1891b). "Sul Concetto Di Numero". Nota I. *Rivista di Matematica*, 1: 87–102

———. (1894). *Notations de Logique Mathématique (Introduction Au Formulaire de Mathématiques)*. Torino: Guadagnini.

———. (1895). "Recens.: *Dr. Gottlob Frege, Grundgesetze der Arithmetik, begriffsschriftlich abgeleitet. Erster Band*, Jena, 1893, Pag. XXXII+254". *Rivista Di Matematica*, 122–128.

———. (1898). *Formulaire de Mathématiques, t. II, §2 [Aritmetica]*. Torino: Bocca.

———. (1906). *Formulario Mathematico. Editio V. Formulario Mathematico. Editio V*. Torino: Bocca.

———. (1913). "Alfred North Whitehead, Bertrand Russell, Principia Mathematica". *Bollettino Di Bibliografia e Storia Delle Scienze Matematiche*, 15: 47–53, 75–81.

———. (1915). "Le definizioni per astrazione". *Bollettino della Mathesis*, 7: 106–120.

———. (2000). *Geometric Calculus, According to the Ausdehnungslehre of H. Grassmann*, trans. Lloyd C. Kannenberg. Boston, Basel, and Berlin: Birkäuser.

———. (2013). *Geometric calculus: according to the Ausdehnungslehre of H. Grassmann*, edited by L.C. Kannenberg. Boston: Birkhäuser.

Pieri, M. (1901). "Sur la géométrie envisagée comme un système purement logique". In *Bibliothèque Du Congrès International de Philosophie*, Vol. 3, 367–404. Paris: Colin.

———. (1906). "Sur la compatibilité des axiomes de l'arithmétique". *Revue de Métaphysique et de Morale*, 14(2): 196–207.

Plato, J.v. (2014). *Elements of Logical Reasoning*. Cambridge: Cambridge University Press.

———. (2017). *The Great Formal Machinery Works: Theories of Deduction and Computation at the Origins of the Digital Age*. Princeton: Princeton University Press.

———. (2018). "The development of proof theory". In *The Stanford Encyclopedia of Philosophy*, edited by Edward N Zalta. Stanford: Metaphysics Research Lab, Stanford University.

———. (2019). "What are the axioms for numbers and who invented them?". In *Philosophy of Logic and Mathematics: Proceedings of the 41st International Ludwig Wittgenstein Symposium*, edited by G. M. Mras, P. Weingartner, and B. Ritter, 343–356. Berlin and Boston: De Gruyter.

———. (2021). "Logic as calculus and logic as language: Too suggestive to be truthful?". *Philosophia Scientiae*, Special Issue Giuseppe Peano and his School: logic, epistemology and didactics, ed. by P. Cantù and E. Luciano, 25(1).

Quine, W.O. (1987). "Peano as logician". *History and Philosophy of Logic*, 8(1): 15–24.

Reck, E.H. (2013a). "Frege, Dedekind, and the origins of logicism". *History & Philosophy of Logic*, 34(3): 242–265.

———. (2013b). "Frege or Dedekind? Towards a reevaluation of their legacies". In *The Historical Turn in Analytic Philosophy*, edited by E. H. Reck, 139–170. London and New York: Palgrave Macmillan.

Roero, S.C., editor. (2010). *Peano e la sua scuola fra matematica, logica e interlingua. Atti del Congresso Internazionale di Studi (Torino, 6–7 Ottobre 2008)*. Torino: Università degli Studi di Torino.

———. (2011). "The *Formulario* between mathematics and history". In *Giuseppe Peano between Mathematics and Logic*, edited by F. Skof, 83–133. Milano: Springer.

Schlimm, D. (2021). "Peano on symbolization, design principles for notations, and the dot notation". *Philosophia Scientiae*, Special Issue *Giuseppe Peano and his School: logic, epistemology and didactics*, ed. by P. Cantù and E. Luciano, 25(1).

Schröder, E. (1898). "On Pasigraphy". *The Monist*, 9(1): 44–62.

Segre, B. (1955). "Peano e il bourbakismo". In *In memoria di Giuseppe Peano*, edited by A. Terracini, 31–40. Cuneo: Liceo Scientifico Statale.

Tappenden, J. (1997). "Metatheory and mathematical practice in Frege". *Philosophical Topics*, 25(2): 213–264.

Vacca, G. (1896–1899). "Sui Precursori della Logica Matematica". II: J.D. Gergonne. *Revue de Mathématiques*, 6: 183–186.

Vailati, G. (1892). "Sui Principi Fondamentali Della Geometria Della Retta". *Rivista di Matematica*, 2: 71–75.

———. (1893). "Sulla raccolta di formule: Part I". *Rivista di Matematica*, 3: 1.

———. (1906a). "Pragmatism and mathematical logic". *The Monist*, 16(4): 481–491.

———. (1906b). "Pragmatismo e Logica Matematica". *Leonardo*, 4(1): 16–25.

Wilson, M. (1992). "Frege: The royal road from geometry". *Nous*, 26: 149–180.

Zalta, E.N. (2015). "Gottlob Frege". In *The Stanford Encyclopedia of Philosophy*, edited by Edward N. Zalta. Stanford: Metaphysics Research Lab, Stanford University.

Zermelo, E. (1908). "Untersuchungen über die Grundlagen der Mengenlehre I". *Mathematische Annalen*, 65(2): 261–281.

10 Logicism in Logical Empiricism

Georg Schiemer

1. Introduction

Logicism, as developed by Frege and Russell, is the thesis that pure mathematics is part of logic.[1] While the logicist thesis was a central doctrine in the philosophy of mathematics of the late nineteenth and early twentieth century, it did not present a uniform research project. Different scholars used the term "logicism" to describe different practices of reducing mathematical theories to higher-order logic or set theory.[2] This holds true, in particular, of work by philosophers related to modern empiricism. Logicism presents one of the cornerstones of logical empiricism.[3] At the same time, the views defended by Carnap, Hahn, and Hempel (among others) differ significantly from Frege's and Russell's original thesis.

The present chapter will focus on several accounts of logicism developed in the main phase of logical empiricism between 1920 and 1940. The aim here is twofold. The first aim is to survey how Frege's classical thesis was modified during the period in question. As we will show, this concerns not only a radically revised conception of the underlying logic but also a new focus on non-arithmetical mathematical theories to be reduced to logic. More specifically, philosophers such as Carnap aimed to formulate a generalized logicism valid for all branches of pure mathematics, including different theories of geometry, topology, and algebra. As we will see in Section 3, his and related accounts are best described as a form of conditional logicism based on an *if-thenist* reconstruction of mathematics.

---

1 This chapter presents an extended version of Schiemer, G., "Nonstandard Logicism", to appear in Uebel, T. (ed.), *Handbook of Logical Empiricism*, Routledge (forthcoming).

2 A more general study of the historical origins of logicism would also have to consider the "non-Fregean" line of early logicism, including the foundational work of Dedekind and Hilbert. See Reck's chapter in the present volume, as well as Sieg and Schlimm (2005) on Dedekind's logicism. Compare Ferreirós (2009) on Hilbert's early logicism.

3 Compare, for example, Goldfarb (1996), Friedman (1999), and Awodey and Carus (2007).

DOI: 10.4324/9780429277894-12

The second aim in this chapter is to clarify how the contributions to conditional logicism are related to other developments in the foundations of logic and mathematics at the time. One focus here will be Wittgenstein's account of the tautological status of logic and the general significance of this view for the logical empiricists' project. In particular, we will retrace how the shift from Frege's and Russell's "universalist" conception of logic to the view of logic as a system of tautologies led to a reformulation of the logicist thesis in work by Carnap. A second issue addressed here concerns Carnap's continued attempts to reconcile classical logicism with a structuralist account of mathematical theories related to the rise of modern axiomatics. A third focus in Section 4 will be on the question of how the logicist thesis was reformulated in his subsequent work on the foundations of mathematics from the late 1930s (and thus after Gödel's incompleteness results).

2.   Classical Logicism and the Type-Theoretic Tradition

The history of classical logicism is well studied. The position is rooted in work on the foundations of mathematics in the nineteenth century, in particular on the rigorization of number theory and analysis by Cantor, Weierstrass, and Dedekind (among others).[4] Frege's logicist project is often described as a direct continuation of this foundational work (see, e.g., Giaquinto 2002). As is well known, Frege developed his program in several steps. He first introduced quantificational logic as the basis for the logicist reduction in his *Begriffsschrift* of 1879. Some years later, an informal characterization of the logicist thesis for arithmetic is outlined in *Grundlagen der Arithmetik* (1884). Based on a critical discussion of Mill's and Kant's respective views on the epistemological status of arithmetic, Frege presents here a new definition of the concept of natural numbers as well as the thesis that arithmetical notions are definable in pure higher-order logic. His main motivation for this reduction of arithmetic to logic is clearly an epistemological one: for Kant, all forms of mathematical knowledge, including arithmetic and geometry, consist of synthetic a priori truths and are thus grounded in *pure* intuition. Whereas Frege still upholds a version of this Kantian view in the case of geometry, his logicist reduction was to show, contra Kant, that the laws of arithmetic have an altogether different, namely purely analytic status.

The technical details of the logicist program are eventually presented in *Grundgesetze der Arithmetik* (1893/1903). In particular, Frege introduces here a higher-order logic together with a naive set theory describing

---

4 Compare, in particular, Ferreirós (1999) and Grattan-Guinness (2000) for detailed historical studies of these developments.

concept extensions. Frege's central axiom on the logical behavior of such extensions of concepts is his notorious Basic Law V:

$$\hat{x}Px = \hat{x}Qx \leftrightarrow \forall x(Px \leftrightarrow Qx)$$

where 'P' and 'Q' are second-order variables ranging over concepts and '$\hat{x}Px$' and '$\hat{x}Qx$' are the extensions of concepts 'P' and 'Q', respectively. This principle is essentially an axiom for unrestricted set abstraction: it states that any two concepts have same extension if they are equivalent. Together with the basic logical laws stated in the *Begriffsschrift* of 1879 as well as a rule of substitution (equivalent to a modern principle of second-order property comprehension), these axioms form Frege's logical system.[5]

Frege's main objective in *Grundgesetze* is to present the technical details of the logicist reduction of arithmetic first outlined in *Grundlagen*. Specifically, based on his explicit definitions of the natural numbers and the successor relation between numbers, it is shown how one can derive each axiom of the Dedekind-Peano axiom system from his system of basic logical laws. Unfortunately, as Russell first pointed out in 1902, Frege's naive theory of classes based on Basic Law V turned out to be inconsistent. In particular, Russell's famous paradox follows from the following instance of naive comprehension:

$$\exists z\forall x(x \in z \leftrightarrow x \notin x)$$

which stipulates the existence of a set that contains as members all sets that do not contain themselves as members.

Contributions to logicism after the discovery of Russell's and related paradoxes were usually based on the logical theory of types, a logical system first introduced in Appendix B of Russell's *Principles of Mathematics* (1903) and then developed systematically in Russell and Whitehead's landmark *Principia Mathematica* (1910–1913). The type theory presented in the three volumes is a higher-order logic describing a rich universe that is stratified into distinct types of objects. Moreover, the logical system presents an intensional logic given that each type is further ramified into different orders, where the order of an object is determined by the kind of formula defining it.[6] Now, Russell and Whitehead's system of ramified type theory was simplified significantly in subsequent work by Carnap, Tarski, Ramsey, and Gödel (among many others). The simplification meant primarily that the original predicativist approach of partitioning type domains into objects of different orders was eventually dropped.

---

5 See, in particular, Heck (2012) for a detailed presentation of Frege's logical system.
6 Compare, for example, Giaquinto (2002), Ferreirós (1999), and Schiemer (forthcoming).

Consequently, Russell's "primacy of intensions" was given up in favor of a purely extensional account of logic.

A second, equally important modification of Russell's original framework concerned the proper formalization of type-theoretic logic: for instance, the clear distinction between the syntax and the semantics of the logic, that is, between the grammatical rules for a type-theoretic language on the one hand and its semantic interpretation on the other hand. While the specific interpretation of type-theoretic languages varied from author to author, the general picture emerging in the late 1920s and early 1930s is that of type theory as a formal set theory, that is, a theory describing a rich universe of sets. Thus, far from being an ontologically neutral theory like first-order logic (as later suggested by Quine), type theory was conceived as a strong logic that describes a rich ontology of sets.[7]

Simple type theory was arguably the standard logical system in the main period of logical empiricism. As Ferreirós put it, before the consolidation of first-order logic, and "as late as 1930 type theory was still regarded by mathematical logicians as the most important and natural system of logic" (Ferreirós 2001, p. 445). In cases where the logical principles of this system (in addition to the standard laws of propositional logic) were explicitly discussed, these usually include an axiom scheme for typed comprehension:

$$\exists x^{i+1} \forall x^i (x^{i+1}(x^i) \leftrightarrow \varphi(x^i))$$

for formulae not containing x^{i+1} free, for all types $i \in \omega$.[8] Informally speaking, the axiom states that every well-formed formula with variable x^i determines a property or set of the objects the formula is true of. The second principle usually mentioned (by Tarski, Gödel, and others) is an axiom scheme of extensionality:

$$\forall x^i (x^{i+1}(x^i) \leftrightarrow y^{i+1}(x^i)) \rightarrow x^{i+1} = y^{i+1}$$

This axiom scheme states that properties or sets of a given type $i + 1$ are identical if co-extensional. In Russell and Whitehead's original presentation of ramified type theory in *Principia Mathematica*, three other axioms were taken to belong to the logical theory: a multiplicative axiom equivalent to the axiom of choice in set theory, an axiom of infinity, and the axiom of reducibility.

---

7 There are other modifications of type theory in the 1930s not discussed here, for instance, the extension of (finite) type theory to theories of transfinite types. See again Ferreirós (1999) for further details. Compare also Mancosu et al. (2009).

8 See, for example, Gödel (1931).

How can the logicist project of representing arithmetic in logic be developed in simple type theory? One way to specify the reduction relation is in terms of the formal notion of interpretability.[9] The interpretation of one theory into another is based on the notion of translation of one formal language into another formal language, usually defined as follows:

Definition 1 A *translation* τ of a language L_S into a language L_T consists of (i) an L_T-formula $\delta(x)$ and (ii) formulas $\varphi_{R_i}(x_1,\ldots x_n)$ (for each primitive n-ary predicate R_i in the language L_S) such that:

1. $(R_i x_1 \ldots, x_n)^{\tau} = \varphi_{R_i}(x_1,\ldots,x_n)$

2. $(x = y)^{\tau} = (x = y)$

3. $(\neg\varphi)^{\tau} = \neg\varphi^{\tau}$

4. $(\varphi \wedge \psi)^{\tau} = \varphi^{\tau} \wedge \psi^{\tau}$

5. $(\forall x\varphi)^{\tau} = \forall x(\delta(x) \to \varphi^{\tau})$

The formula $\delta(x)$ presents a 'domain formula' in language L_T for the variables occurring in L_S-formulas. Formulas $\varphi_{R_i}(x_1,\ldots,x_n)$ provide 'interpretations', within the language L_T, of the non-logical terminology of the primitive terms of L_S. Given this notion of a translation, one can then define the notion of an interpretation as follows:

Definition 2 A translation τ is an *interpretation* of theory S in theory T if for every formula φ such that $S \vdash \varphi$, we have $T \vdash \varphi^{\tau}$.

Frege's project of reducing arithmetic to higher-order logic is often described in the literature as an interpretability result of this form. Roughly speaking, it expresses the fact that (second-order) Dedekind-Peano arithmetic is interpretable in logical type theory. In particular, one can show that all arithmetical statements can be translated into logical statements based on Frege's definitions of the primitive vocabulary '0','successor', and 'being a natural number' of Peano arithmetic. The translation of arithmetical statements into purely logical ones gives an interpretation in the previous sense: for any statement in the language of Peano arithmetic in symbols $\varphi \in L_{PA}$, if $PA \vdash \varphi$ holds, we can show that $TT \vdash [\varphi(0,s,N)]^{\tau}$ also holds.[10] Arithmetic is thus reducible to type theory (possibly including

---

9 See, for example, Burgess (2005), Walsh (2014), and Schiemer (forthcoming). As we will see in the next two chapters, this approach to treat the logicist thesis as an interpretability result is closely connected to Carnap's account of logicism.
10 Compare again Burgess (2005, pp. 50–51).

Russell's axioms of choice and infinity) if the former is interpretable in the latter.[11]

3. Logical Empiricism and Conditional Logicism

Logical empiricism in the 1920s and 1930s was strongly shaped by debates on the epistemological status of mathematics and by logicism in particular. Kant's traditional conception of mathematical principles as synthetic *a priori* truths was generally considered incompatible with a purely empiricist account of scientific knowledge. Logicism, in turn, provided twentieth-century empiricists with an alternative picture of the nature of mathematics which did not conflict with their general philosophical view. Compare Carnap in his "Intellectual Autobiography" on the general significance of the logicist thesis for the philosophers of the Vienna Circle:

> But to the members of the Circle there did not seem to be a fundamental difference between elementary logic and higher logic, including mathematics. Thus we arrived at the conception that all valid statements of mathematics are analytic in the specific sense that they hold in all possible cases and therefore do not have any factual content. What was important in this conception from our point of view was the fact that it became possible for the first time to combine the basic tenet of empiricism with a satisfactory explanation of the nature of logic and mathematics.
>
> (Carnap 1963, p. 46)

Frege's logicism thus provided the logical empiricists with a strategy to establish the purely analytic status of mathematical knowledge. Nevertheless, the previous passage already indicates that the understanding of Carnap and others was in several ways different from the classical program outlined previously.

To characterize the logical empiricists' account of logicism, two aspects should be mentioned here. First, given the discussion in the previous section, it is not surprising that the standard logical system used by philosophers working in Vienna at the time was also a version of Russell and Whitehead's type-theoretic logic of *Principia Mathematica*. Carnap's work from the 1920s and early 1930s contains important contributions to the simplification of type theory. In fact, his *Abriss der Logistik* (1929) can be viewed as one of the first textbooks of modern logic where a purely extensional version of type theory is presented in full detail. A second figure to mention in this respect is Hans Hahn, then head of the mathematics

---

11 We will return to a slightly different notion of interpretability in Section 5.

department at the University of Vienna and one of the founders of the Vienna Circle. Like Carnap, Hahn was also an active proponent of Russell's type-theoretic logic.[12]

The second point to mention here is that the very understanding of logic changed radically in the period in question, mainly in reaction to Wittgenstein's *Tractatus Logico-Philosophicus* (1922). Wittgenstein's new conception of logic (already mentioned by Carnap in the previous passage) can be characterized roughly as follows: logical laws are tautological in nature, that is, statements without factual content. Tautologies do not express facts about the world but are statements true simply by virtue of their logical form. In Wittgenstein's own words:

> The propositions of logic are tautologies. Therefore, the propositions of logic say nothing. (They are the analytic propositions.) All theories that make a proposition of logic appear to have content are false.
>
> (1922, 6.1–6.111)

Philosophers such as Schlick, Carnap, and Hahn fully embraced Wittgenstein's new account of logic in their work from the 1920s and early 1930s.[13] How did this fact impact their view of logicism? In several published texts from the time, logicism is officially described in the classical Fregean sense. Compare, for instance, Carnap's presentation of the thesis in his paper "Die Mathematik als Zweig der Logik":

> The basic idea of logicism can be formulated as [the claim that] mathematics is a branch of logic. That means: there are no specifically mathematical, extra-logical basic concepts and basic propositions. The concepts of mathematics can be derived from the logical concepts, i.e., from concepts which are indispensable for the development of logic even in the ordinary, non-mathematical sense; the propositions of mathematics form a part of the logical propositions.
>
> (Carnap 1930a, p. 298)

---

12 Hahn taught a seminar on the logic of *Principia Mathematica* in Vienna in 1924/1925. See Uebel (2005) for a detailed study of Hahn's work on logic and philosophy of mathematics. See also Hahn (1929), (1933).

13 Compare again Carnap in his autobiography on the significance of the *Tractatus* for his own work: "The most important insight I gained from his work was the conception that the truth of logical statements is based only on their logical structure and on the meaning of the terms. Logical statements are true under all conceivable circumstances; thus their truth is independent of the contingent facts of the world. On the other hand, it follows that these statements do not say anything about the world and thus have no factual content" (Carnap 1963, p. 25).

The general description of the program given here sounds very similar to Frege's original position. However, Carnap's account of the status of logical principles as tautologies clearly differs from Frege's and Russell's respective views.

This fact is particularly interesting given that both Hahn and Carnap were aware that the attempt to extend the tautological character of elementary (i.e., propositional) logic to higher mathematics was in itself deeply problematic.[14] Moreover, Wittgenstein himself not only rejected the logicist thesis in the *Tractatus*, there is also strong textual evidence that he did not take a higher-order system such as type theory to be properly logical in nature. For instance, he is clear on the point that Russell's "existential" axioms of infinity, choice, and reducibility as well as set theory more generally should not be seen as a part of logic (see, in particular, his 5.535, 6.031, and 6.1232 in the *Tractatus*).[15]

One approach adopted by logical empiricists to defend type-theoretic logicism against this objection consists in a form of logical *if-thenism*. Roughly speaking, if-thenism is based on the reformulation of the theorems of a given mathematical theory as universally quantified conditional statements. Such statements contain a ramsified version of the relevant axioms in the antecedent and the ramsified theorem in the consequent. The general approach goes back to Russell and was first formulated systematically in *Principia Mathematica*. Compare Russell and Whitehead on the axiom of choice in volume I of the book:

> We have not assumed its truth in the general [non-finite] case where it cannot be proved, but have included it in the hypotheses of all propositions which depend upon it.
>
> (Russell and Whitehead 1910–13, vol. 1, p. 504)

More generally, in his *Introduction to Mathematical Philosophy*, Russell famously claims that the if-thenist maneuver must be applied to any axiom which is problematic from a logical point of view:

---

14 Tarski, in his logical work from the 1930s, was also critical of the logical positivists' use of the notion of tautology. See, for instance, his article "On the Concept of Logical Consequence" of 1936: "the concept of *tautology* (i.e., of a statement which 'says nothing about reality'), a concept which to me personally seems rather vague, but which has been of fundamental importance for the philosophical discussions of L. Wittgenstein and the whole Vienna Circle" (Tarski 1983, pp. 419–420).

15 A similar view is also expressed in Carnap's work. For instance, in his *Abriss der Logistik*, he holds that: "The axiom of choice should not be included among the basic principles of logic, since its admissibility has been problematic. This is connected with its character as an existential assertion. However, the axiom is required in the proofs of certain theorems of set theory on transfinite powers (infinite cardinal numbers)" (1929, §24b). A similar verdict is adopted in the case of the axiom of infinity.

no principle of logic can assert "existence" except under a hypothesis. . . . Propositions of this form, when they occur in logic, will have to occur as hypotheses or consequences of hypotheses, not as complete asserted propositions.

(Russell 1919, p. 204)

The original motivation for the if-thenist reconstruction was thus to find a way of reducing mathematics to logic in a way in which one does not have to assert the logical truth of Russell's problematic axioms. The resulting conditional logicism was also embraced by several logical empiricists in order to address the problem of the non-tautological character of these axioms. In Carnap's *Abriss*, following a discussion of the status of the axiom of choice, he holds that:

If the axiom is not taken as a basic principle, these theorems can be formulated only as conditional propositions, as implications whose implicans is the axiom of choice.

(1929, §24b)

This Russellian approach is presented more explicitly in Carnap's "The Logicist Foundations of Mathematics":

[Russell] . . . transformed a mathematical sentence, say S, the proof of which required the axiom of infinity, I, or the axiom of choice, C, into a conditional sentence; hence S is taken to assert not S, but $I \supset S$ or $C \supset S$, respectively. This conditional sentence is then derivable from the axioms of logic.

(1931, p. 96)

Interestingly, if-thenism was adopted by Carnap not only as a way to deal with the non-tautological character of several existential axioms of type theory. It was also used as a way to formulate a logicist thesis for non-arithmetical branches of pure mathematics. This is true, in particular, of Carnap's attempts from the 1920s to reconcile classical logicism with the structural approach underlying modern Hilbertian axiomatics.[16] Consider, for instance, Carnap's early monograph *Der Raum* of 1922. Carnap distinguishes between three concepts of spaces in the book, namely between formal, intuitive, and physical space. Formal space, that is, the subject matter of pure geometry in the sense studied also by Russell, is

---

16 Carnap's approach is again strongly influenced by Russell's preceding work on an if-thenist reconstruction of geometrical theorems first presented in *Principles of Mathematics* (1903). See, in particular, Musgrave (1977) Coffa (1981), and Gandon (2012) for further details of Russell's position. Compare also Putnam (1967).

described here as an abstract "relational system" that can be specified in two ways. The first one is in terms of an axiom system in the style of Hilbert. Compare Carnap on this account of axiomatic geometry as a "pure theory of relations or order theory":

> The object of this discipline is not space, i.e., the system of points, lines, and planes determined by *geometrical* axioms . . ., but a "relational or structural system" determined by the *formal* axioms. As this represents the formal design of the spatial system, and turns into the spatial system again when spatial elements are substituted for indeterminate relata, it too will be called "space": "*formal space*".
>
> (1922, p. 8)

An axiomatic theory does not describe a particular and independently accessible domain but rather an abstract structure shared by all systems that satisfy its axioms.

The second approach discussed in the book is more closely connected to the logicist approach. This is the idea to explicitly define a geometrical space in purely logical terms or based on a "logical construction". A formal space, according to Carnap, is what is definable in higher-order logic:

> The construction of formal space can also be undertaken by a different path, however, not just by the above way of setting up certain axioms about classes and relations: by deriving (ordered) series and, as a special case, continuous series from *formal logic*, the general theory of classes and relations.
>
> (Carnap 1922, p. 8)

Thus, Carnap envisages two distinct but equivalent approaches to characterize the subject matter of theories of pure geometry, namely (i) in terms of implicit definitions through an axiom system and (ii) in terms of explicit constructions in a logical system. This duality of methods clearly reflects Carnap's previously mentioned attempt to synthesize a Fregean (or Russellian) foundational stance with Hilbert's modern axiomatic approach (see, in particular, Awodey and Carus 2001 and Reck 2004).

It should be noted here that the logical background system used for such constructions in geometry was not yet made precise in 1922. Nevertheless, the two approaches to studying pure mathematics are still present in Carnap's later work on "general axiomatics" from the late 1920s. Here, the formalization of mathematical theories is expressed in a fully specified logical type theory. For instance, Carnap gives a type-theoretic formalization of several axiomatic theories in his *Abriss der Logistik*. He argues here that there are two ways in which axiom systems can be understood, namely as fully interpreted or as schematic. In the first case, mathematical primitives should be treated as non-logical constants with a fixed semantic interpretation. In the second reading, axiomatic theories are to be treated as formal (in roughly the modern sense of the term). Its primitive terms are thus non-interpreted and can be expressed by higher-order variables. More specifically, according to Carnap, axiom systems can be formalized in the language of simple type theory in the following way: the primitive terms of a theory are expressed as variables (of a given arity and type). The axioms, axiom systems, and theorems, in turn, are expressed as sentential functions.

Given this approach, Carnap argues that an axiomatic theory gives an explicit definition of a higher-level concept, the *"Explizitbegriff"* of an axiom system. Put in modern terms, this is simply the class of models satisfying the theory. In Carnap's own terms:

> For instance, if $x, y, \ldots \alpha, \beta, \ldots P, Q, \ldots$ are the primitive variables of the AS and if we name the conjunction of axioms (that is a propositional function) $AS(x, y, \ldots \alpha, \beta, \ldots P, Q, \ldots)$, then the definition of the explicit concept of this AS is:
>
> $$\hat{x}, \hat{y}, \ldots \hat{\alpha}, \hat{\beta}, \ldots \hat{P}, \hat{Q}, \ldots \{AS(x, y, \ldots \alpha, \beta, \ldots P, Q, \ldots)\}$$
> (Carnap 1929, p. 72)[17]

Two comments are in order here. First, Carnap's idea to describe the content of an axiom system in terms of an explicit concept (or rather its extension) is clearly motivated by Frege's critical analysis of Hilbert's axiomatics. Frege, in his correspondence with Hilbert and in subsequent writings, famously argued that an axiom system cannot be understood as an implicit definition of first-level concepts but rather as an explicit definition of second-level or higher-level concepts. Precisely this idea, with which Carnap was well acquainted with from his time as a student of Frege's lectures in Jena, is presented in *Abriss* in the framework of a type-theoretic logic.[18]

Second, Carnap's formalization of axiomatic theories gives rise to an alternative, weakened form of the logicist thesis. In particular, given the ramsification of theories, the explicit concept corresponding to any axiom

---

17 Carnap discusses a number of mathematical axiom systems in his book to illustrate this account. His examples include formalizations of different projective geometries, of set theory, and of Peano arithmetic, as well as of Hausdorff's topological neighborhood axioms.

18 Compare, for example, Carnap's notes based on Frege's lecture "Logic in Mathematics" of 1914, in particular (Awodey and Reck 2004, pp. 164–166). Carnap has also read and commented on Frege's two articles titled "The Foundations of Geometry" (1903) in which the understanding of axiom systems as definitions of second-level concepts is developed in closer detail. This can be seen from two shorthand notes with comments on Frege's papers written by Carnap in 1921 and documented in Carnap's *Nachlass* (ASP/RC 081-28-01).

system is clearly a logical concept, since it can be expressed in purely logical terms. It follows from this that any theory reconstructed in this way turns out to be logical in character. Compare again Carnap on this point:

> the explicit concept of a geometrical AS, e.g. an AS of projective geometry presents the logical concept of the relevant type of space (e.g. the concept "projective space"). In this sense geometry can also be represented as a branch of logistic itself (as arithmetic) instead of being a case of application of logistics to a nonlogical domain.
>
> (Carnap 1929, p. 72)

In "Proper and Improper Concepts" (1927), Carnap further discusses how his understanding of mathematical theories is related to the conditional logicism described previously. Given that the mathematical primitives of a theory are not defined explicitly but only implicitly through an axiom system, they refer to so-called "improper" concepts and should therefore be symbolized by (free) variables. Mathematical axioms and theorems containing them are hence open formulas and not statements. However, again following Russell in this respect, Carnap argues that the real content of a theorem can be expressed in terms a quantified conditional statement that contains the "explicit concept" of an axiom system in antecedent:

> Are the propositions of (Peano) arithmetic or (Hilbert) geometry then not sentences? After all, they contain symbols for improper concepts, thus variables. As they stand, indeed, they are not sentences, but rather functional expressions. But they serve as very effective abbreviations for proper sentences on the basis of an implicit convention. A sentence-like expression of this kind, in which variable symbols of a given AS occur, is to be taken as short for the sentence that looks like this . . .: first comes a universal prefix containing all the variables of the AS and which applies to the entire implication, then comes the symbol for the logical product of the axioms of the AS as antecedent, and finally comes the sentence-like expression at issue as the consequent. The variables thus occur here only as apparent variables.
>
> (Carnap 1927, p. 371)

The "implicit convention" described here is precisely the if-thenist translation of mathematical statements indicated previously. Theorems of mathematical theories such as Peano arithmetic or Hilbert's Euclidean geometry can thus be translated into purely logical statements of the following form:

$$\forall x, y, \ldots \alpha, \beta, \ldots P, Q, \ldots [AS(x, y, \ldots \alpha, \beta, \ldots P, Q, \ldots) \rightarrow \varphi(x, y, \ldots \alpha, \beta, \ldots P, Q, \ldots)]$$

The variables $x, y, \ldots \alpha, \beta, \ldots P, \ldots Q, \ldots$ present the "primitive signs" of the theory in question, AS presents the conjunction of the universally ramsified axioms, and φ the ramsified theorem considered.[19]

Generally speaking, there are two important reasons for Carnap and other logical empiricists to adopt this form of if-thenism. The first one is to express in purely logical terms the structural character of axiomatic theories. As is well known, the development of modern axiomatics brought with it a model-theoretic conception of theories: axioms and mathematical theorems deducible from them are not merely true in an intended interpretation, but they hold in *any* model or structure that satisfies the primitive structural properties expressed in the axioms. Moreover, from a mathematical point of view, neither one of these structures is preferable to another one. This model-theoretic generality (i.e., the generalization over all possible interpretations of a theory) is certainly characteristic of modern axiomatic mathematics. In Carnap's work, it is expressed logically in terms of the symbolic representation of primitive mathematical signs in terms of variables.

Carnap's second motivation for his if-thenism was to develop an alternative to classical (i.e., arithmetical) logicism. This is established by the fact that a mathematical statement can always be translated into a purely logical statement by the methods of universal ramsification and conditionalization.[20] Generally speaking, we can understand this approach to reduce mathematical theories (including non-arithmetical ones) to a logical system as a kind of "if-thenist logicism". The position is aptly characterized by Musgrave in terms of two conditions:

---

19 Interestingly, the same if-thenist reconstruction of mathematics is also present in philosophical writings of other logical empiricists. Compare, for instance, Hahn (1930) on a general "logization of geometry": "Every theorem of geometry thus appears as a (tautological) implication $P \rightarrow Q$ whose antecedent is the logical product of the axioms and whose consequent Q is the theorem in question. The axioms no longer appear here as self-evident but non-provable truths, but as stipulations from which one can deduce: the primitive concepts no longer appear as elements that cannot be further reduced by definition, but as immediately perceivable trough intuition, but rather as logical variables. Given that every single axiom is a relation between the variables representing the primitive concepts, it follows that geometry appears as a special chapter of the theory of relations, as an investigation of certain special relational systems" (p. 102) A similar discussion of this form of if-thenist reconstruction of mathematical theories is discussed in Hempel's classic article "On the Nature of Mathematical Truth" from 1945.

20 Consider another example given in Carnap's *Abriss*, namely Hausdorff's theory of topological spaces. As he shows, any theorem of that theory is thus best translated into purely logical statements of the form $\forall X(hausd(X) \rightarrow \varphi(X))$ (where $hausd(X)$ presents the logical product of the topological neighborhood axioms). Since both the concept *hausd* and the statement φ are universally ramsified here, the conditional statement is a purely logical statement in the language of type theory.

(1*)All mathematical statements can be translated into purely logical ones, namely as quantified conditional statements with a conjunction of "mathematical axioms" as antecedent and a 'mathematical theorem' as consequent.

(2*)All true mathematical statements can be deduced from logical axioms.

(Musgrave 1977, pp. 117–118)

Taken for itself, the first condition expresses a kind of "language logicism" in the sense that all "mathematical sentences can be paraphrased in such a way that they contain no non-logical vocabulary" (Rayo 2005). Notice that this form of the logicist reduction can also be expressed in terms of the notion of translation presented in Section 2. Thus, condition (1*) suggests a syntactic translation τ that maps each theorem φ of a mathematical theory A expressible in a mathematical language L (with a given mathematical signature $\vec{R}$) to a purely logical statement of the form:

$$[\varphi(\vec{R})]^\tau = \forall \vec{X}(A(\vec{X}) \to \varphi(\vec{X}))$$

The translation $[\varphi]^\tau$ presents the universal ramsification, that is, the result of uniformly substituting variables of the appropriate type for all non-logical primitives in conditional formula $(A \to \varphi)$.

Now, conditional logicism is usually considered a stronger thesis than language logicism. As pointed out by Musgrave and others, it is based also on a second thesis, namely that all mathematical statements so reconstructed are derivable within the logical system. This can be viewed as a form of "consequence logicism" (Rayo 2005). Thesis (2*) thus states that the if-thenist translation is also theorem-preserving, that is, that $[\varphi]^\tau$ is provable from the logical axioms if φ is deducible from theory A. The translation τ thus forms an *interpretation* of a mathematical theory A in type theory TT in the sense specified in Section 2: for every φ such that $A \vdash \varphi$, we have $TT \vdash \varphi^\tau$.

Returning to Carnap's work: while a clear exposition of condition (2*) is missing in his published work from the time, one can find indirect textual evidence in related unpublished work that he understood the logicist thesis precisely in this form of conditional logicism. In particular, Carnap's *Untersuchungen* manuscript (2000) contains his most systematic discussion of the modern axiomatic method. Carnap presents here a general method of formalizing axiomatic theories as well as several metatheoretical concepts in a logical "basic system" ("*Grunddisziplin*"). What is relevant in the present context is that the manuscript contains also a version of if-thenism in the previous sense. More specifically, the notions of theorems ("*Lehrsätze*") and of "logical consequence" are introduced here in the following way: a sentence g is a consequence of an axiom system f if the purely logical statement $\forall X(f(X) \to g(X))$ holds in the type-theoretic basic system. Logical consequence is thus specified here in terms of the material conditional or, more precisely, in terms of a quantified conditional statement expressible in the purely logical language of the basic system. Notice that this logical reconstruction clearly suggests the kind of conditional logicism specified previously. Not only are mathematical theorems expressible in purely logical terms. These logical translations should also be valid in the underlying logical system. This latter condition is clearly a version of condition (2*) stated previously.

It is a matter of scholarly debate how the primitive notion of truth in the basic (type-theoretic) system was understood by Carnap *anno* 1928. One understanding explicitly mentioned by him in a related paper refers to Wittgenstein's notion of a tautology:

> "*consequence*" of f, if f generally implies g: $\forall R(fR \rightarrow gR)$, abbreviated: $f \rightarrow g$. The consequence is, as is the AS, not a sentence, but a propositional function; only the associated implication $f \rightarrow g$ is a sentence, namely a purely logical sentence, thus a tautology, since no nonlogical constants occur.
>
> (Carnap 1930b, p. 304)

As we saw previously, it is notoriously unclear how Wittgenstein's notion of a tautological truth can be extended to apply also to statements of logical type theory. A second, more promising approach is to interpret the notion of "holding in type theory" purely syntactically, namely as being derivable from the logical axioms of the logical system. There is again indirect textual evidence for such as reading as well. In particular, after presenting his notion of logical consequence in (Carnap 2000), Carnap goes on to argue that the notion of logical consequence should not be conflated with Hilbert's notion of "derivability in a formal system". More specifically, Carnap holds that while "g follows from f" and "g is derivable from f in the basic system" are not identical, they are equivalent notions (2000, p. 92). Now, in light of the incompleteness of higher-order logic, Carnap's argument and the intended equivalence result turned out to be false. However, his discussion is based on a correct version of the deduction theorem for such systems, which is also relevant in the present context. In his own terminology, if a proposition g can be formally deduced from the axiom system f, then the statement $\forall X(fX \rightarrow gX)$ is deducible from the principle of type theory, and conversely.

This result allowed Carnap to assume a version of "consequence logicism" that is comparable to the standard logicist thesis that all arithmetical theorems are deducible from purely logical ones. At the same time, this account is certainly weaker than Frege's original logicism. There are two central differences: first, consequence logicism does not require the logicist definitions of the primitive terms of a mathematical theory in a pure higher-order language. As Carnap pointed out in his *Abriss*, the only thing that is explicitly defined by an axiomatic theory is a higher-level property, namely the property (or class) of its models. Second, what is also missing in the account is the requirement that mathematical axioms can be derived from purely logical principles. In contrast, what the present account of conditional logicism effectively shows is that all proofs of theorems can be expressed in a type-theoretic framework. Thus, for any mathematical theory S and every statement φ in the language L_S, the following equivalence can be established:

$$TT \cup \{S\} \vdash \varphi \Leftrightarrow TT \vdash \forall \vec{X}(S(\vec{X}) \to \varphi(\vec{X}))$$

Thus, for every theorem φ of theory S, the universal ramsification of the conditional statement $(S \to \varphi)$ can be derived from the logical principles alone.

4.   The Logicist Thesis and Semantic Interpretability

While classical logicism lost much of its popularity as a foundational approach in the 1930s (mainly as a consequence of Gödel's incompleteness results), it remained an important topic in subsequent work by philosophers affiliated with logical empiricism. This is true, in particular, of Carnap's contributions from the late 1930s.[21] His monograph *Foundations of Logic and Mathematics* (1939) contains a detailed discussion of the logicist reduction of mathematics to higher-order logic. The book is particularly interesting, since it marks the starting point of Carnap's work on formal semantics, eventually culminating in three *Series in Semantics* volumes published in the course of the 1940s. What is characteristic of Carnap's account in 1939 is that the purely syntactic approach of *Logical Syntax of Language* (1934) is complemented by a semantic analysis of the languages of logic and mathematics (as well as of theoretical languages used in the physical sciences). Specifically, the scope of the metatheoretic study of mathematical languages is extended here from a pure "syntax theory" (as developed in detail in *Logical Syntax*) to a systematic exposition of different semantic systems used for the interpretation of such languages. Consequently, Carnap's central notion of analyticity (or *L*-truth) is defined now in a semantic way, based on the notion of truth relative to a semantic system.[22]

It is against the background of this new framework that the logicist thesis is addressed again by Carnap, now in a decidedly semantic form. Sections 16 and 17 of the book deal with so-called "*Non-logical Calculi (Axiom systems)*", that is, mathematical theories presented in axiomatic

---

21 See, in particular, Bohnert (1975) for a detailed study of the different forms of logicism defended by Carnap throughout his intellectual career. See also Reck (2007).

22 Very roughly, a statement of a given semantical system S is logically true (or *L*-true) if its truth can be determined solely on the basis of the semantical rules of S. See Carnap (1939, §1.7). Compare also Koellner (unpublished) for more detailed discussions of Carnap's book. See also Goldfarb & Ricketts (1992).

form. As Carnap argues, such theories usually consist of two parts, namely a logical calculus and a "specific" mathematical calculus. The logical calculus presented here is a higher-order logic similar to the system of simple type theory discussed previously. An example of an "elementary" mathematical calculus that Carnap discusses is a version of second-order Peano arithmetic with a full induction (henceforth PA). As he points out, this theory has an intended or "customary" mathematical interpretation, which he describes as follows:

> The *customary interpretation* of the Peano system may first be formulated in this way: "*b*" designates the cardinal number 0; if ". . ." designates a cardinal number *n*, then ". . ."designates the next one, i.e., *n* + 1; "*N*" designates the class of finite cardinal numbers. Hence in this interpretation the system concerns the progression of finite cardinal numbers, ordered according to magnitude.
>
> (Carnap 1939, p. 40)

Carnap's central contribution in the section is to show how Peano arithmetic can be reduced to the higher-order logic in a roughly Fregean sense. Interestingly, the logicist reduction is described here as an *interpretability* result (in the technical sense of the term) that is comparable to the kind of type-theoretic logicism outlined in Section 2. However, in Carnap's *Foundations*, the notion of interpretability is not merely understood *syntactically*, that is, in terms of the notion of formal provability, but also *semantically*, in terms of the construction of an interpretation (understood as a semantic system) based on the "translation" of the arithmetical calculus in the higher logical calculus.

The notion of a translation between calculi presented in Carnap (1939) corresponds roughly to the modern definition of the interpretation of a theory into another one specified previously (see ibid., p. 40). As we saw, the latter notion is based on a translation function between the formulas of two formal languages that preserves their logical structure as well as the theorems of the interpreted theory. Interestingly, Carnap argues that such a theorem-preserving (in his terminology, a C-true) translation of this form also allows one to construct new "interpretations" for the calculi in question. Compare his description of this translation-based method of model construction:

> If we have an interpretation I_1, for the calculus K_1, then the translation of K_2 into K_1 determines in connection with I_1 an interpretation I_2 for K_2. I_2 may be called a *secondary interpretation*. If the translation is C-true and the (primary) interpretation I_1 is true, I_2 is also true.
>
> (ibid, p. 40)[23]

---

23 To say that a translation is C-true means for Carnap that the translation determines an interpretation (in the modern sense of the term) of K_2 in K_1. See ibid., p. 40.

Paraphrased in modern terms, Carnap's idea seems to be roughly this: an interpretation (in the modern sense of the term) of theory T in theory S based on a translation of language L_T in language L_S allows one to construct a model of T based on a given model of S. More specifically, one can say that an L_T structure M is interpretable in a L_S-structure N in this sense if M is *definable* (in the model-theoretic sense) in N, that is, if the domain, relations, functions, and distinguished individuals of M are definable in N.[24]

Given this general method of model construction, Carnap shows how it can be applied to the program of reducing the arithmetic to higher-order logic. The logicist reduction is presented here in terms of the construction of a purely logical interpretation of PA based on the translation of the language of PA into a pure higher-order calculus. Compare again Carnap on this approach:

> We shall now state rules of translation for the Peano system into the higher functional calculus and thereby give a secondary interpretation for that system. The logical basic calculus is translated into itself; thus we have to state the correlation only for the specific primitive signs. As correlates for "b", "$'$", "N", we take "0", "$+$", "finite cardinal number", for any variable, a variable of two levels higher.
>
> (ibid., pp. 40–41)

The translation mentioned previously is based on the well-known logicist definitions of the primitive arithmetical vocabulary in purely logical terms. These bridge definitions allow one to represent the axioms of arithmetic as purely logical statements. Moreover, as Carnap shows, the same syntactic translation also allows one to construct a purely logical version of the "customary interpretation" of PA as a subsystem of the standard or "normal interpretation" of the logical calculus. Given this genuinely semantic approach of reinterpreting arithmetic in a "secondary", purely logical interpretation, Carnap then specifies the logicist thesis in the following way:

> If we assume that the normal interpretation of the logical calculus is true, the given secondary interpretation for the Peano system is shown to be true by showing that the correlates of the axioms are C-true. And it can indeed be shown that the sentences $P1$–5 are provable in the higher functional calculus, provided suitable rules of transformation are established. As the normal interpretation of the logical

---

24 See, for example, Walsh (2014) for a more detailed discussion of the notion of semantic interpretability.

calculus is logical and L-true, the given interpretation of the Peano system is also logical and L-true.

(ibid., p. 41)

Carnap's line of reasoning can be recast in modern terms as follows: a mathematical theory can be shown to be reducible to logic if it is true in a purely logical model. Given that the higher-order logical calculus (henceforth HOC) is true in the intended logical universe, say, V, and that there exists an interpretation of PA into HOC that allows the construction of a logical model M of PA as a model interpretable in V, it follows that PA is reducible to logic. More generally, what is shown here is an interpretability argument of the following form: if a mathematical theory T is interpretable in HOC, then the underlying translation function allows one to construct a model of T *within* the intended universe of HOC. Since this universe is purely logical and theory T is interpretable in HOC, it follows that T also has a purely logical interpretation.

Given Carnap's arithmetical logicism in *Foundations* of 1939, two further points of commentary should be made here. First, his technical presentation of the thesis obviously differs in several respects from the classical thesis of Frege and Russell. In particular, Carnap explicitly describes the logicist reduction of arithmetic to higher-order logic (or type theory) as an interpretability result, based on the notion of a theorem-preserving translation of one calculus into another one. Moreover, as we saw, his account is decidedly semantic in nature: as he points out, the interpretation of one axiomatic theory in another one is complemented by an additional semantic constraint, namely the fact that this interpretation also gives a uniform way to construct a model of the interpreted theory. Thus, according to this particular logicist thesis, a theory like PA is reducible to a logical theory such as HOC if (i) PA is interpretable in HOC and (ii) the standard model of PA is *semantically* interpretable in the logical universe of HOC.[25]

This form of "interpretational" logicism clearly echoes Carnap's general focus on axiomatic mathematics and can thus be traced back to his work on general axiomatics from the 1920s. In fact, one can identify a similar (however less explicit) form of interpretational logicism in his pre-semantical work from the time. For instance, in the discussion on the foundations of mathematics at the famous Königsberg meeting in 1930, based on his talk on the logicist foundations of mathematics, Carnap gives the following well-known remark on the "logical analysis of the formalistic system(s)":

---

25 Compare again Walsh (2014) for a more systematic discussion of arithmetical logicism and different versions of interpretability.

(1) For every mathematical sign one or more interpretations are found, and in fact purely logical interpretations.

(2) If the axiom system is consistent, then upon replacing each mathematical sign by its logical interpretation (or one of its various interpretations), every mathematical formula becomes a tautology.

(3) If the axiom system is complete . . ., then the interpretation is unique; every sign has exactly one interpretation, and with that the formalist construction is transformed into a logicist one.

(Hahn et al. 1931, pp. 143–144)

This characterization of the logicist thesis based on the interpretation of axiomatically defined primitive terms already anticipates Carnap's position in *Foundations* of 1939. The basic idea already expressed in 1930 is that a mathematical theory, presented in axiomatic form, can be described as a branch of logic if one can construct a "purely logical interpretation" of it.

The second point to mention here concerns the scope of Carnap's logicism. While his discussion in *Foundations* is restricted primarily to the case of elementary arithmetic, he makes clear in later sections of the book that all other theories of pure mathematics can also be reduced to higher-order logic. In particular, he explicitly mentions in §18 that different "higher mathematical calculi", for example, that of real analysis, can be reduced to Peano arithmetic and hence also to the logical calculus in question.[26] In §21, Carnap turns to a detailed discussion of "geometrical calculi and their interpretations". Geometrical theories are usually presented in axiomatic form, according to him. While the customary interpretation of such systems is "descriptive" and thus empirical, Carnap points out that also purely logical interpretations can be constructed for them, based on the translation of geometrical terms into terms of real analysis. Given that real analysis can be reduced to arithmetic and thus to logic in the sense outlined previously, it follows that a purely "logico-mathematical interpretation" can be given for geometry as well. Compare Carnap on this point:

> Of especial importance for the development of geometry in the past few centuries has been a certain translation of the geometrical calculus into the mathematical calculus. This leads, in combination with the customary interpretation of the mathematical calculus, to a logical interpretation of the geometrical calculus. The translation was found by Descartes and is known as analytic geometry or geometry of

---

26 It should be noted here that Carnap's claim that real analysis can also be reduced to type-theoretic logic is contentious and currently under discussion. I would like to thank a reviewer for emphasizing this point.

coordinates. "P_1" (or, in ordinary formulation, "point") is translated into "ordered triple of real numbers"; "P_3" ("plane") into "class of ordered triples of real numbers fulfilling a linear equation", etc. The axioms, translated in this way, become *C*-true sentences of the mathematical calculus; hence the translation is *C*-true. On the basis of the customary interpretation of the mathematical calculus, the axioms and theorems of geometry become *L*-true propositions.

(ibid., pp. 53–54)

Thus, as was originally shown in Hilbert's *Grundlagen der Geometrie* (1899), axiomatic Euclidean geometry can be interpreted in a purely analytic model. Given that analysis can be reduced to logic, it follows that a purely "logico-mathematical interpretation" can be given for geometry as well.

These remarks clearly show that Carnap's approach in *Foundations* is closely connected to his pre-*Syntax* work on logicism and general axiomatics. Specifically, his version of the logicist thesis given in 1939 corresponds closely to classical type-theoretic logicism, complemented by a semantic claim, namely that the logicist translation of the language of arithmetic into a purely logical language also allows one to construct a purely logical model of PA. Now, one can view the semantic version of his interpretational logicism as Carnap's most systematic attempt to "reconcile" the traditional logicist thesis with formalism or with the axiomatic approach in mathematics. At the same time, it is also evident that he upheld a more deflationist account of logicism in 1939 (that is also in spirit with his scattered remarks on the topic in his *Logical Syntax* of 1934). In particular, in §20 of the book, he argues that the former controversy between the foundational doctrines logicism and formalism "has at present lost much of its former appearance of importance" (ibid., p. 49). This is mainly due to the fact that both the axiomatic and the logicist approach are compatible with each other and should thus no longer be subject to philosophical dispute.

5. Conclusion

This chapter showed that Frege's classical logicism was subject to a number of transformations in the work of logical empiricists throughout the 1920s and 1930s. The focus here was on three contributions. Based on a brief account of the development of the classical type-theoretic logicism, we first surveyed how the logicist thesis was understood by Carnap in the course of the 1920s. As we saw, his contributions to the reduction of mathematics to logic were strongly influenced by Wittgenstein's *Tractatus*-view of the tautological nature of logical truths. This new conception of logic, paired with a sustained critique of the existential and thus non-logical character of the type-theoretic axioms of choice, infinity, and reducibility, led Carnap and others to develop a version of "conditional" logicism, first outlined in Russell's and Whitehead's *Principia Mathematica*.

The second development analyzed in the chapter concerns the application of conditional logicism to non-arithmetical fields of mathematics. Carnap, in his work on general axiomatics, presented a precise account of a generalized logicist thesis based on two steps: (i) the type-theoretic formalization of axiomatic theories of different branches of mathematics and (ii) an *if-thenist* reconstruction of mathematical theorems. Concerning the latter, he argued that all theorems can be translated into quantified conditional statements where the mathematical primitives are substituted by variables of the correct type. As we saw, Carnap's if-thenism can also be viewed as a form of conditional logicism that aims to reconcile Frege's original thesis with a structuralist account of modern axiomatics.[27]

Finally, we surveyed how logicism was further developed in Carnap's work from the 1930s, that is, after his involvement in the Vienna Circle. Our focus here was on his re-adoption of classical logicism in *Foundations of Logic and Mathematics*, the first book belonging to his post-syntactic or semantic period. As we saw, Carnap explicitly formulated the reduction of arithmetic to higher-order logic in terms of an interpretability result in this book. Classical arithmetical logicism is usually expressed in terms of the syntactic interpretability of arithmetic in higher-order logic. However, as Carnap first showed in 1939, it can also be recast as a genuinely semantic result: arithmetic is reducible to logic if the standard model of the natural number can be constructed within the type theoretic universe.

6.   Acknowledgments

The author wishes to thank Francesca Boccuni, Andrea Sereni, Erich Reck, and an anonymous reviewer, as well as the participants of the workshop *Origins and Varieties of Logicism* at IUSS Pavia (2015) for their helpful remarks on topics addressed in this chapter. Research on this project has received funding by the *European Research Council* (ERC) under the European Union's Horizon 2020 research and innovation program (grant agreement No. 715222).

References

Awodey, S. and Carus, A. (2001). "Carnap, completeness, and categoricity: the Gabelbarkeitssatz of 1928". *Erkenntnis*, 54:145–172.

Awodey, S. and Carus, A. (2007). "The turning point and the revolution: philosophy of mathematics in logical empiricism from *Tractatus* to *Logical Syntax*". In: Richardson, A. and Uebel, T. (eds.), *The Cambridge Companion to Logical Empiricism*, Cambridge: Cambridge University Press, 165–192

---

27 It would be interesting to further investigate the relation between Carnap's if-thenism and Hilbert's preceding views on logicism and axiomatics from the 1890s. See Ferreirós (2009) on Hilbert's early logicism.

Awodey, S. and Reck, E. (2004). *Frege's Lectures on Logic: Carnap's Student Notes 1910–1914*. La Salle, IL: Open Court.

Bohnert, H. G. (1975). "Carnap's logicism". In: Hintikka, J. (ed.), *Rudolf Carnap, Logical Empiricist*. Dodrecht: Reidel, 183–216.

Burgess, J. (2005). *Fixing Frege*. Princeton: Princeton University Press.

Carnap, R. (1922). *Der Raum: Ein Beitrag zur Wissenschaftslehre*. Berlin: Reuther & Richard.

Carnap, R. (1927). "Eigentliche und uneigentliche Begriffe". *Symposion*, 1:355–374.

Carnap, R. (1929). *Abriss der Logistik*. Wien: Springer.

Carnap, R. (1930a). "Die Mathematik als Zweig der Logik". *Blätter für deutsche Philosophie*, 4:298–310.

Carnap, R. (1930b). "Bericht über Untersuchungen zur allgemeinen Axiomatik". *Erkenntnis*, 1:303–307.

Carnap, R. (1931). "Die logizistische Grundlegung der Mathematik". *Erkenntnis*, 2:91–105.

Carnap, R. (1934). *Logische Syntax der Sprache*. Wien: Springer.

Carnap, R. (1939). *Foundations of Logic and Mathematics, International Encyclopedia*, vol. 1, nr. 3. Chicago: University of Chicago Press.

Carnap, R. (1963). "Intellectual autobiography". In: Schilpp, P. (ed.), *The Philosophy of Rudolf Carnap*, vol. 11, The Library of Living Philosophers. La Salle, IL: Open Court, 3–84.

Carnap, R. (2000). *Untersuchungen zur allgemeinen Axiomatik*. Darmstadt: Wissenschaftliche Buchgesellschaft.

Coffa, A. (1981). "Kant and Russell". *Synthese*, 46:247–263.

Ferreirós, J. (1999). *Labyrinth of Thought—A History of Set Theory and Its Role in Modern Mathematics*. Basel: Birkhäuser.

Ferreirós, J. (2001). "The road to modern logic—an interpretation". *Bulletin of Symbolic Logic*, 7:441–484.

Ferreirós, J. (2009). "Hilbert, logicism, and mathematical existence". *Synthese*, 170:33–70.

Frege, G. (1893). *Grundgesetze der Arithmetik. I. Band*. Hildesheim: Georg Olms Verlagsbuchhandlung.

Friedman, M. (1999). *Reconsidering Logical Positivism*. Cambridge: Cambridge University Press.

Gandon, S. (2012). *Russell's Unknown Logicism: A Study in the History and Philosophy of Mathematics*. Basingstoke: Palgrave Macmillan.

Giaquinto, M. (2002). *The Search for Certainty—A Philosophical Account of Foundations of Mathematics*. Oxford: Oxford University Press.

Goldfarb, W. (1996). "Philosophy of mathematics in early positivism". In: Giere, R. N. and Richardson, A. (eds.), *Origins of Logical Empiricism*. Minneapolis, MN: University of Minnesota Press, 213–230.

Goldfarb, W. and Ricketts, T. (1992). "Carnap and the philosophy of mathematics". In: Bell, D. and Vossenkuhl, W. (eds.), *Science and Subjectivity: The Vienna Circle and 20th Century Philosophy*. Berlin: Akademieverlag, 61–78.

Grattan-Guinness, I. (2000). *The Search for Mathematical Roots 1870–1940. Logics, Set Theories and the Foundations of Mathematics from Cantor through Russell to Gödel*. Princeton: Princeton University Press.

Hahn, H. (1929). "Empirismus, Mathematik, Logik," in: *Forschungen und Fortschritte 5*. Trans. "Empiricism, Mathematics, Logic," in H. Hahn,

Empiricism, Logic and Mathematics (ed. by B. McGuinness), Dordrecht: Reidel, 1980, pp. 20–30.

Hahn, H. (1933). *Logik, Mathematik, Naturerkennen*, Vienna: Gerold. Trans. "Logic, Mathematics and Knowledge of Nature," in: B. McGuinness (ed.), *Unified Science*, Dordrecht: Reidel, 1987, pp. 24–45.

Hahn, H. 1930. "Die Bedeutung der wissenschaftlichen Weltauffassung insbesondere für Mathematik und Physik", *Erkenntnis* 1, 96–105, English translation in: H. Hahn, *Empiricism, Logic and Mathematics*, in B. McGuinness, ed., Dordrecht: Reidel, pp. 20–30.

Hahn, H., et al. (1931). "Diskussion zur Grundlegung der Mathematik". *Erkenntnis*, 2:135–151.

Heck, R. (2012). *Reading Frege's Grundgesetze*. Oxford: Clarendon Press.

Hempel, G. (1945). "On the nature of mathematical truth". *The American Mathematical Monthly*, 52:543–556.

Koellner, P. (2009). *Carnap on the Foundations of Logic and Mathematics*, unpublished manuscript.

Mancosu, P., Zach, R. and Badesa, C. (2009). "The development of mathematical logic from Russell to Tarski". In: Haaparanta, L. (ed.), *The Development of Modern Logic*. Oxford: Oxford University Press, 318–347.

Musgrave, A. (1977). "Logicism revisited". *British Journal of Philosophy of Science*, 28:99–127.

Putnam, H. (1967). "The thesis that mathematics is logic". In: Schoenman, R. (ed.), *Bertrand Russell, Philosopher of the Century*. London: Allen & Unwin, 273–303.

Rayo, A. (2005). "Logicism reconsidered". In: Shapiro, S. (ed.), *The Oxford Handbook of Philosophy of Mathematics and Logic*. Oxford: Oxford University Press.

Reck, E. (2004). "From Frege and Russell to Carnap: Logic and logicism in the 1920s". In: Awodey, S. and Klein, C. (eds.), *Carnap Brought Home: The View from Jena*. Chicago: Open Court, 151–180.

Reck, E. (2007). "Carnap and modern logic". In: Friedman, M. and Creath, R. (eds.), *The Cambridge Companion to Carnap*. Cambridge: Cambridge University Press, 176–199.

Russell, B. (1903). *The Principles of Mathematics*. London: Routledge.

Russell, B. (1919). *Introduction to Mathematical Philosophy*. London: George Allen & Unwin.

Russell, B. and Whitehead, N. A. (1910–13). *Principia Mathematica*. Cambridge: Cambridge University Press. 3 volumes.

Schiemer, G. (forthcoming). "Nonstandard logicism, to appear in: Uebel, T. (ed.)," *Handbook of Logical Empiricism*. Routledge (forthcoming).

Sieg, W. and Schlimm, D. (2005). "Dedekind's analysis of number: systems and axioms". *Synthese*, 147:121–170.

Tarski, A. (1983). *Logic, Semantics, Metamathematics*. Indianapolis: Hackett, 2 edition.

Uebel, T. E. (2005). "Learning logical tolerance: Hans Hahn on the foundations of mathematics". *History and Philosophy of Logic*, 26(3):175–209.

Walsh, S. (2014). "Logicism, interpretability and knowledge of arithmetic". *The Review of Symbolic Logic*, 7(1):84–119.

Wittgenstein, L. (1922). *Tractatus Logico-Philosophicus*. London: Routledge and Kegan Paul.

Part III
The Roads Ahead

11 Ordinals by Abstraction[1]

Bob Hale

The neo-Fregean programme in the philosophy of mathematics seeks to provide foundations for fundamental mathematical theories in abstraction principles—that is, principles of the form $\forall\alpha\forall\beta(f(\alpha) = f(\beta) \leftrightarrow Eq(\alpha, \beta))$, understood as implicit definitions of the function f in terms of the equivalence relation Eq. The best known and most discussed such principle is, of course, Hume's principle:

(HP) $\forall F\forall G(NxFx = NxGx \leftrightarrow F \approx G)$

which seeks to define the cardinal number operator $Nx \ldots x \ldots$ in terms of one-one correlation ($\approx$). Abstractions for real numbers and, with less conspicuous success, for sets have also been proposed. Until quite recently, however, very little has been published on introducing *ordinal* numbers by abstraction. In a welcome addition to that little,[2] Ian Rumfitt (Rumfitt 2018) proposes to introduce ordinal numbers by means of an abstraction principle, (ORD), which says, roughly, that 'the ordinal number attaching to one well-ordered series is identical with that attaching to another if, and only if, the two series are isomorphic' (Rumfitt 2018, p. 195). Although I shall myself reject (ORD) in favour of an alternative form of ordinal abstraction, I shall discuss Rumfitt's proposal in some detail, because it poses a sharp and serious challenge to those of us seeking to advance the neo-Fregean programme. As Rumfitt explains, (ORD) runs pretty directly into trouble, as one might expect, in the shape of

---

1 This chapter originally appeared as Essay 14 in: Bob Hale, *Essence and Existence: Selected Essays*, edited by Jessica Leech, Oxford University Press, Oxford, 2020, pp. 240–255.

2 Roy Cook (Cook 2003) discusses a simple order-type abstraction principle (OAP): $\forall R\forall S(OT(R) = OT(S) \leftrightarrow R \cong S$, which runs into Burali-Forti's paradox, and proposes a Size-Restricted Abstraction Principle (SOAP): $\forall R\forall S[OT(R) = OT(S) \leftrightarrow (((\neg WO(R) \vee Big(R)) \wedge (\neg WO(S) \vee Big(S)) \vee (WO(R) \wedge WO(S) \wedge R \cong S \wedge \neg Big(R) \wedge \neg Big(S))))]$, but his main focus in this paper lies elsewhere, in developing an abstractionist version of the iterative approach to set theory.

DOI: 10.4324/9780429277894-14

Burali-Forti's paradox. But, Rumfitt argues, we may and should uphold (ORD)—in his view the best abstractionist treatment of the ordinals—by blocking the derivation of Burali-Forti's contradiction in another place, viz. its reliance upon an impredicative comprehension principle. Instead, we should adopt a significantly weaker second-order logic, allowing no more than Δ_1^1-comprehension. However, as Rumfitt himself emphasizes, this remedy ill-suits the neo-Fregean enterprise, at least as it has been conceived by its principal proponents. For Δ_1^1-comprehension is insufficient for the derivation of arithmetic on the basis of (HP), which requires a stronger form of comprehension—either Σ_1^1 or Π_1^1. Thus if neo-Fregean foundations for elementary arithmetic are to be saved, we must explain why we do not have to block the derivation of Burali-Forti's contradiction from an ordinal abstraction principle by restricting comprehension in the way Rumfitt proposes, but can obstruct it elsewhere. I shall be exploring the prospects for doing so.

1.   Rumfitt's Proposal

I begin with a summary of Rumfitt's approach and his diagnosis of the Burali-Forti paradox.

1.1.   *Series, Relations-in Extension, and Ordinal Abstraction*

On Rumfitt's approach, ordinal numbers are identified with *order-types* of *well-ordered series*.

> An ordinal number, then, is something which isomorphic serial relations, i.e. isomorphic well-ordered series, have in common. Thus one well-ordered series consists of the Mozart-Da Ponte operas, arranged in order of composition: *Le Nozze di Figaro, Don Giovanni, Così Fan Tutte*. Another consists of the Norman Kings of England, arranged in order of succession to the throne: William the Conqueror, William Rufus, Henry I. These two series are isomorphic, so both are instances of a common order-type, viz. the ordinal number 3.
>
> (Rumfitt 2018, pp. 190–191)

Whilst Rumfitt's identification of ordinals with order-types is more or less standard, his conception of series is not. A *series*, Rumfitt insists, is not a set-theoretic entity but a certain kind of plurality:

> I understand a series to be some things (plural) in a particular order. . . . The series is not constituted by its terms alone, but by them in tandem with a two-place relation between terms.
>
> (Rumfitt 2018, p. 192)

However, he argues, we do not have to think of series as *composite* entities, comprising *both* a relation *and* a plurality:

> Logicians achieve a useful economy by identifying a series with the relevant binary relation, ... The identification is legitimate, but only if the relation with which the series is identified is what we may call a *relation-in-extension*. By this, I mean a plurality of pairs of objects which stand in the universal or pure relation. Thus the series of Mozart-da Ponte operas may be identified with the relation-in-extension *<Figaro, DonGiovanni>*, *<Figaro, Così>*, *<DonGiovanni, Così>*. The series, then, comprises those pairs of Mozart-da Ponte operas which stand in the pure relation of *being composed before*. In general, a relation-in-extension is a plurality of pairs—specifically, the pairs of objects in a domain which stand in the relevant pure (or universal) relation.
>
> (Rumfitt 2018, p. 192)

A relation-in-extension R is (or forms) a series if it is a *total order*, i.e. R is irreflexive, transitive and connected in the relevant domain,[3] and is *well-ordered* if it is a total order and for every subdomain X of its field (i.e. plurality of objects which either bear R to some object or to which some object bears R), some object is R-minimal among the Xs (i.e. one of the Xs has no other X bearing R to it).[4]

Two series R and S are *isomorphic*—briefly $R \cong S$—if and only if there is an order-preserving bijection f from the field of R onto the field of S (i.e. f is such that: $\forall x \forall y (Rxy \leftrightarrow Sf(x)f(y))$).

Ordinal numbers, identified with order-types, are then introduced by the abstraction:

$$(\text{ORD}) \quad \forall R \forall S \big(W(R) \wedge W(S) \rightarrow \big(ord(R) = ord(S) \leftrightarrow R \cong S\big)\big)$$

On the basis of (ORD), we may define a predicate $On(x)$ (read: 'x is an ordinal'):

$$(\text{ON}) \quad On(x) =_{def} \exists R (W(R) \wedge x = ord(R))$$

---

3   N.b. for Rumfitt, domains are pluralities, not sets.
4   More formally (cf. Rumfitt 2018, p. 194) a binary relation-in-extension R ($=\langle x_1, y_1 \rangle$, $\langle x_2, y_2 \rangle$, . . ., $\langle x_i, y_i \rangle$, . . .) is a *series* iff R is a total order, i.e.

$\forall x \neg Rxx$
$\forall x \forall y \forall z (Rxy \wedge Ryz \rightarrow Rxz)$
$\forall x \forall y (Rxy \vee Ryx \vee x = y)$

R is well-ordered, written $W(R)$, iff R is a total order and
(WO) $\forall X (\exists xx \propto X \wedge \forall x (x \propto X \rightarrow \exists y((Rxy \vee Ryx) \rightarrow \exists y(y \propto X \wedge \forall z(z \propto X \rightarrow z = y \vee Rzy)))$
where X varies over pluralities of objects in the domain and $z \propto X$ means that z is one of the Xs.

1.2.   *Burali-Forti*

In this setting, Rumfitt obtains Burali-Forti's contradiction as follows. Let α and β be ordinals, and suppose $\alpha = ord(R)$ and $\beta = ord(S)$. Rumfitt defines:

$(<)$ $\alpha < \beta$ iff $R \cong S'$ where S' is a *proper initial segment* of S [5]

$(<_\alpha)$ $<_\alpha$ is the restriction of $<$ to the ordinals strictly less than α, i.e.

$$\beta <_\alpha \gamma \leftrightarrow \left(On(\alpha) \wedge On(\beta) \wedge On(\gamma) \wedge \beta < \gamma < \alpha \right)$$

Cantor proved that if α is the order-type of a relation R, then $<_\alpha$ is isomorphic to R (i.e. $\alpha = ord(R) \rightarrow <_\alpha \cong R$). On certain assumptions, $<$ is a well-ordering on the ordinals (Rumfitt 2018, p. 196). Let $\leq$ denote the relation-in-extension got by applying $<$ to the domain of all ordinals. Then it can be shown that the series of all ordinals is well-ordered, i.e. $W(\leq)$. Since (ORD) entails

$$\forall R \left(W(R) \rightarrow \left(ord(R) = ord(R) \leftrightarrow R \cong R \right) \right)$$

we may instantiate R by $\leq$ and so, given $W(\leq)$, detach to obtain

$$ord(\leq) = ord(\leq) \leftrightarrow \leq \; \cong \; \leq$$

The identity mapping gives us the right-hand component, so we may infer the left, and thence

$$\exists x(x = ord(\leq))$$

That is, there exists an order-type of the series of all ordinals under $<$. Denote this by Ω. Since Ω is the order-type of a well-ordered series, Ω is itself an ordinal, i.e. $On(\Omega)$.

Consider $<_\Omega$. This is a well-ordered series. By the result of Cantor noted above, its order-type $b = \Omega$. But $<_\Omega$ is a *proper* initial segment of the series of all ordinals, so that $b < \Omega$. But then $\Omega < \Omega$, contradicting the irreflexivity of $<$.

---

5   S' is a proper initial segment of (the relation-in-extension) S iff there some object z in S's field such that z bears S' to no y in S's field, but bears S to some y in the field of S. Rumfitt denotes the proper initial segment of S up to such an element z by S_z, as I shall also do.

1.3. Rumfitt's Resolution

As Rumfitt observes, the derivation of the contradiction requires a proof that $W(\leq)$. The first step in proving this is to show that the whole series of ordinals exists, and this, Rumfitt contends, can be accomplished by an application of the relational comprehension scheme $\exists R \forall x \forall y (Rxy \leftrightarrow \varphi(x, y))$, provided that this is assumed to permit the occurrence of second-order quantifiers in the comprehension formula $\varphi(x, y)$. Specifically, Σ_1^1-comprehension is required:

> As applied to two-place predicates, the unrestricted Comprehension Schema takes as a logical axiom any instance of
>
> $$(CS) \quad \exists R \forall x \forall y (< x, y > \propto R \leftrightarrow \phi(x, y))$$
>
> where 'x' and 'y' are free in '$\phi(x,y)$'. To legitimate the series of all *ordinals* (i.e. $\leq$), one replaces '$\phi(x,y)$' by '$x < y$', i.e. by '$\exists R \exists S(x = ord(R) \wedge y = ord(S) \wedge \exists z(R \cong S_z))$'. If there are no such things as the ordinals, there are also no such things as the pairs of ordinals, and hence there is no such thing as the series of all ordinals. The proposed solution to the Burali-Forti Paradox, then, requires finding an appropriate restriction on (CS).
>
> (Rumfitt 2018, pp. 203–204)[6]

If Rumfitt is right, then we may—and he contends that we should—block the derivation of Burali-Forti's contradiction by disallowing even the quite modest degree of impredicativity permitted by Σ_1^1-comprehension. However, there is, he argues, no need to retreat to a fully predicative second-order logic; a restriction to Δ_1^1-comprehension—that is comprehension which allows Σ_1^1-formulae provided that they are equivalent to Π_1^1-formulae and vice-versa—is enough to block the derivation.

2. Saving Impredicative Comprehension

The fact that an impredicative instance of comprehension is needed to generate Burali-Forti's contradiction does not, of course, mean that there are not other ways, besides restricting comprehension, in which the derivation may be blocked. Most obviously, Rumfitt's derivation relies upon an ordinal abstraction principle which, as it stands, is completely unrestricted—and while he does not himself even mention the possibility of restricting it, that is clearly the obvious course for the neo-Fregean

---

6 *[Note by Jessica Leech] Editor's note: The cited passage in Hale's original version was changed in the published version of Rumfitt's paper. I have replaced the passage with the closest available.*

abstractionist.[7] Rumfitt's proposed abstraction, (ORD), in which the only bound second-order variables are *relation* variables, tends to obscure this possibility. And it is in part for that reason that I shall myself propose an alternative form of ordinal abstraction, in which the scope for relevant restriction is more apparent. But first I want to challenge the impression, which Rumfitt's paper certainly encourages, that it is impredicative comprehension which is mainly, if not wholly, to blame for the contradiction.

As we have seen, Rumfitt's abstraction introduces ordinals as order-types of *well-ordered series*, which are to be identified with what he calls *relations-in-extension*. His way of generating Burali-Forti's contradiction accordingly requires securing the existence of the order-type of the relation-in-extension he denotes by $\leq$, and this requires establishing the existence of $\leq$ itself. As the last passage I quoted makes clear, he thinks that the existence of this relation-in-extension can be established by appeal to an instance of comprehension—specifically, Σ_1^1-comprehension. But this belief, it seems to me, is not obviously correct, and I shall argue that neo-Fregeans are free to reject it.

The relevant instance of the second-order comprehension scheme is:

$$\exists R \forall x \forall y (xRy \leftrightarrow \exists S \exists T(x = ord(S) \wedge y = ord(T) \wedge \exists z(S \cong T_z)))$$

It bears emphasis that this must be taken to justify us in claiming that there exists a *relation-in-extension* defined by the open sentence $\exists S \exists T(x = ord(S) \wedge y = ord(T) \wedge \exists z(S \cong T_z))$—that is, a *plurality of ordered pairs* $\langle x, y \rangle$ whose first terms bear the *pure* relation < to their second terms, and which collectively covers *all* the ordinal numbers.

It is, of course, true that *if* the second-order relation variables are interpreted as ranging over sets of ordered-pairs, as they are in the standard semantics for second-order languages, then this instance of comprehension *will* indeed guarantee the existence of a set of ordered pairs meeting that condition. However, it is obvious that to appeal to the standard semantics at this point would be simply to beg the question against the neo-Fregean abstractionist, whose aim is to provide a foundation for mathematical theories, *including* some form of *set-theory*, by means of abstraction principles which, adjoined to a suitable second-order logic, will permit the derivation of the fundamental laws of the target theories. Clearly an abstractionist must reject the idea that second-order logic either *is* a form of set theory ('set theory in sheep's clothing', as its most celebrated critic once put it) or that it must receive its canonical interpretation in

---

7 [Note by Jessica Leech] *Editor's note: In the published version of his paper, Rumfitt does briefly discuss Roy Cook's proposal for solving the Paradox by restricting the ordinal abstraction principle (Rumfitt 2018, pp. 200–201). However, this discussion does not feature in the earlier version of the paper from which Hale was working.*

set-theoretic terms. In my view, he should instead take the second-order variables to range over properties and relations, where these are intensionally individuated functions from objects to truth-values, and so not to be confused with sets (of objects or *n*-tuples objects).

Now while by taking relations-in-extension to be *pluralities* of ordered pairs rather than *sets* of such, Rumfitt avoids endorsing the full standard, set-theoretic, interpretation of second-order logic, it remains the case that he tacitly relies upon an interpretation of the relation-variables of second-order logic which the abstractionist is entirely free to reject. Just as 1-*place predicate variables* are, in the abstractionist's view, to be understood as ranging over monadic *properties* of objects, and *not* over *sets* of objects, *nor* over *pluralities* of objects, so 2-*place* predicate variables should be understood as ranging over *pure* dyadic *relations*, as opposed not only to *sets* of ordered pairs, but equally to *pluralities* of such.

The key point, then, is that ≤ is not a relation in the sense in which the abstractionist will understand it—what Rumfitt calls a 'pure' or 'universal' relation—but a relation-in-extension; it is the relation-in-extension got by applying the pure relation < to the domain of all ordinals. However—or so the abstractionist may object—what Σ_1^1-comprehension, applied to the comprehension formula $\exists S \exists T (x = ord(S) \wedge y = ord(T) \wedge \exists z(S \cong T_z))$, delivers is not the relation-in-extension, but only the corresponding pure relation. Neither this, nor any other instance of comprehension with this comprehension formula, guarantees the existence of any particular domain, such as the domain comprising all ordinals, for the relation it allows us to define. However, to ensure the existence of ≤, we require an *exhaustive* plurality of pairs of ordinals related by <. Comprehension, as properly understood on the abstractionist approach, affords no guarantee that there is any such exhaustive plurality, and so a definite relation-in-extension, in which every ordinal occurs as a member of a pair ordered by <.

In sum, the abstractionist may confront Rumfitt with a simple dilemma. Either comprehension is to be understood, as the abstractionist understands it, as a means of introducing only *pure* relations, or it is to be understood—as Rumfitt's argument requires and as he appears to understand it—as directly introducing relations-in-extension. Understood in the first way, there is a gap to be closed—between, on the one side, the pure relation < which comprehension gives us, and on the other, the relation-in-extension whose constituent pairs of ordinals cover *all the ordinals*, which is needed to obtain the contradiction. But understanding comprehension in the second way—i.e. as a means of *directly* introducing relations-in-extension—simply begs the question against the abstractionist.

Rumfitt himself, of course, while insisting that there are ordinals, denies that there are any such things as *the* ordinals. Such, he argues, is the conclusion we should draw from Burali-Forti's paradox. With that much, we may well wish to agree. But that obliges us to restrict Σ_1^1-comprehension only if that is what, without any further questionable assumptions, commits us to the existence of the ordinals—and as we have seen, it does not.

3.   Ordinal Abstraction—A Better Way

Some relations possess some of their properties absolutely, in the sense that they have those properties no matter what is taken to be their field. Most obviously, perhaps, identity is reflexive, symmetric, and transitive throughout the whole universe of objects. More commonly, however, relations have various properties, including their ordering properties, only in certain restricted fields. For one example, consider the relation of *predeceasing*: this well-orders Vivaldi, Bach, and Handel, but not Prokoviev and Stalin. For another, contrast the relation < on the natural numbers with the same relation on the integers—we have a well-ordering in the first case, but not in the second.[8]

This suggests that if ordinals are to be identified, as usual, with order-types, a suitable and perspicuous form of ordinal abstraction should abstract—not over relations alone but—over relations taken as relating some things, rather than others.[9] This might be done by taking the relations relative to some set, as is standardly done in set theory texts, where a relation R is said to be *connected* (or *transitive*, etc.) *in a set X*. But it is not necessary to bring sets into the account in this way, and an abstractionist has strong reason to avoid doing so. Although an abstractionist foundation for set theory is likely to be *one* of his goals, he may aim to provide foundations for *other* mathematical theories—arithmetic and analysis, to take the obvious examples—without presupposing set theory. If he is to do so, he must not only reject the usual set-theoretic identification of relations with sets of *n*-tuples, but also avoid relativizing their properties to sets. So I shall, instead, take the other parameter to be a *property*, where this is to be understood non-extensionally, as a condition—the application or satisfaction condition—associated with a predicate which determines its extension. Thus the order-type of the ordering of Mozart's da Ponte operas got by taking the relation *being composed before* should be seen as obtained by abstracting over that relation *together with* the property of *being an opera composed by Mozart in collaboration with*

---

8 Rumfitt himself gives another nice example: the relation of *being composed before* well-orders Mozart's da Ponte operas, since *Figaro* was composed before *Don Giovanni*, and *Don Giovanni* before *Così Fan Tutte*; but it does not so order Beethoven's symphonies, because those we know as the fifth and the sixth were composed concomitantly.

9 Although Rumfitt does not advertise it, he is, in effect, taking care of this point in his construction of the ordinals by insisting that the relation variables R and S in (ORD) are to be interpreted as ranging over relations-in-extension—the specification of a relation-in-extension fixes the relevant domain and range of the pure relation.

Lorenzo da Ponte. More generally, ordinal abstraction should take something like the shape:

$$Ord(R, F) = Ord(S, G) \leftrightarrow R \upharpoonright F \cong S \upharpoonright G$$

where R, S denote binary relations, while F, G denote properties defined on the things in the fields of those relations, and $R \upharpoonright F$, $S \upharpoonright G$ denote the restrictions of R and S to F and G.

I say 'something like' that shape, because we shall certainly need some restrictions. Most obviously, we shall require that R and S well-order F and G, and we shall want a restriction to block the Burali-Forti paradox, and there is some latitude over how they might be implemented. I shall suppose the first requirement implemented by conditionalizing our abstraction, much as Rumfitt conditionalizes his version:

$$(Ord) \ W(R, F) \wedge W(S, G) \rightarrow Ord(R, F) = Ord(S, G) \leftrightarrow R \upharpoonright F \cong S \upharpoonright G^{10}$$

One obvious advantage of formulating our principle so that the ordinal abstraction operator operates on a relation *and* a property is that this allows us to impose restrictions in a more selective way. In particular, we can impose restrictions on F, G even when R, S meet the conditions necessary for ordinal abstraction (such as well-ordering the Fs and the Gs (assuming F, G are in good standing)), and even when the pure relations and properties R, S, F and G may all be introduced by perfectly acceptable (perhaps impredicative) instances of comprehension.

Burali-Forti might then be blocked by imposing a restriction, not on comprehension but on the properties F, G which can figure in the basis for ordinal abstraction. Specifically, we might reject the property of *being an ordinal* as an inadmissible value of F and G, whilst allowing properties of the form *being an ordinal strictly less than a*. The application of (*Ord*) required to generate Burali-Forti—one in which we abstract over $<\upharpoonright ordinal$ to obtain $Ord(<, ordinal)$—would then be blocked; the relational component is in perfectly good standing, and in particular, there is nothing amiss in the instance of comprehension which justifies its definition. The trouble lies rather with the property *ordinal* as opposed to properties of the form *ordinal* $< \beta$.

4. Restricting (*Ord*)

How exactly should (*Ord*) be restricted so as to block the derivation of Burali-Forti's contradiction? One well-known diagnosis of this and various other paradoxes sees them as having their source in what Russell

---

10 An alternative would be to incorporate the well-ordering condition into the abstraction's righthand side, as Cook does with (SOAP)—see footnote 2.

called *self-reproductive processes and classes* and Dummett subsequently described as *indefinitely extensible concepts*. Thus Russell wrote:

> [T]here are what we may call *self-reproductive* processes and classes. That is, there are some properties such that, given any class of terms all having such a property, we can always find a new term also having the property in question. Hence we can never collect all the terms having the said property into a whole; because, whenever we hope to have them all, the collection which we have immediately proceeds to generate a new term also having the said property.
>
> (Russell 1906, in Russell 1973, p. 144)

Dummett characterizes indefinite extensibility in somewhat similar terms as follows:

> [an] indefinitely extensible concept is one such that, if we can form a definite conception of a totality all of whose members fall under that concept, we can, by reference to that totality, characterize a larger totality all of whose members fall under it.
>
> (Dummett 1993, p. 441)

An obvious—and in my view, plausible—thought is that we might simply require that the properties *F* and *G* should be *definite*, where being definite requires (at least) *not* being *indefinitely extensible* (in Dummett's sense) or *self-reproductive* (in Russell's). There is, however, an immediate difficulty to be confronted.

For, as Shapiro and Wright observe (Shapiro and Wright 2006, p. 236), Dummett's characterization makes apparently essential use of the notion of *definiteness*, but it is unclear how this is to be explained save as, or at least in terms of, the *complement* of indefinite extensibility. Russell's explanation suffers from a similar problem; by 'any class of terms all having such a property' he presumably means us to take it that classes are definite wholes. We appear to be in a very small and seemingly vicious circle.

4.1. *Defining Indefinite Extensibility*

Shapiro and Wright (Shapiro and Wright 2006, p. 264ff) offer a promising way past this problem. In essence, their proposal is that we may first define, without circularity, a *relativized* form of indefinite extensibility; having done so, we may then use it as the basis for defining a non-relative or *absolute* kind of indefinite extensibility.

Here is their explanation of the relativized concept:

> Let *P* be a concept of items of a certain type τ. Typically, τ will be the (or a) type of individual objects. Let Π be a concept of concepts of

type τ items. Let us say that P is *indefinitely extensible with respect to* Π if and only if there is a function F from items of the same type as P to items of type τ such that if X is any sub-concept of P such that ΠX then

(1) FX falls under the concept P,
(2) it is not the case that FX falls under the concept X, and
(3) $\Pi X'$, where X' is the concept instantiated just by FX and every item which instantiates X (i.e. $\forall x[X'x \equiv (Xx \lor x = FX)]$; in set-theoretic terms, X' is $(X \cup \{FX\})$.

(Shapiro and Wright 2006, p. 266)

The intuitive idea, Shapiro and Wright explain, is

that the sub-concepts of P of which Π holds have no maximal member. For any sub-concept X of P such that ΠX, there is a proper extension X' of X such that $\Pi X'$.

(*ibid*)

They illustrate the idea with a variety of examples, of which I reproduce just two:

> *Shapiro-Wright 1.* Px iff x is a finite ordinal number; ΠX iff there are only finitely many Xs; FX is the successor of the largest X. So being a finite ordinal is indefinitely extensible with respect to 'finite'.
> *Shapiro-Wright 11.* Px iff x is a set that does not contain itself; ΠX iff the Xs form a set, i.e. $\exists y \forall x(x \in y \equiv Xx)$. FX is just the set of Xs: $\{x : Xx\}$. So being a set that does not contain itself is indefinitely extensible with respect to the property of being, or constituting, a set.

Shapiro and Wright's next move starts from the observation that their examples—they give twelve—fall into two groups. Thus in the case of *Shapiro-Wright 1*, and the majority of their other examples, we can—'helping ourselves to the classical ordinals', as they put it—cite an ordinal which places a *limit* on the length of the series of Π-preserving applications of the function F to any X such that ΠX; but with *Shapiro-Wright 11*, along with two others they give, there is *no* ordinal limit we can place on the Π-preserving iterations of F. The promising thought is then that we may define a de-relativized notion of indefinitely extensibility in terms of the absence of limits on F-iteration. Here is their proposal in more detail:

First, they define a more refined form of relative indefinite extensibility. Where λ varies over ordinals,

> *P* is *up-to-λ-extensible with respect to* Π just in case *P* and Π meet the conditions for the relativized notion as originally defined but λ places a limit on the series of Π-preserving iterations of *F* to any sub-concept *X* of *P* such that Π*X*.
>
> (Shapiro and Wright 2006, p. 269)

This notion is then contrasted with *proper indefinite extensibility* with respect to Π, defined by:

> *P* is *properly indefinitely extensible with respect to* Π just in case *P* meets the conditions for the relativized notion as originally defined and there is no λ such that *P* is up-to-λ extensible with respect to Π.

Finally, their de-relativized notion of indefinite extensibility is defined:

> *P* is *indefinitely extensible (simpliciter)* just in case there is a Π such that *P* is *properly indefinitely extensible with respect to* Π.

4.2.   *Two Problems*

Can we use the notion of indefinite extensibility, as explained by Shapiro and Wright, to restrict (*Ord*)?[11] There are two immediate and fairly obvious obstacles. To abbreviate their presentation and subsequent discussion, I shall henceforth write *IE* for 'indefinitely extensible simpliciter', *IE*$^\Pi$ for 'indefinitely extensible with respect to Π', and *IE*$^{\lambda\Pi}$ for 'indefinitely extensible up to λ with respect to Π' and *Prop IE*$^\Pi$ for 'properly indefinitely extensible with respect to Π'. I write $P \subseteq Q$ for '*P* is a sub-property of *Q*', defined as $\forall x(Px \rightarrow Qx)$.

Problem 1: The first difficulty appears as soon as we notice that if a concept/property *P* is to qualify as *IE*—and so must qualify as *Prop IE*$^\Pi$ for some Π—it must not itself be Π (i.e. it must be that ¬Π*P*). For $P \subseteq P$, so that if Π*P*, then *P* must satisfy conditions (1)-(3) in the definiens for *IE*$^\Pi$*P*. But it cannot do so, since condition (1)—that *FX* falls under the concept *P*—requires in this case that *P*(*F*(*P*)), while condition (2)—that *FX* does *not* fall under the concept *X*—requires precisely the opposite. We get a paradox of indefinite extensibility when a concept *P* appears to be *Prop IE*$^\Pi$ for some Π but also appears itself to be Π, as in Shapiro and Wright's example 10, where *Px* is '*x* is an ordinal' and Π*X* is 'each *X* is

---

11 I should emphasize that Shapiro and Wright are not themselves concerned to explain indefinite extensibility in such a way that it may be used to restrict abstraction principles. Their leading question is whether the usual paradoxes have their source in indefinite extensibility.

an ordinal and the Xs exemplify a well-ordering type'.[12] But now it seems that, having restricted (*Ord*) to non-*IE* concepts, we cannot exempt *ordinal* from its scope on the ground that it is *IE*, because we cannot establish that *ordinal* is *IE*, since that very claim lands us in a contradiction.

Problem 2: The second, equally serious and perhaps more obvious, obstacle to using the notion of indefinite extensibility to restrict ordinal abstraction is that Shapiro and Wright's explanation *presupposes* that ordinals are *already* in place—as they put it, they 'help themselves to the classical ordinals' in defining $IE^{\lambda\Pi}$. We cannot, therefore, use their notion to restrict (*Ord*) without falling into a vicious circle.

Can we circumvent these problems? If we assume, as I shall, that Shapiro and Wright's basic idea, of explaining indefinite extensibility by de-relativizing a relative notion, offers the best prospect of a satisfactory definition, and if we further assume that the relative notion should be IE^{Π}, or something close to it, then we shall need, if we are to overcome the first problem, to further restrict the sub-concepts of P for which conditions (1)-(3) must be met, if P is to qualify as IE^{Π}. In particular, we shall need to ensure, somehow, that *ordinal* itself does not qualify as one of the sub-concepts of *ordinal* which must meet those conditions, if *ordinal* is to be IE^{Π}. As for the second problem, we shall need to modify Shapiro and Wright's sequence of definitions, either so as to avoid presupposing the *ordinals* in explaining bounded extensibility, or perhaps by avoiding bounded extensibility altogether.

5. Solving Problem 1: Re-Defining IE^{Π}

The most obvious and straightforward way to exempt *ordinal* from the sub-concepts of X which are required to satisfy conditions (1)-(3) (in the definition of IE^{Π}) is to weaken Shapiro and Wright's definition so as to require only that those conditions be met by *proper* sub-concepts of P. Then, when P is *ordinal* and ΠX is *the Xs are ordinals and collectively exemplify a*

---

12 In full, Shapiro-Wright example 10 is: '*Px* iff *x* is an ordinal (or von Neumann ordinal); ΠX iff each of the Xs is an ordinal and the Xs are themselves isomorphic to an ordinal (under the natural ordering). In other words, ΠX iff each X is an ordinal and the Xs have (or exemplify) a well-ordering type. . . . F(X) is the successor of the union of the Xs. So being an ordinal is indefinitely extensible with respect to the property of being isomorphic to an ordinal (or exemplifying a well-ordering type).' (Shapiro and Wright 2006, p. 268) But as Shapiro and Wright subsequently observe: '. . . in case P is *ordinal* and ΠX holds just if the Xs *exemplify a well-ordering type*, it seems irresistable to say that *ordinal* itself falls under Π. After all, the ordinals are well-ordered. But then the relevant principle of extension kicks in and dumps a new object on us that both must and cannot be an ordinal—must because it corresponds, it seems, to a determinate order-type, cannot because the principle of extension always generates a non-instance of the concept to which it is applied. So we have the Burali-Forti paradox'. (Shapiro and Wright 2006, pp. 269–270)

well-ordering type (as in Shapiro-Wright example 10), even if *ordinal* is Π (as it certainly appears to be), then the fact that conditions (1) and (2) cannot both be met with $X = ordinal$ no longer poses an obstacle, because *ordinal* is improper and so is not itself required to meet those conditions.

Unfortunately, this simple remedy is insufficient to dispose of the problem. Consider, for example, the concept *non-zero ordinal*, briefly *ordinal⁺*. Since *ordinal⁺* is a proper sub-concept of *ordinal* which meets the higher-order condition Π as above, it is required, if *ordinal* is to qualify as IE^Π, to meet conditions (1)-(3). It appears clearly to meet condition (1) just as *ordinal* itself does. That is, the successor of the union of the positive ordinals is itself an ordinal—*ordinal*($F(ordinal⁺)$). The problem lies with condition (2). For $F(ordinal⁺)$ must clearly be a non-zero ordinal, so that we have *ordinal⁺*($F(ordinal⁺)$) and *not* its negation as (2) requires.[13]

It is clear that the difficulty generalizes. For exactly the same problem will arise from taking *any* proper sub-concept of *ordinal* whose instances are all and only the ordinals belonging to a non-initial segment of the ordinals which is not bounded above—for any F-extension of the ordinals greater than λ, for any non-zero ordinal λ, must itself be an ordinal greater than λ, so that condition (2) must fail. The difficulty does not arise, however, for those proper sub-concepts whose instances are just the terms of some proper *initial* segment of the ordinals in their natural well-ordering by $<$. This suggests a more radical revision of Shapiro and Wright's definition, along the following lines:

> P is *indefinitely extensible with respect to* Π if and only if there is a well-ordering of the Ps and a function F such that if (*a*) X is any proper sub-concept of P whose instances are all and only the terms of a proper initial segment of the Ps so ordered, and (*b*) ΠX, then
>
> (1) FX falls under the concept P,
> (2) it is not the case that FX falls under the concept X, and
> (3) $\Pi X'$, where X' is the concept instantiated just by FX and every item which instantiates X (i.e. $\forall x[X'x \equiv (Xx \lor x = FX)]$).

If *up-to-λ-extensibility with respect to* Π, *proper extensibility with respect to* Π, and *indefinite extensibility simpliciter* are defined as before, the intuitive idea of indefinitely extensibility thereby captured is of a concept whose extension is unbounded because any proper initial segment of its instances (under some well-ordering) can be extended to a longer, but still proper, initial segment which can therefore be further extended . . . and so on, indefinitely.

---

13 I am indebted to Hannes Leitgeb for drawing attention to this difficulty in discussion of an earlier version of this paper.

This characterization assumes that successful candidates for indefinite extensibility will always be concepts whose instances can be well-ordered. Whilst this assumption clearly holds for such paradigms as *ordinal* and *cardinal*, it is obvious neither that it holds for all intuitively plausible candidates, nor that so strong an assumption is needed—what is strictly required appears to be only that the instances may be so ordered that we can take larger and larger initial segments under their ordering.

As an example, consider the concept *set*. According to the iterative conception, sets are formed in *stages*. We start with whatever individuals (non-sets) there may be. At the first stage (stage 0) are formed all the sets which may be formed from these. If there are no individuals, the only set formed at this stage is the null set. But we may assume that there are many individuals, and so many sets—2^n, where n is the number of individuals—formed at this stage. At stage 1 are formed all the sets having individuals and stage 0 sets as members. At stage 2 are formed all the sets having individuals, stage 0 and stage 1 sets as members. And so on. At each stage are formed all sets that can be formed from individuals and sets formed at earlier stages.

We thus have a cumulative hierarchy, the start of which looks like this:

stage 0 $\varnothing$,{a}, {b}, {c}, . . ., {a, b}, {a, c}, . . ., {a, b, c}, . . .
stage 1 $\varnothing$,{a}, {b}, {c}, . . ., {a, b}, {a, c}, . . ., {a, b, c}, . . ., {$\varnothing$},
{{a}}, {{b}}, . . .
stage 2 $\varnothing$,{a}, {b}, {c}, . . ., {a, b}, {a, c}, . . ., {a, b, c}, . . ., {$\varnothing$},
{{a}}, {{b}}, . . ., {{$\varnothing$}}, {{{a}}}, {{{b}}}, ...

The stages themselves are denoted by ordinals, and so are well-ordered as usual by $<$. The sets at any given stage are partially ordered by the stages at which they are first formed, so that at stage n we have first the sets formed at stage 0, followed by the sets first formed at stage 1, followed by the sets first formed . . . at stage $n - 1$ and finally the sets first formed at stage n. But—assuming the existence of individuals—the sets formed at any given stage—and *a fortiori* the sets in general—are not thereby *well*-ordered, since many sets are first formed at the same stage.[14] They do, however, possess a kind of order under which we can discern larger and larger initial segments of the hierarchy, corresponding to the stages. Thus the first initial segment comprises the sets of stage 0, the second those of stage 1, and so on. If we denote the *rank* of the sets formed at stage 0 as *rank* 0, that of the new sets formed at stage 1 *rank* 1, and so on, then we may say the sets are ordered by rank. The general concept here may be fixed as follows:

---

14 Of course, if we assume the Axiom of Choice, it follows that the sets generated at any given stage can be well-ordered, and if a global form of Choice is assumed, the sets as a whole can be well-ordered. But the usual informal argument to show that *set* is indefinitely extensible does not rely on any such strong Choice principle.

A binary relation R rank-orders the Ps if

(i)   some Ps—the P's—bear R to all other Ps
(ii) $R \upharpoonright P$ is asymmetric and transitive
(iii) any two Ps neither of which bears R to the other bear the same R-relations to all other Ps.

Clause (iii) ensures that any two Ps neither of which bears R to the other belong to the same rank.

We may now redefine IE^{Π} along the lines previously suggested, replacing the requirement that the Ps be well-ordered by the weaker requirement that they be rank-ordered.[15] That is:

> P is *indefinitely extensible with respect to* Π if and only if there is a rank-ordering of the Ps and a function F such that if X is any sub-concept of P whose instances all belong to a proper initial segment of the Ps so ordered and ΠX then

> (1) FX falls under the concept P
> (2)   it is not the case that FX falls under the concept X, and
> (3) $\Pi X'$, where X' is the concept instantiated just by FX and every item which instantiates X.[16]

6.   Solving Problem 2: Doing Without Bounds

6.1.   *Doing Without Bounds Altogether*

Can we define a non-relative notion of indefinite extensibility *without* going through an intermediate definition of bounded extensibility? Fairly obviously we *could* directly define a de-relativization of our amended relative notion by simply existentially generalizing it:

> (IE^*) P is IE iff $\exists \Pi (P$ is $IE^{\Pi})$

---

15 Note that any well-ordering is a rank-ordering. In particular, if the Ps are well-ordered by R, there is an R-least P, and this bears R to every other P. That $R \upharpoonright P$ is asymmetric and transitive holds by the definition of well-ordering, and since $R \upharpoonright P$ is connected, it holds vacuously that any two Ps neither of which bears R to the other bear the same R-relations to all other Ps.

16 Condition (i) for R to rank-order the Ps will be vacuously met if there are some Ps but none of them bears R to any P. In this case, conditions (ii) and (iii) will likewise be vacuously met. But then there will be no sub-concept of P whose instances all belong to a proper initial segment of the Ps rank-ordered by R, so that the conditions for P to be indefinitely extensible with respect to Π will be vacuously met. This is harmless. In fact, P will in this case be *boundedly* indefinitely extensible with respect to Π, with bound 1, since there are *no* Π-preserving iterations of F.

But would this notion be suitable for our purpose, i.e. to restrict (*Ord*) and perhaps other abstraction principles? It is easy to see that it would not.

It is straightforward to verify that *ordinal* is *IE** by verifying that it is *IE*$^{\Pi}$ with Π taken, as above, as 'exemplifies a well-ordering type'. We take the extending function F as Shapiro and Wright do, so that for any proper sub-concept X of *ordinal*, $F(X)$ is the successor of the union of the Xs, which must be an ordinal greater than each of the Xs, so that $P(F(X))$ but $\neg X(F(X))$, satisfying conditions (1) and (2). And clearly $\Pi X'$ (where $X'x \leftrightarrow (Xx \lor x\,F(X))$ since each X' is an ordinal and the X's collectively exemplify a well-ordering type.

However, it is equally straightforward to verify that, for example, *finite ordinal* qualifies as *IE** by verifying that it is *IE*$^{\Pi}$ with Π taken as *finitely exemplified* (as in Shapiro-Wright Example 1). Let G be any finitely instantiated proper sub-property of *finite ordinal*, so that ΠG. And let $F(G)$ be the successor of the largest G. Then clearly $F(G)$ is a finite ordinal—so that condition (1) is met, but it is not one of the Gs, so that $\neg G(F(G))$, so that condition (2) is met. And clearly where $G'x \leftrightarrow Gx \lor x\,F\,G$, G' is finitely instantiated (i.e. $\Pi G'$). So condition (3) is met, and *finite ordinal* is *IE* with Π taken as *finitely exemplified*, and so *IE**.

Thus a blanket restriction on (*Ord*) to non-*IE** properties would exclude properties which are *merely* boundedly extensible (in Shapiro and Wright's sense), such as *finite ordinal*, along with those which, like *ordinal* itself, are *IE* simpliciter. But there appears to be no good reason to exclude boundedly extensible properties from the scope of ordinal and other abstraction principles—in particular, including them does not appear to lead to paradox. Indeed, there appears to be good reason *not* to exclude them—for doing so threatens to render abstraction principles *too weak*. For example, restricting set-abstraction to non-*IE** properties would mean that there can be no set of finite ordinals, i.e. natural numbers.

In short, *IE** is too blunt an instrument. The prospects of sharpening it to make it suitable for our purpose do not look promising, and I shall not pursue them here. Instead, I want to explore an alternative, and in one way more radical, solution.

6.2. *Doing Without Ordinal Bounds*

Although Shapiro and Wright do in fact define bounded extensibility by reference to *ordinal* bounds, it is not obvious that they had to do so. Consideration of their examples of bounded extensibility suggests that they might just as well have taken bounds to be set by *cardinals*, rather than ordinals.[17] But if that is right, then there is, it may seem, a simple

---

17 Shapiro and Wright's examples of bounded extensibility are their examples 1-9. For some of these (examples 1,2,6,7,8,9) a cardinal bound on iterated applications of the function F which are Π-preserving can be set using $\aleph_0$ or $2^{\aleph_0}$. For the others (examples

way around the immediate problem—i.e. how to restrict ordinal abstraction so as to escape Burali-Forti. For provided that we can first introduce *cardinal* numbers independently of ordinals—as neo-Fregeans have all along proposed to do, by means of Hume's principle—they may be used, without circularity, in defining bounded extensibility and hence indefinite extensibility proper and indefinite extensibility simpliciter, leaving us free to restrict *F* and *G* in (*Ord*) to *definite* properties.

6.2.1.   *An Apparently Decisive Objection*

To this simple proposal there is an obvious, and at first glance devastating, objection. Far from solving the problem, it may be claimed, the proposed remedy merely causes it to come up in another place, with the definition of cardinal numbers via Hume's principle. The principle cannot, on pain of re-introducing vicious circularity, be restricted so as to apply only to concepts which are not indefinitely extensible. But without such a restriction, it will deliver cardinal numbers corresponding to such paradigm indefinitely extensible concepts as *set, ordinal number* and *cardinal number* itself.[18] But that, surely, is a disaster. How can there possibly be a cardinal number of the *F*s, where *F* is indefinitely extensible?[19] Must not the assumption that there exist such 'monster' cardinals as *the cardinal number of ordinal numbers, the cardinal number of sets*, and *the cardinal number of cardinal numbers* land us in contradictions?

6.2.2.   *A Radical Response*

Devastating as it may appear, I am not sure that this objection *is* unanswerable. It may be argued that the consequences of dispensing with a restriction of (HP) to definite (and so not indefinitely extensible) concepts as admissible values of its second-order variables, though dramatically at odds with orthodoxy, are neither as dire, nor as unpalatable, as the

---

3-5) larger cardinals are required. It is true that for some examples (8,9), a *more precise* bound can be set using ordinals than using cardinals (in these cases, the first non-recursive ordinal, as opposed to $2^{\aleph_0}$), but for the purposes of the explanation of indefinite extensibility by de-relativization of proper indefinite extensibilty, this loss of information is not crucial; the existence of a cruder cardinal bound is enough.

18 For example, there will be nothing to obstruct reasoning as follows: first, instantiate the bound variables *F* and *G* in (HP) to obtain:

$$Nx \; cardinal(x) = Nx \; cardinal(x) \leftrightarrow cardinal \approx cardinal$$

where the predicate '*x* is a cardinal number' is itself explicitly defined in terms of the number operator by setting $cardinal(x) =_{def} \exists F(x = NyFy)$. Then, since the right-hand side can be established by using the identity mapping, we may detach the left-hand side and existentially generalize, whence $\exists y(Nx \; cardinal(x) = y)$.

19 cf Wright (1999, pp. 13–14): 'If there is anything at all in the notion of an indefinitely extensible totality . . . one principled restriction on Hume's Principle will surely be that [cardinal numbers] *not* be associated with such totalities.'

objection claims. The crucial questions are whether indefinitely extensible concepts must be numberless (i.e. whether, when F is indefinitely extensible, there can be no such thing as the number of Fs), and whether the supposition that there are such numbers as the number of cardinals, the number of ordinals, and the number of sets, must be rejected, because it lands us in contradictions. To anyone who is not prepared to take an affirmative answer to the first question as simply obvious, the second is the more important, since an affirmative answer to it would provide the best possible grounds for answering the first the same way. Nevertheless, we may begin with a response to the first question.

We should agree straightaway that an indefinitely extensible property cannot have a *definite* number of instances. So *if* all cardinal numbers are definite, such properties cannot be assigned cardinal numbers. But *are* all cardinals definite? Obviously *finite* cardinals are. It is anything but obvious that the same goes for *transfinite* cardinals. Entries for 'finite' and 'infinite' in my dictionary (SOED 2007) point to a negative answer. Thus the primary meanings of 'finite' are given as 'having bounds, ends, or limits' and 'fixed, determined, definite', while those for 'infinite' speak of indeterminacy or indefiniteness, and 'infinite' in mathematics is said to mean 'having no limit; greater than any assignable number or magnitude'. Of course, it is open to question how much weight we should place on these entries. However accurately they may record ordinary usage of the terms, it seems clear that the lexicographers have not consulted the set theorists! In particular, their entries seem to leave no room for distinctions among transfinite cardinals or for their use, as proposed here, in setting bounds. But the idea that the standard set-theoretic treatment of transfinite cardinals does not deal in definite numbers may find support elsewhere. For one thing, there is the striking contrast between the cardinals and the ordinals—assuming the Continuum Hypothesis, there are no cardinals between $\aleph_0$ and $2^{\aleph_0}$, but there are uncountably many ordinals $\omega + 1$, $\omega + 2$, . . ., $\omega.2$, $\omega.2 + 1$, . . ., each cardinally indistinguishable from ω. Further, there are $\aleph_0$ natural numbers. Leaving out the first fifteen natural numbers, or the even numbers, say, there are still $\aleph_0$ natural numbers. Similarly, there are $2^{\aleph_0}$ real numbers. Leaving out the natural numbers, we still have $2^{\aleph_0}$ reals. There are, as we might put it, $\aleph_0$ natural numbers and $2^{\aleph_0}$ reals, *give or take* fifteen, or even $\aleph_0$. 100 give or take 5 isn't a definite number. Why should we count $2^{\aleph_0}$ give or take $\aleph_0$ as one?

A necessary, but insufficient, condition for an affirmative answer to the second question is that the assumption of numbers of cardinals, ordinals, etc., does actually lead to contradictions. If the assumption of cardinal numbers for *IE* properties such as *ordinal*, *set*, or *cardinal* itself gave rise, *by itself*, to contradictions, that would settle the matter. But it is as good as known that it doesn't. For as George Boolos proved, the addition of unrestricted *Hume's principle* to second-order logic (the system he calls *Frege Arithmetic*) is consistent if and only if second-order arithmetic (PA^2) is—and few, if any, seriously doubt the consistency of the latter.

Contradictions are forthcoming, of course, in incautious extensions of Frege Arithmetic. To take an obvious example, if we add a further—unrestricted—abstraction principle for sets, we shall be able to derive Cantor's paradox: by unrestricted set-abstraction, there will be a set of all sets, S, whose powerset, $\mathcal{P}(S)$, must be one of its subsets, so that the number of members of S can be no smaller than the number of members of $\mathcal{P}(S)$; but by Cantor's Theorem, the number of members of $\mathcal{P}(S)$ is strictly greater than the number of members of S. There is an obvious remedy—for whilst we cannot, without circularity, restrict Hume's Principle to definite properties, we can so restrict set-abstraction, and that blocks the derivation at its first step. Further, there is independent reason to blame naive set-abstraction (rather than Hume's Principle), since it lands us in trouble—Russell's paradox—without any play with cardinal abstraction.

7. Concluding Remark: Sets and Numbers

On this radical response, on which set-abstraction is restricted to definite properties but cardinal abstraction is not, the higher-order properties of *having a cardinal number* and of *determining a set*, which are co-extensive on the standard set-theoretic account, come dramatically apart—indefinitely extensible concepts have cardinal numbers, but they fail to determine sets. This is heterodox but defensible. The essence of cardinality and sethood alike lies in the identity-conditions for cardinal numbers and those for sets. Properties have the same cardinal number if and only if they are *one-one correspondent*, for which it is required that they be *sortal*, but *not* that they be *definite*. But sets are the same if and only if they have the same members, and for this condition to be met, a property that determines the set must be *not only* sortal *but also* possess a *definite* extension.

References

Cook, R. (2003). "Iteration one more time". *Notre Dame Journal of Formal Logic*, 44(2):63–92.

Dummett, M. (1993). *The Seas of Language*. Oxford: Clarendon Press.

Rumfitt, I. (2018). "Neo-Fregeanism and the Burali-Forti Paradox". In Fred, I. and Leech, J., editors, *Being Necessary: Themes of Ontology and Modality from the Work of Bob Hale*, pages 188–223. Oxford: Clarendon Press.

Russell, B. (1906). "On some difficulties in the theory of transfinite numbers and order types". *Proceedings of the London Mathematical Society*, 4:29–53.

Russell, B. (1973). *Essays in Analysis*. London: George Allen and Unwin.

Shapiro, S. and Wright, C. (2006). "All things indefinitely extensible". In Rayo, A. and Uzquiano, G., editors, *Absolute Generality*, chapter 10, pages 255–304. Oxford: Clarendon Press.

SOED (2007). *Shorter Oxford English Dictionary*. Oxford: Oxford University Press.

Wright, C. (1999). "Is Hume's Principle analytic?". *Notre Dame Journal of Formal Logic*, 40(1):6–30.

12 Logicism, Separation, and Complement

Roy T. Cook

1. Introduction

According to neo-logicism, we can provide definitions of mathematical concepts, and provide epistemological access to the abstract objects falling under those mathematical concepts, via laying down abstraction principles:[1]

$$A_E : (\forall\alpha)(\forall\beta)[@_E(\alpha) = @_E(\beta) \leftrightarrow E(\alpha, \beta)]$$

where $@_E$ denotes a non-logical function mapping arguments of type α to objects and where $E(\alpha, \beta)$ is an equivalence relation on entities of the type indicated by α and β. Here we shall restrict our attention to the case that has received the most attention in the literature on abstraction principles, where α and β range over unary concepts. Thus, we can rewrite our schema for abstraction principles as:

$$A_E : (\forall X)(\forall Y)[@_E(X) = @_E(Y) \leftrightarrow E(X, Y)]$$

Abstraction principles serve to implicitly define the relevant mathematical concepts via providing identity conditions on the abstract objects falling under those concepts (i.e., the range of $@_E$) in terms of the equivalence relation $E(X, Y)$.

The two most well-known and well-studied abstraction principles are Hume's Principle and Basic Law V. Hume's Principle:

$$HP : (\forall X)(\forall Y)[\#(X) = \#(Y) \leftrightarrow X \approx Y]$$

provides a neo-logicist definition of the concept Cardinal Number and is the paradigm instance of an acceptable abstraction principle. Basic Law V:

$$BLV : (\forall X)(\forall Y)[\S(X) = \S(Y) \leftrightarrow (\forall x)(X(z) \leftrightarrow Y(z))]$$

---

1 For more details and defenses of neo-logicism, see (Hale & Wright 2001) and the essays in (Cook 2007), (Cook 2009), and (Linnebo 2018).

DOI: 10.4324/9780429277894-15

on the other hand, is inconsistent (in second-order logics with sufficiently strong comprehension principles). The failure of Basic Law V to provide an acceptable neo-logicist account of extensions of concepts (or sets) highlights two interrelated problems that neo-logicism faces:

1.   To provide a philosophically principled account of the line that distinguishes the acceptable abstraction principles (e.g., HP) from the unacceptable (e.g., BLV). This is the Bad Company problem.
2.   To provide a neo-logicist account of the concept SET.

Here our explicit focus shall be on the second issue. In particular, I will show that there are good reasons to think that any acceptable neo-logicist account of sets will satisfy the axiom of complement and as a result will be forced to reject the axiom of separation. As a result, the set theory obtained within the neo-logicist framework will look very different from traditional Zermelo Fraenkel–style set theories (such as e.g. ZFC).

We will not neglect the first issue entirely, however. In fact, the argument for the axiom of complement within the neo-logicist framework will depend essentially on the kinds of strategies that have been adopted by neo-logicists to address the Bad Company problem. Our argument will proceed as follows. First, in §2, we will review a particularly interesting version of the Russell Paradox—one we shall call the Comp-Sep Russell Paradox—which shows that we cannot accept both the unrestricted axiom of separation and the unrestricted axiom of complement. We will then, in §3, survey all of the various conditions on abstraction principles that have been proposed as desiderata that acceptable abstraction principles must satisfy and show that they all are closed under a condition we shall call *strong cardinality equivalence*. Then, in §4, we will develop a novel strategy for reconstructing set theory within the neo-logicist framework—the *global* approach to neo-logicist sets. In §5, we will then prove the main result of this chapter: that the global approach to sets entails the axiom of complement (and hence requires a rejection of the axiom of separation), regardless of which of the various accounts of acceptability canvassed in §3 turns out to be correct. Finally, in §6, we shall tie up a few loose ends.

2.   The Comp-Sep Russell Paradox

Our first task is to review the tension between the axiom of separation and the axiom of complement. Since, once we move to neo-logicist accounts of set theory, we will have no guarantee that the abstraction operator is onto (that is, we have no guarantee that every object is a set), we shall carry out our examination of separation versus complement in a language for set theory that allows for urelements and hence includes a primitive set predicate $\mathsf{Set}(x)$. The axiom of separation, familiar from second-order ZFCU, is thus formulated in the present context as:

$$\mathsf{Sep} : (\forall x)(\mathsf{Set}(x) \rightarrow (\forall Y)(\exists z)(\mathsf{Set}(z) \wedge (\forall w)(w \in z \leftrightarrow (w \in x \wedge Y(w)))))$$

and the (similarly familiar but less widely accepted) axiom of complement is formulated as:

$$\mathsf{Comp} : (\forall x)(\mathsf{Set}(x) \rightarrow (\exists y)(\mathsf{Set}(y) \wedge (\forall z)(z \in y \leftrightarrow z \notin x)))$$

The important point, for present purposes, is that the axiom of separation and the axiom of complement are jointly incompatible with the existence of any sets:[2]

Theorem 2.1. $\mathsf{Sep}, \mathsf{Comp}, (\exists x)(\mathsf{Set}(x)) \vdash \bot$

Proof. Assume some set exists. Then, by the axiom of separation, $\varnothing$ exists. By the axiom of complement, the universal set U exists. By the axiom of separation, the Russell set:

$$\{x : x \notin x\}$$

exists. Proceed as usual. □

Thus, no consistent (classical) set theory can accept both the axiom of separation and the axiom of complement—we must choose one or the other.

Admittedly, most of us are conditioned to opt for the axiom of separation. But I would like to suggest that this is genuinely a conditioned response—the majority of us are corrupted by theory due to the

---

2 This argument depends on there being concepts corresponding to non-self-identity ($x \neq x$) and non-self-membership ($x \notin x$). Since throughout this chapter we are assuming full second-order comprehension for concepts, both assumptions are legitimate. But it should be noted that the latter assumption is illegitimate in predicative second-order logic, since membership is defined, in the neo-logicist context, as:

$$x \in y =_{df} (\exists Z)(y = \S(Z) \wedge Z(x))$$

Thus, the predicativist approach is immune to the **Comp-Sep** Russell Paradox, and both the axiom of complement and the axiom of separation (as formulated previously) hold in the theory consisting of second-order logic with predicative comprehension plus Basic Law V [see (Heck 1996) for details]. Of course, the predicativist might have qualms about other versions of separation, such as the schematic [for any formula $\Phi(x)$]:

$$(\forall x)(\mathsf{Set}(x) \rightarrow (\exists y)(\mathsf{Set}(y) \wedge (\forall z)(z \in y \leftrightarrow (z \in x \wedge \Phi(z)))))$$

Thanks are owed to an anonymous referee for emphasizing this point.

prominence of ZFC and ZFC-like set theories. The axiom of separation and the axiom of complement, although incompatible, strike me as equally intuitively plausible insofar as I can divorce my own intuitions from the weight of ZFC-dominated history. The axiom of separation and the axiom of complement codify distinct, and arguably equally plausible, intuitions regarding the process of forming collections of objects. We can, rather simplistically but I think helpfully, get at these intuitions via the following analogues of the two axioms in question, formulated in terms of building fences:

> Fence Separation: If you have built a fence that separates the objects that satisfy $\Phi(x)$ from all other objects, then you can build a second fence separating those objects inside the first fence that satisfy some second condition $\Psi(x)$ from those objects inside the first fence that don't satisfy condition $\Psi(x)$.
>
> Fence Complement: If you have built a fence that separates the objects that satisfy $\Phi(x)$ from all other objects, then you have built a fence that separates the objects that satisfy $\neg\Phi(x)$ from all other objects. In short, there is no privileging one side of a fence over the other.

Although the analogy between set formation and the building of fences is imperfect (and becomes rather strained once we consider fence-theoretic analogues of sets of sets), these informal fence-building-theoretic principles nevertheless get at important, and I think plausible, intuitions regarding collection-formation. In short, if sets are those collections that are determinate or definite enough to be reified—that is, to be somehow turned into objects—then the axiom of separation and the axiom of complement codify different conceptions of DEFINITENESS or DETERMINACY. The Sep-Comp Russell Paradox then shows that, surprisingly, we cannot have a (classical) set theory that supports both of these intuitions.

So what is a neo-logicist to do? One option, of course, would be to bow down to the pressures of history and accept the axiom of separation because everyone else does. Selecting one's set theory based on wanting to get along with the in-crowd isn't a particularly compelling philosophical methodology, however. Instead, we should remember that neo-logicism is being proposed as an alternative foundation for mathematics—one that is meant to replace, in important ways, set theory. The point is not that neo-logicists will abandon set theory altogether. On the contrary, I take it that one desideratum of a successful development and defense of neo-logicism is that it somehow account for contemporary work in set theory and hence provide some way to understand that work as concerned with abstracts of some sort. Rather, the point is that there are certain *foundational* roles that have traditionally been played by set theory (or, more recently, by category theory, or homotopy type theory, etc.) that will, for the neo-logicist, be played by a theory consisting of acceptable abstraction principles. Thus, although the neo-logicist must provide an account of set theory, since set theory is a legitimate mathematical theory in need of a foundation, the neo-logicist need feel no pressure to reconstruct set theory in a manner in which it could continue to play the foundational role traditionally attributed to it.

Thus, the neo-logicist should approach the question of separation versus complement unburdened by the past and instead assess this question solely in terms of how the axioms fare from within the neo-logicist framework. And, as we shall see in the following, adopting such an approach allows us to recognize that, from the perspective of neo-logicism, the axiom of complement is nearly forced upon us, along with the corresponding rejection of the axiom of separation.

3. Bad Company and Candidates for Acceptability

Our next task is to review various criteria that have been proposed as constraints on acceptable abstraction principles. The first, and most obvious, requirement that must be imposed on abstraction principles is that they be satisfiable. But, as we will see, we can (and have) asked for much more than mere satisfiability.

We begin with a number of definitions. First, we assume that we are working in a language $\mathcal{L}$ that contains the full resources of second-order logic and also contains a distinct abstraction operator $@_E$ for each equivalence relation on concepts $E(X, Y)$. We can now formulate a number of criteria that abstraction principles in $\mathcal{L}$ might satisfy in terms of whether they are satisfiable "at" particular cardinals:

> **Definition 3.1.** *Given any cardinal κ, an abstraction principle A_E is κ-satisfiable if and only if A_E is satisfiable in a (and hence any) domain of cardinality κ.*
>
> **Definition 3.2.** *An abstraction principle A_E is satisfiable if and only if there is a cardinal κ such that A_E is κ-satisfiable. SAT = the class of satisfiable abstraction principles.*
>
> **Definition 3.3.** *An abstraction principle A_E is unbounded if and only if, for any cardinal γ, there is a cardinal $\kappa \geq \gamma$ such that A_E is κ-satisfiable. UNB = the class of unbounded principles.*
>
> **Definition 3.4.** *An abstraction principle A_E is stable if and only if there is a cardinal γ such that, for all cardinals $\kappa \geq \gamma$, A_E is κ-satisfiable. STB = the class of stable principles.*
>
> **Definition 3.5.** *An abstraction principle A_E is strongly stable if and only if there is a cardinal γ such that A_E is κ-satisfiable if and only if $\kappa \geq \gamma$. S-STB = the class of strongly stable principles.*

Definition 3.6. *An abstraction principle* A_E *is conservative if and only if, for every cardinal* κ, A_E *is* κ-*satisfiable.* CON = *the class of conservative principles.*

The following diagram sums up the relationships between these classes of abstraction principles, where arrows represent inclusion:

$$\text{CON} \to \text{S-STB} \to \text{STB} \to \text{UNB} \to \text{SAT}$$

Here, and in the following, we will not provide proofs of the inclusion or distinctness of various classes of abstraction principles, since all results reported in this section are either familiar from the literature or are easy corollaries of such results. The reader interested in a relatively comprehensive treatment of this material should consult (Cook & Linnebo 2018).[3]

Of these four classes of abstraction principles, only **S-STB** and **STB** are "technically feasible" candidates for the class of acceptable abstraction principles: On the one hand, **UNB** (hence **SAT**) are too broad, since there exist abstraction principles A_{E_1} and A_{E_2} where both A_{E_1} and A_{E_2} are unbounded (hence satisfiable), yet A_{E_1} and A_{E_2} are not jointly satisfiable. On the other hand, **CON** is too narrow, since not even Hume's Principle is in **CON**.

In addition to the cardinality-based criteria, various notions of *conservativeness* have been proposed as additional criteria that acceptable abstraction principles should meet. Crispin Wright suggests that:

> A legitimate abstraction, in short, ought to do no more than introduce a concept by fixing truth conditions for statements concerning instances of that concept. . . . How many sometime, someplace zebras there are is a matter between that concept and the world. No principle which merely assigns truth conditions to statements concerning objects of a quite unrelated, abstract kind—and no legitimate second-order abstraction can do any more than that—can possibly have any bearing on the matter. What is at stake . . . is, in effect, conservativeness in (something close to) the sense of that notion deployed in Hartry Field's exposition of his nominalism.
>
> (Wright 1997, 296)

The first observation to make along these lines is to note that there is no point in requiring that abstraction principles be conservative on the standard understanding of this notion, which we will call *inferentially conservative*:

---

3 The literature on the various conditions proposed as candidates for delineating the Bad Company divide is rife with errors. These errors are identified and cleaned up in (Cook & Linnebo 2018).

Definition 3.7. $\mathcal{L} \backslash @_E$ *is the language obtained by removing from $\mathcal{L}$ all formulas containing $@_E$. Similarly, if S is a set of abstraction operators, then $\mathcal{L} \backslash S$ is the language obtained by removing from $\mathcal{L}$ all formulas containing any abstraction operator $@_E \in S$.*

Definition 3.8. *An abstraction principle A_E is inferentially conservative if and only if, for any theory T and formula Φ in $\mathcal{L} \backslash \{@_E\}$, if $T \cup \{A_E\} \vDash \Phi$, then $T \vDash \Phi$. I-CON = the class of conservative abstraction principles.*

The reason is simple: the class of inferentially conservative abstraction principles is identical to the class of conservative abstraction principles:

Proposition 3.9. CON = I-CON.

Thus, not even Hume's Principle is conservative in this sense.[4]

With this in mind (and, in part, inspired by Wright's mention of Field's notion quoted previously), (Weir 2003) proposes two weaker notions of conservativeness that Hume's Principle, and many other abstraction principles, do in fact meet:

Definition 3.10. *Given any formula Φ and unary predicate Ψ (where Ψ might be a second-order variable), Φ^Ψ is the relativization of (the quantifiers of) Φ to Ψ.*

Definition 3.11. *An abstraction principle A_E is Field conservative if and only if, for any theory T and formula Φ in $\mathcal{L} \backslash \{@_E\}$, if $T^{\neg(\exists Y)(x = @_E(Y))} \cup \{A_E\} \vDash \phi^{\neg(\exists Y)(x = @_E(Y))}$, then $T \vDash \phi$. F-CON = the class of Field conservative principles.*

Definition 3.12. *An abstraction principle A_E is Caesar-neutral conservative if and only if, for any theory T and formula Φ in $\mathcal{L} \backslash \{@_E\}$ and primitive predicate $\Psi(x)$ not occurring in T or Φ, if $T^{\Psi(x)} \cup \{A_E\} \vDash \phi^{\Psi(x)}$, then $T \vDash \phi$. C-CON = the class of Caesar-neutral conservative principles.*

Clearly, Field conservativeness entails Caesar-neutral conservativeness, hence:

Proposition 3.13. F-CON $\subset$ C-CON.

We note the following fact:

Proposition 3.14. C-CON = UNB.

---

4 A simpler way to see that Hume's Principle is not inferentially conservative is to note that it entails the purely logical formula expressing that the domain contains infinitely many objects.

F-CON, however, is not equivalent to any of the notions introduced thus far (nor any combination of them), since whether an abstraction principle is Field conservative depends not only on the cardinals κ for which it is κ-satisfiable but also on how many abstracts it "generates" on domains on which it is satsifiable. Thus, some further definitions:

> **Definition 3.15.** A_E is κ-full *if and only if, for any model* $\mathcal{M} = \langle \Delta, \mathcal{I} \rangle$ *where* $|\Delta| = \kappa$, *we have:*

$$|\{x \in \Delta : (\exists Y \subseteq \Delta)(\mathcal{I}(@_E(Y)) = x\}| = \kappa$$

Loosely put, an abstraction principle A_E is κ-full, for some cardinal κ, just in case any model of A_E whose domain is κ-sized contains κ-many abstracts.

> **Definition 3.16.** A_E is full *if and only if:* A_E *is* κ-full *if and only if* A_E *is* κ-satisfiable. FULL = *the class of full abstraction principles.*

We should pause here to note that FULL, like CON (= I-CON introduced earlier), is not a viable candidate for the class of acceptable abstraction principles, since Hume's Principle is not full. There are, however, weaker versions of fullness that apply to Hume's Principle:

> **Definition 3.17.** κ *is a critical point of* A_E *if and only if* A_E *is* κ-satisfiable, *and there is a* $\gamma < \kappa$ *such that* A_E *is not* λ-satisfiable *for all* $\gamma \leq \lambda < \kappa$. $\operatorname{crit}(A_E)$ *is the set of critical points of* A_E.
> **Definition 3.18.** A_E *is critically full if and only if* A_E *is* κ-full *for any* $\kappa \in \operatorname{crit}(A_E)$. C-FULL = *the class of critically full abstraction principles.*
> **Definition 3.19.** A_E *is weakly critically full if and only if for any* $\kappa_1 \in \operatorname{crit}(A_E)$, *there is a* $\kappa_2 \geq \kappa_1$ *such that* A_E *is* κ_2-full. WC-FULL = *the class of weakly critically full abstraction principles.*

Clearly, the following inclusions hold between the various notions of fullness:

FULL → C-FULL → WC-FULL

We now have an equivalent to Field conservativeness formulated in terms of cardinality considerations:

> **Proposition 3.20.** F-CON = WC-FULL ∩ UNB

(Weir 2003) also introduces the idea that an abstraction principle be *irenic*, which in the present context can be divided into two distinct notions corresponding to our two distinct notions of conservativeness:

Definition 3.21. *An abstraction principle* A_{E_1} *is Field irenic if and only if it is Field conservative and, for any Field-conservative abstraction principle* A_{E_2}, *there is a cardinal* κ *such that* A_{E_1} *and* A_{E_2} *are both* κ-*satisfiable.* **F-IRN** = *the set of Field irenic abstraction principles.*

Definition 3.22. *An abstraction principle* A_{E_1} *is Caesar-neutral irenic if and only if it is Caesar-neutral conservative and, for any Caesar-neutral conservative abstraction principle* A_{E_2}, *there is a cardinal* κ *such that* A_{E_1} *and* A_{E_2} *are both* κ-*satisfiable.* **C-IRN** = *the set of Caesar-neutral irenic abstraction principles.*

The Caesar-neutral version of irenicity is equivalent to stability:

Proposition 3.23. C-IRN = STB

and the Field version of irenicity is equivalent to stability plus weak critical fullness:

Proposition 3.24. F-IRN = WC-FULL ∩ STB

We can sum up the implications between various combinations of these notions in the following diagram:

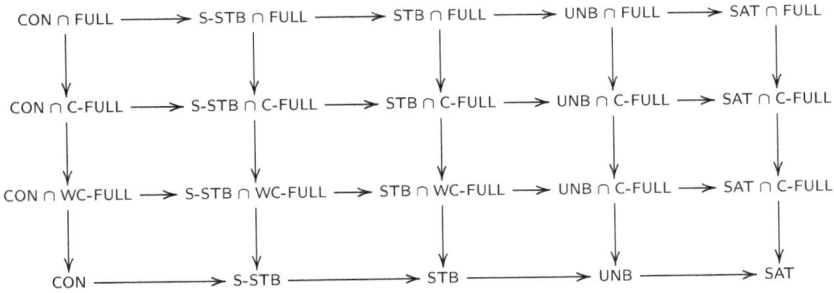

Finally, Richard Kimberly Heck [in a postscript to (Heck 2012a), included in (Heck 2012b)] has suggested that some form of monotonicity should hold of abstraction principles. One way to capture that idea is as follows:

Definition 3.25. A_E *is monotonic if and only if, given any two models* $\mathcal{M}_1 = \langle \Delta_1, \mathcal{I}_1 \rangle$ *and* $\mathcal{M}_2 = \langle \Delta_2, \mathcal{I}_2 \rangle$ *where:*

$$\mathcal{M}_1 \vDash A_E$$
$$\mathcal{M}_2 \vDash A_E$$

if $|\Delta_1| < |\Delta_2|$ *then:*

$$| \{ x \in \Delta_1 : (\exists Y \subseteq \Delta_1)(\mathcal{I}_1(@_E(Y)) = x\} |$$
$$\leq |\{ x \in \Delta_2 : (\exists Y \subseteq \Delta_2)(\mathcal{I}_2(@_E(Y)) = x\} |$$

MONO = *the class of monotonic abstraction principles.*

Clearly, every full abstraction principle is monotonic (although not vice versa):

Proposition 3.26. FULL ⊂ MONO

In addition, although none of the classes appearing in these equivalences contain Hume's Principle and hence are not viable candidates for the class of acceptable abstraction principles, the following are worth noting:

Proposition 3.27.

C-FULL ∩ MONO ∩ CON = WC-FULL ∩ MONO ∩ CON = MONO ∩ CON C-FULL ∩ CON = WC-FULL ∩ CON = CON

No other implications or equivalences hold between the various combinations of these conditions, however [again, for details, see (Cook & Linnebo 2018)]. Ignoring those classes that either (i) fail to contain Hume's Principle (i.e., any subset of CON and any subset of FULL ∩ SAT) or (ii) are not satisfiable (i.e., any superset of FULL ∩ UNB), we obtain the following diagram:

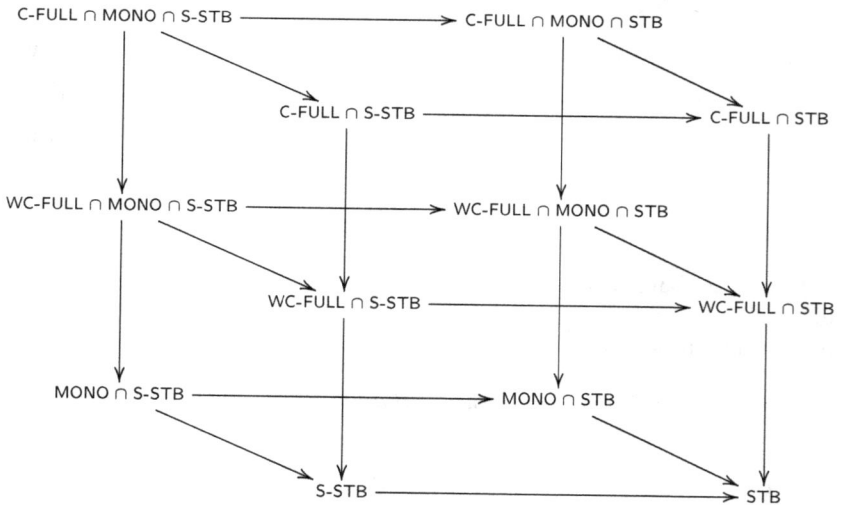

We need not, for the purposes of this chapter, decide which of these is the class of all and only the acceptable abstraction principles. The argument we will give in the following will apply to any choice from among these twelve options. All we need in order to carry out the argument given in the following sections is that the class of acceptable abstraction principles be closed under a certain kind of equivalence. We now define the kind of equivalence in question:[5]

Definition 3.28. A_{E_1} *is strongly cardinality equivalent to* A_{E_2}—*that is:*

$$A_{E_1} \Leftrightarrow_{SC} A_{E_2}$$

if and only if, for any cardinal κ, A_{E_1} *is* κ-*satisfiable if and only if* A_{E_2} *is* κ-*satisfiable, and, for any model* $\mathcal{M} = \langle \Delta, \mathcal{I} \rangle$ *such that:*

$$\mathcal{M} \vDash A_{E_1}$$
$$\mathcal{M} \vDash A_{E_2}$$

we have:

$$| \{ x \in \Delta : (\exists Y \subseteq \Delta)(\mathcal{I}(@_{E_1}(Y)) = x) \} |$$
$$= | \{ x \in \Delta : (\exists Y \subseteq \Delta)(\mathcal{I}(@_{E_2}(Y)) = x) \} |$$

Less formally put, two abstraction principles are strongly cardinality equivalent if and only if they are κ-satisfiable at the same κs, and, in addition, for any such κ, the collections of abstracts given by each principle on a domain of size κ are equinumerous.

Closure under cardinality equivalence is defined as one would expect:

Definition 3.29. *A set of abstraction principles* S *is closed under strong cardinality equivalence if and only if, for any* A_{E_1} *and* A_{E_2}, *if* $A_{E_1} \in S$ *and* $A_{E_1} \Leftrightarrow_{SC} A_{E_2}$, *then* $A_{E_2} \in S$.

We now note the following fact:

Theorem 3.30. *Let* $\mathcal{A}$ *be any subset of:*

{CON, STB, W-STB, UNB, SAT, FULL, C-FULL, WC-FULL, MONO}

---

5 There is a weaker notion of cardinality equivalence, which we can call *weak cardinality equivalence* and which can be defined as follows:

A_{E_1} *is weakly cardinality equivalent to* A_{E_2}—*that is:*

$$A_{E_1} \Leftrightarrow_{WC} A_{E_2}$$

if and only if, for any cardinal κ, A_{E_1} is κ-satisfiable if and only if A_{E_2} is κ-satisfiable. A set of abstraction principles S is *closed under weak cardinality equivalence* if and only if, for any A_{E_1} and A_{E_2}, if $A_{E_1} \in S$ and $A_{E_1} \Leftrightarrow_{WC} A_{E_2}$, then $A_{E_2} \in S$.

Note, however, that, of the eight notions in our final diagram, only W-STB and STB are closed under weak cardinality entailment.

Then ∩𝒜 *is closed under strong cardinality equivalence.*

Proof. Straightforward consequence of definitions, left to the reader. □

Simply put, any class of abstraction principles we might obtain by restricting our attention to those principles that meet some list of conditions considered previously will be closed under strong cardinality equivalence.[6] None of these conditions (nor any combination of them) "carve up" the space of abstraction principles more finely than that. This result will be a central ingredient in our argument that any viable neo-logicist set theory must accept complement and reject separation.

4.   Local and Global Neo-Logicist Set Theories

In order to determine whether the neo-logicist should opt for the axiom of separation or the axiom of complement, it is not enough merely to look at some acceptable abstraction principle (whichever criterion of acceptability we opt for) and ascertain whether the principle in question entails separation or complement (or neither). The reason is simple: we can easily construct individual abstraction principles that satisfy all of the "technically viable" criteria surveyed in the previous two sections that give us a set theory with separation but without complement, and we can equally easily construct individual abstraction principles that likewise satisfy all of these criteria and give us complement but not separation.

Before looking at examples, we need the following notions of *exponential bigness* and *exponential smallness*, due to (Fine 2002):[7]

Definition 4.1.

$$^\mathrm{E}\mathsf{Big}(X) =_{df} |\{Y \subseteq \mathrm{U} : |Y| \leq |X|\}| > |\mathrm{U}|$$
$$^\mathrm{E}\mathsf{Sm}(X) =_{df} \neg^\mathrm{E}\mathsf{Big}(X)$$

Now, if we want separation but not complement, we can adopt the following *separation abstraction principle*:[8]

$$\mathsf{BLV}_\mathsf{Sep} : (\forall X)(\forall Y)[\S_\mathsf{Sep}(X) = \S_\mathsf{Sep}(Y)$$
$$\leftrightarrow ((\forall z)(X(z) \leftrightarrow Y(z)) \vee (^\mathrm{E}\mathsf{Big}(X) \wedge {}^\mathrm{E}\mathsf{Big}(Y)))]$$

---

6 Note that this result applies not only to the twelve "technically viable" classes of abstraction principles contained in our final diagram but also to any class of abstraction principles definable via conjoining any combination of the conditions surveyed in this section.

7 U here is the domain in question. Although expressed here in terms of intuitive set-theoretic notation in the definition, both exponential bigness and exponential smallness are expressible in the language of second-order logic [see (Fine 2002) for details].

8 Note that these are merely minor modifications of NewV and DualV (Boolos 1989).

Adopting the following definitions of set and membership:

$$\text{Set}_{\text{Sep}}(x) = (\exists Y)(^{\text{E}}\text{Sm}(Y) \wedge x = \S_{\text{Sep}}(Y))$$
$$x \in_{\text{Sep}} y = (\exists Z)(y = \S_{\text{Sep}}(Z) \wedge {}^{\text{E}}\text{Sm}(Z) \wedge Z(x))$$

we can easily derive the axiom of separation. If, however, we desire an abstraction principle for sets that delivers complement but not separation, we can instead consider the following *complement abstraction principle*:

$$\text{BLV}_{\text{Comp}} : (\forall X)(\forall Y)[\S_{\text{Comp}}(X) = \S_{\text{Comp}}(Y) \leftrightarrow ((\forall z)(X(z) \leftrightarrow Y(z))$$
$$\vee ({}^{\text{E}}\text{Big}(X) \wedge {}^{\text{E}}\text{Big}(\neg X) \wedge {}^{\text{E}}\text{Big}(Y) \wedge {}^{\text{E}}\text{Big}(\neg Y)))]$$

Adopting slightly different definitions of set and membership (due to the fact that we have imposed slightly different requirements for an object to be assigned to a single concept and hence be a set):

$$\text{Set}_{\text{Comp}}(x) = (\exists Y)(({}^{\text{E}}\text{Sm}(Y) \vee {}^{\text{E}}\text{Sm}(\neg Y)) \wedge x = \S_{\text{Comp}}(Y))$$
$$x \in_{\text{Comp}} y = (\exists Z)(y = \S_{\text{Comp}}(Z) \wedge ({}^{\text{E}}\text{Sm}(Z) \vee {}^{\text{E}}\text{Sm}(\neg Z)) \wedge Z(x))$$

we can easily derive the axiom of complement. Both BLV_{Sep} and BLV_{Comp} are strongly stable, critically full, and monotonic—that is:

$$\text{BLV}_{\textit{Sep}} \in \text{C-FULL} \cap \text{MONO} \cap \text{S-STB}$$
$$\text{BLV}_{\textit{Comp}} \in \text{C-FULL} \cap \text{MONO} \cap \text{S-STB}$$

hence, since **C-FULL** ∩ **MONO** ∩ **S-STB** is a subset of every class of abstraction principles identified previously as viable candidates for the class of acceptable abstraction principles, none of the criteria surveyed previously will select one rather than the other of BLV_{Sep} and BLV_{Comp} as a better (much less the *correct*) account of neo-logicist set theory.

Thus, if we want to locate genuine neo-logicist reasons for preferring one of separation or complement to the other, piecemeal consideration of this or that abstraction principle won't do the job. Fortunately, there is another strategy.

Instead of looking *locally*, at individual abstraction principles on their own, we can instead consider neo-logicist set theory from a *global* perspective, where we consider the entire structure of abstracts generated by acceptable abstraction principles all at once.[9] Viewed this way, we

---

9 Although the view proposed in this chapter is quite different from (Fine 2002) in its details, the general methodology adopted here—to consider all abstraction principles at once as a single unified theory rather than considering each individually in turn—is lifted straight from Fine's insightful analysis of abstraction.

can think of both BLV_{Sep} and BLV_{Comp} as providing us with genuine sets, since it is natural to consider any abstract that is the abstract of a unique concept to be a set. Thus, BLV_{Sep} governs some but not all of the neo-logicist sets, and separation holds when restricted to this substructure of the realm of sets provided by BLV_{Sep}. BLV_{Comp} governs a different, over-lapping but not identical, substructure of the entire realm of neo-logicist sets—a substructure that, when viewed in isolation, does satisfy the axiom of complement.

With this in mind, it is natural to define the *global sets* as follows:[10]

Definition 4.2. *x is a global set, that is:*

$Set_{Glob}(x)$

if and only if there is an acceptable abstraction principle:

$A_E : (\forall X)(\forall Y)[@_E(X) = @_E(Y) \leftrightarrow E(X,Y)]$

such that:

$(\exists Y)(x = @_E(Y) \wedge (\forall Z)(x = @_E(Z) \rightarrow (\forall w)(Y(w) \leftrightarrow Z(w))))$

Informally, an object is a global set just in case there is some acceptable abstraction principle such that the object in question is the abstract of a unique concept according to that abstraction principle. A corresponding definition of global membership is as follows:

Definition 4.3. $x \in_{Glob} y$ *iff there an acceptable abstraction principle:*

$A_E : (\forall X)(\forall Y)[@_E(X) = @_E(Y) \leftrightarrow E(X,Y)]$

such that:

$(\exists Z)(y = @_E(Z) \wedge (\forall W)(y = @_E(W) \rightarrow (\forall v)(Z(v) \leftrightarrow W(v))) \wedge Z(x))$

Informally, an object α is a global member of an object β just in case there is an abstraction principle such that β is the abstract of a single concept according to that abstraction principle, and α is an instance of the single concept whose abstract is β according to that abstraction principle.

---

10 Note that the notions "global set" and "global membership" must be formulated in a metatheory that allows one to quantify over all acceptable abstraction operators.

In order to obtain a well-behaved set theory via the global approach, we will need one more ingredient. As it stands so far, we have not ruled out the possibility that there are two abstraction principles:[11]

$$A_{E_1} : (\forall X)(\forall Y)[@_{E_1}(X) = @_{E_1}(Y) \leftrightarrow E_1(X,Y)]$$
$$A_{E_2} : (\forall X)(\forall Y)[@_{E_2}(X) = @_{E_2}(Y) \leftrightarrow E_2(X,Y)]$$

and a concept X such that X corresponds to a set according to both A_{E_1} and A_{E_2} but not the *same* set. In short, we have as of yet said nothing to rule out the following situation:

$$(\exists Y)[(\forall Z)(@_{E_1}(Z) = @_{E_1}(Y) \rightarrow (\forall w)(Y(w) \leftrightarrow Z(w)))$$
$$\wedge (\forall Z)(@_{E_2}(Z) = @_{E_2}(Y) \rightarrow (\forall w)(Y(w) \leftrightarrow Z(w)))$$
$$\wedge @_{E_1}(Y) \neq @_{E_2}(Y)]$$

Equally worrisome, we have also not ruled out the possibility that there is a single object that is a set corresponding to distinct concepts according to different abstraction principles—that is, we have said nothing to rule out the following situation:

$$(\exists Y_1)(\exists Y_2)[(\forall Z)(@_{E_1}(Z) = @_{E_1}(Y_1) \rightarrow (\forall w)(Y_1(w) \leftrightarrow Z(w)))$$
$$\wedge (\forall Z)(@_{E_2}(Z) = @_{E_2}(Y_2) \rightarrow (\forall w)(Y_2(w) \leftrightarrow Z(w)))$$
$$\wedge @_{E_1}(Y_1) = @_{E_2}(Y_2)$$
$$\wedge \neg(\forall z)(Y_1(z) \leftrightarrow Y_2(z))]$$

The solution to both of these problems is straightforward. These two problems are just special cases of what has come to be called the $\mathbb{C}$-$\mathbb{R}$ problem. This problem arises when we consider cross-abstraction identities—that is, when we ask whether the abstracts generated by one abstraction principle are identical to or distinct from the abstracts given to us by a second abstraction principle [see (Cook & Ebert 2005) for a good discussion of

---

11 A variant of this worry is the following: We have not ruled out the possibility that there are two distinct abstraction principles:

$$A_{E_1} : (\forall X)(\forall Y)[@_{E_1}(X) = @_{E_1}(Y) \leftrightarrow E_1(X,Y)]$$
$$A_{E_2} : (\forall X)(\forall Y)[@_{E_2}(X) = @_{E_2}(Y) \leftrightarrow E_2(X,Y)]$$

where the equivalence relations in question are equivalent—that is:

$$(\forall X)(\forall Y)[E_1(X,Y) \leftrightarrow E_2(X,Y)]$$

yet some (or even all!) of the abstracts corresponding to particular concepts according to one of the abstraction principles are distinct from the abstracts corresponding to those same concepts according to the second abstraction principle. Thanks are owed to an anonymous referee for pointing out this possibility and pointing out the fact that Fine's Principle also prevents it.

the $\mathbb{C}$-$\mathbb{R}$ problem]. The simplest solution to the $\mathbb{C}$-$\mathbb{R}$ problem is also the obvious one to apply if we wish to obtain a viable global neo-logicist set theory: We should adopt Kit Fine's *identity principle* (2002), which states that two abstracts given by two distinct abstraction principles are identical if and only if they correspond to the same class of concepts. More carefully:

Identity Principle: Given two abstraction principles:

$$A_{E_1} : (\forall X)(\forall Y)[@_{E_1}(X) = @_{E_1}(Y) \leftrightarrow E_1(X,Y)]$$
$$A_{E_2} : (\forall X)(\forall Y)[@_{E_2}(X) = @_{E_2}(Y) \leftrightarrow E_2(X,Y)]$$

cross-abstraction identity conditions are given by:

$$(\forall X)(\forall Y)(@_{E1}(X) = @_{E2}(Y) \leftrightarrow (\forall Z)(E_1(X,Z) \leftrightarrow E_2(Y,Z)))$$

Adopting Fine's identity principle guarantees that, in our global approach, each concept will be mapped to at most one set, and no two distinct concepts will ever be mapped to the same set. Hence it rules out both of the problematic possibilities just considered. With this additional machinery in place, we are now in a position to ask what sort of set theory the neo-logicist will obtain on the global approach to set theory and, in particular, whether this approach to set theory will provide us with separation or complement.

5.   Global Set Theory and Comprehension

In this section, we will prove that, if the class of acceptable abstraction principles is closed under strong cardinality equivalence (and we saw good reasons for thinking that it must be in §3), then the axiom of complement is true (and the axiom of separation is false) within the global neo-logicist set theory developed in the previous section. We begin by rehearsing some simple facts about equivalence relations on concepts:[12]

Definition 5.1. *Given any domain Δ and any equivalence relation $E(X,Y)$ on $\mathcal{P}(\Delta)$:*

$$E_{Comp}(X,Y) = E(\neg X, \neg Y)$$

Theorem 5.2. *If Δ is a domain and $E(X,Y)$ is an equivalence relation on $\mathcal{P}(\Delta)$, then $E_{Comp}(X,Y)$ is also an equivalence relation on $\mathcal{P}(\Delta)$.*

Proof. Trivial. $\square$

---

12 $E(\neg X, \neg Y)$ abbreviates:

$$(\forall Z)(\forall W)(((\forall v)(Z(v) \leftrightarrow \neg X(v)) \wedge (\forall v)(W(v) \leftrightarrow \neg Y(v))) \rightarrow E(Z,W))$$

More informally, $E(\neg X, \neg Y)$ says that the equivalence relation E holds of the complements of X and Y.

Theorem 5.3. *Given any domain Δ and equivalence relation $E(X,Y)$ on $\mathcal{P}(\Delta)$, if $E(X,Y)$ partitions $\mathcal{P}(\Delta)$ into κ equivalence classes, then $E_{Comp}(X,Y)$ partitions $\mathcal{P}(\Delta)$ into κ-many equivalence classes.*

Proof. Trivial. ☐

We now prove our main result:

Theorem 5.4. *If the class of acceptable abstraction principles is closed under cardinality equivalence, then the axiom of (global) complement:*

$$(\forall x)(Set_{Glob}(x) \rightarrow (\exists y)(Set_{Glob}(y) \wedge (\forall z)(z \in_{Glob} y \leftrightarrow z \notin_{Glob} x)$$

is true of the universe of abstracts.

Proof. Assume α is a global set. Then there is an acceptable abstraction principle:

$$A_E : (\forall X)(\forall Y)[@_E(X) = @_E(Y) \leftrightarrow E(X,Y)]$$

such that:

$$(\exists Y)(\alpha = @_E(Y) \wedge (\forall Z)(\alpha = @_E(Z) \rightarrow (\forall w)(Y(w) \leftrightarrow Z(w))))$$

Let us call the unique concept witnessing this formula M. Then:

$$(\forall Z)(@_E(M) = @_E(Z) \rightarrow (\forall w)(M(w) \leftrightarrow Z(w)))$$

Since the class of acceptable abstraction principles is closed under cardinality equivalence, **Theorem 5.3** entails that:

$$A_{E_{Comp}} : (\forall X)(\forall Y)[@_{E_{Comp}}(X) = @_{E_{Comp}}(Y) \leftrightarrow E_{Comp}(X,Y)]$$

is an acceptable abstraction principle. Consider $@_{E_{Comp}}(\neg M)$. Let Y be any concept such that:

$$@_{E_{Comp}}(\neg M) = @_{E_{Comp}}(Y)$$

Then:

$$E_{Comp}(\neg M, Y)$$

Hence

$$E(M, \neg Y)$$

So:

$$@_E(M) = @_E(\neg Y)$$

and thus, since $@_E(M)$ is a set, we have:

$$(\forall w)(M(w) \leftrightarrow \neg Y(w))$$

which is equivalent to:

$$(\forall w)(\neg M(w) \leftrightarrow Y(w))$$

Since Y was arbitrary, we obtain:

$$(\forall Z)(@_{EComp}(\neg M) = @_{EComp}(Z) \rightarrow (\forall w)(\neg M(w) \leftrightarrow Z(w))))$$

Hence $@_{E_{Comp}}(\neg M)$ is a set. Furthermore, since

$$(\forall z)(M(z) \leftrightarrow \neg(\neg(M(z))))$$

we have:

$$(\forall z)(z \in_{Glob} @_E(M)) \leftrightarrow z \notin_{Glob} @_{E_{Comp}}(\neg M)$$

So $@_{E_{Comp}}(\neg M)$ is the complement of $@_E(M)$. □

> **Corollary 5.5.** *If the class of acceptable abstraction principles is closed under cardinality equivalence, then the axiom of (global) separation:*
>
> $$(\forall x)(Set_{Glob}(x) \rightarrow (\forall Y)(\exists z)(Set_{Glob}(z) \wedge (\forall w)(w \in z \leftrightarrow (w \in x \wedge Y(w)))))$$
>
> *is false of the universe of abstracts.*

Of course, as we already saw, all extant proposals for the class of acceptable abstraction principles are closed under strong cardinality equivalence. Hence (unless some other approach to the Bad Company problem is adopted), it seems like we have no choice but to accept that the axiom of complement is true, and the axiom of separation is false, from a neologicist perspective.

6. Conclusions

Thus, if we attend to the global perspective and consider a set to be any object that is the abstract of a unique concept according to some acceptable abstraction principle, then the axiom of complement is true, and the axiom of separation is false. This, however, cannot be the whole story.

As we noted at the outset, if neo-logicism is to be a viable account of the foundations of mathematics, then it must be able to provide some account of actual set theoretic practice. As a result, it needs to be able to account for the work of set theorists working within set theories, such as first- or second-order ZFC, that accept the axiom of separation and reject the axiom of complement. But making sense of such practices within the current framework is straightforward—we need only point out that, from this perspective, the working set theorist is studying some, but not all, of the (global) sets.

We can make this a bit more precise, as follows. First, we define the notion of a concept being closed under set formation:

Definition 6.1.

$$\mathsf{Closed}_{Glob}(X) = (\forall y)((\mathsf{Set}_{Glob}(y) \wedge (\forall z)(z \in_{Glob} y \rightarrow X(z)) \rightarrow X(y))$$

Loosely put, a concept X is closed if and only if, for any set y, if X holds of all of the members of y, then X holds of y. We can then use this notion of closure to define the pure sets:

Definition 6.2.

$$Pure(x) = (\forall Y)(\mathsf{Closed}_{Glob}(Y) \rightarrow Y(x))$$

The pure sets, intuitively, are those sets that can be "built up" from the empty set.

It is straightforward to prove that the pure sets satisfy the axiom of separation.[13] Thus, in order to make sense of standard set-theoretic practice, we need merely interpret the set theorist as working with the pure (global, neo-logicist) sets rather than with *all* of the sets that neo-logicism provides. Thus, from a neo-logicist perspective, there is nothing wrong with current set-theoretic practice. The set theorist working with, for example, ZFC, is just not taking advantage of all of the sets that are at their disposal.[14]

---

13 Which of the other axioms of ZFC are satisfied by the pure sets depends on which abstraction principles turn out to be acceptable. Thus, the previous comments are meant only to reassure the reader that the neo-logicist can make sense of set theoretic practices that include acceptance of separation. Defending the whole of ZFC from such a perspective, while possible, is beyond the scope of this chapter.

14 This material was presented at the 2012 New Foundations 75th Anniversary Conference at the Department of Mathematics, Cambridge University; at the 2011 Truth, Paradox, and Abstraction Workshop at the Northern Institute of Philosophy, University of Aberdeen; and at the 2011 Midwest Philosophy of Mathematics Workshop at the University of Notre Dame. Thanks are owed for generous feedback from these audiences and to a helpful anonymous referee.

References

Beuno, O. & Ø. Linnebo (eds.) (2009). *New Waves in Philosophy of Mathematics*, Basingstoke, UK: Palgrave.

Boolos, G. (1989). "Iteration Again", *Philosophical Topics* 7: 5–21.

Cook, R. (ed.) (2007). *The Arché Papers on the Mathematics of Abstraction*, Dordrecht: Springer.

Cook, R. (2009). "New Waves on an Old Beach: Neo-Fregean Philosophy of Mathematics Today", in (Bueno & Linnebo 2009): 13–34.

Cook, R. & P. Ebert (2005). "Abstraction and Identity", *Dialectica* 59(2): 121–139.

Cook, R. & Ø. Linnebo (2018). "Cardinality and Acceptable Abstraction", *Notre Dame Journal of Formal Logic* 59: 61–74.

Fine, K. (2002). *The Limits of Abstraction*, Oxford: Oxford University Press.

Hale, B. & C. Wright (2001). *The Reason's Proper Study*, Oxford: Oxford University Press.

Heck, R. (1996). "The Consistency of Predicative Fragments of Frege's *Grundgesetze der Arithmetik*", *History and Philosophy of Logic*, 17: 209–220.

Heck, R. (ed.) (1997). *Language, Thought and Logic: Essays in Honour of Michael Dummett*, Oxford: Oxford University Press.

Heck, R. (2012a). "On the Consistency of Second-Order Abstraction Principles", in (Heck 2012b): 230–236.

Heck, R. (2012b). *Frege's Theorem*, Oxford: Oxford University Press.

Linnebo, Ø. (2018), *Thin Objects: An Abstractionist Account*, Oxford: Oxford University Press.

Weir, A. (2003). "Neo-Fregeanism: An Embarrassment of Riches?", *Notre Dame Journal of Formal Logic* 44: 13–48.

Wright, C. (1997). "On the Philosophical Significance of Frege's Theorem", in (Heck 1997): 201–244.

13 *Principia Mathematica* Redux

Gregory Landini

1. Introduction

No science, empirical or *a priori*, can be more fundamental than the science of mathematics. No notion of "necessity" can be more foundational than that of *mathematical necessity*. But what is the science of mathematics about, and what ontological commitments are indispensable to it? Traditional Empiricists and Rationalists alike imagined it is about abstract particulars. But neither could explain the epistemology of mathematics, and indeed the Rationalist appeal to innate contact with abstract particulars such as numbers seemed nothing short of occult. Dewey's pragmatism would have the question ignored, appealing instead to applied "mathematical" know-how as found, for example, in carpentry, metallurgy, and the building of bridges (Dewey, 1920, p. 116). The center stage is given over to human practices and learning from the trials and tribulations of life and death. Following in the steps of Dewey's pragmatism, Putnam hopes that commitments to mathematical abstract particulars are legitimized by the "success" of an empirical physics at building bridges. Quine's empiricism without the analytic-synthetic distinction would have us accept a doctrine of ontological relativity and believe that there are no absolute facts about what there is—as if the evidentiary basis for believing there are abstract particulars such as numbers and triangles and Quine-sets is no different in kind from the empirical evidence that there are black holes, Higgs particles, and coelacanths living in the south China seas. All empiricisms leave us in a morass today no better than in Kant's day when he offered his *Transcendental Idealism* as a solution to the question "How is synthetic *a priori* possible?" But at the turn of the twentieth century, mathematicians such as Whitehead and Russell thought that they had arrived at the correct, non-Empiricist, non (traditional) Rationalist and non-Kantian answer by following the new revolution in mathematics (pioneered by Cantor) that *relational order is all that matters to mathematics*. Their unique doctrine of Logicism was born from two revolutions: the revolution within mathematics *and* Frege's revolution within logic.

DOI: 10.4324/9780429277894-16

Frege was *not* among the revolutionaries. Whitehead and Russell espouse a *unique* form of Logicism quite antithetical to that of Frege. Its agenda was to reveal that *no* abstract particulars (numbers, classes, spatial figures, etc.), and no special mathematical *necessity* governing them, are involved in any field of mathematics.[1] Their Logicism embraces a transformation of the fields of mathematics (including geometry, Euclidean, and otherwise) into studies of relational structures. These studies, pioneered by Cantor, Weierstrass, Pieri, the non-Euclidean geometers, and many other mathematicians, broadly assume the existence of relations that characterize the structures under study. Logic, according to Whitehead and Russell, is the synthetic *a priori* study of relational structures, and it can be conducted by studying relations independently of whether they are exemplified. It is by means of our Rationalist *acquaintance* with universals (some of which is innate) that it studies relational structures independently of whether the relations are exemplified. Alas, this has largely been lost and forgotten.[2] It won't be found in the scholarly pieces written by Carnap (1931), Church (1976), Quine (1963) or Putnam (1967) or among orthodox metaphysicians committed to abstract particulars and a special kind of invariance notion of (metaphysical) *necessity*. It wants a revival. As kindling, I offer a *Principia Mathematica Redux*.

2.   Revolution *Within* Mathematics

Frege's *Grundgesetze der Arithemtic* (1893, 1903) has served for many years as a *paradigm* of Logicism, and this has tended to obscure the

---

1 The agenda comes fully clear in *Principia Mathematica* if one reads summary sections. In volume 1, we find the summary of Part II, which comes on page 329 directly after the logic of relations of Part I. It concerns a *Prolegomena* to the cardinal arithmetic set out in *Principia's* volume 2. Whitehead and Russell write:

> The objects to be studied in this Part are not sharply distinguished from those studied in Part I. The difference is only of degree. . . . Although cardinal arithmetic is the goal which determines our course in Part II, all the objects studied will be found to be also required in ordinal arithmetic and the theory of series. As this Part advances, the approach to cardinal arithmetic becomes gradually more marked, until at last nothing is lacking except the definition of cardinal numbers with which Part III opens. . . . Section E deals with mathematical induction, not in the special form in which it applies to finite integers (this is considered in Part III, Section C), but in a general form in which it applies to all relations.

Part III of Section C in *Principia's* volume 2 concerning finite and infinite cardinals offers a nice quote, as does page 498 of volume 2, concerning Part V concerning limits and continuity. Similarly, there are comments on page vi of volume 3. We shall quote from these later in the chapter.

2 Traditional as well as contemporary empiricists (holding that whatever is, is *particular*) may well also object to embracing *innate* acquaintance with universals and even to the existence of universals as abstract entities. I'm only concerned here with the oft-forgotten revolution *within* mathematics against abstract particulars.

quite different orientation involved in Whitehead and Russell's *Principia Mathematica*. Frege's logicism endeavored to argue that numbers (natural, signed, rational, real,[3] complex) are abstract particulars of an ontology of logic. Frege never doubted that numbers are abstract particulars ("objects" as opposed to "functions" in his technical use of these words). His *Begriffsschrift* (1879) was better on this point—for it was published prior to his having taken a stand on what abstract particulars numbers are. Its goal was to advance a revolutionary new notion of "logic" as a synthetic *a priori* science embracing *impredicative comprehension*. (I call this "cp-Logic".) No mathematician doubts that functions on numbers are assured to exist. Frege maintained that there is no difference in kind between the functions invoked in logic and those invoked in the mathematics of number. This breakthrough concerning the comprehension of functions was unprecedented. Unfortunately, few mathematicians of his day separated the function *itself* from its intimate relationship with the numbers upon which the equations indicating the functions were comprehended. Mathematicians were comprehending functions not just by equations but by recursion and the technique of *characteristic functions* as well. Thus, in Frege's view, the *whole* of the mathematics of functions resides within the science of logic. Geometry, he thought, is not captured by attention only to functions and must rely upon the ontology of abstract particulars in space which are not logical objects. Numbers as abstract particulars, he held, are not functions either. But numbers, Frege held, are logical objects.

Frege was *not* among the mathematical revolutionaries eschewing abstract particulars. Though he accepted numerical second-level functions, he never wavered from his commitment to numbers as objects.[4] It won't help to say, as Boolos (1994) and Heck (2012) do, that Frege adamantly required numbers as objects because he naturally accepted the arithmetic necessity that there is an infinity of numbers. The impetus goes in the inverse direction. His accepting the necessary infinity of numbers is a consequence of his taking them to be logical objects. Nothing in the revolutionary Cantorian conception of natural number (in terms of relational properties of one-to-one correspondence) entails that there be infinitely many natural numbers. As we shall see in Section 5, *Principia*—which embraced the revolution *within* mathematics—accepted this result. *Grundgesetze* imagined (*per impossible*) that numbers are logical particulars by appealing, in direct violation of Cantor's power theorem, to

---

3 Frege's account of real numbers was incomplete and, because of its connections to geometry, it remains unclear to what extent he embraced them as purely logical objects.

4 Even when confronted with Russell's paradox, Frege couldn't bring himself to retreat to the safety of his cp-Logic of levels of functions. He demanded that natural numbers be objects.

a naïve correlation thesis: logic assures that there is a non-homogeneous function $\acute{z}(\Phi z)$ that correlates each first-level function $f\xi$, one-to-one, with a purely logical object $\acute{z}fz$. These logical objects, which Frege called "value ranges" (*Werthferlauf*), are certainly not classes (or sets). Frege never seems to have brought himself to accept that an analog of Cantor's power-set theorem applies to his non-homogeneous functions. But his naïve correlation thesis, he came to admit, is the source of an analog of Russell's paradox (of the class of all classes not members of themselves) being derivable within his system of value-ranges.

Frege's greatest achievement lies, not as Boolos imagines, in developing arithmetic from the assumption that Hume's Principle governs the objects that are cardinal numbers. Hume's Principle asserts that the number of Fs equals the number of Gs *iff* entities that are Fs are in one-to-one correspondence with entities that are Gs. Frege's greatest achievement lies elsewhere—namely in his discovery of impredicative comprehension logic—cp-Logic. Frege's *Begriffsschrift* of 1879 pioneered a revolution in logic according to which logic, that is, cp-Logic, embodies impredicative comprehension, which, he explained, is the source of (mathematical) inductive proof itself. For all his revolutionary work in accepting impredicative comprehension and the informativity of cp-Logic which it entails, the focus of scholarship on Frege remains focused on fixing on his failed attempt to agree with the metaphysicians that numbers are abstract particulars. The focus among neo-Fregeans, for example, is Hume's Principle itself, with some advocates even rejecting Frege's higher-level non-homogeneous functions and even impredicative comprehension itself.

The history of logicism went a different way. Frege's revolutionary cp-Logic was accepted by Whitehead and Russell (with the impredicative comprehension of functions abandoned in favor of the impredicative comprehension of many-one relations). Frege exacted impredicative comprehension in logic by his higher-level quantification theory together with his rule of uniform substitution of complex function terms for function variables. *Principia* adopts impredicative comprehension outright as an axiom schema (see *12.1 and *12.11 in the following). Impredicative comprehension is at the foundation of the formal apparatus of *Principia*'s Logicism. Indeed, Russell's formal theory of types of relations (and properties) parallels, in many important ways, Frege's hierarchy of levels of functions. And with the revolution within mathematics in place, its thesis is that we have captured the entirety of mathematics, including geometry. Few remember this today. Impredicative comprehension is what makes logic a genuine synthetic *a priori* science. It, not quantification theory (at any level), is the source of the *informativity* of logic. It is unfortunate that this has been lost.

Whitehead and Russell offer *Principia Mathematica* as capturing the revolution within mathematics. They had become convinced that the fields of mathematics concern only relational structures after attending a

1900 Congress in Paris where Peano and the Italian school of mathematicians were presenting work. In Russell (1901) "Mathematics and the Metaphysicians," which largely reports on what they learned from the conference,[5] Russell points out that his Logicism relies on *two* revolutions: the revolution that logic[6] embraces impredicative comprehension of relations and the revolution in mathematics that transforms its into studies of relational structures. He wrote (*MM*, p. 91)

> Mathematics has, in modern times, brought order into greater and greater prominence. In former days, it was supposed (and philosophers are still apt to suppose) that quantity was the fundamental notion of mathematics. But nowadays, quantity is banished altogether, except from one little corner of Geometry, while order more and more reigns supreme. The investigation of different kinds of series and their relations is now a very large part of mathematics, and it has been found that this investigation can be conducted without any reference to quantity, and, for the most part, without any reference to number. All types of series are capable of formal definition, and their properties can be deduced from the principles of symbolic logic.

As we can see, Russell regards Cantor as a leading member of the revolutionary transformation of mathematics that comes from holding that relations of bijection are the heart of the mathematical notion of *number* and that well-ordering relations are the heart of the mathematical notion of *ordinals* and *order type*. Russell simply puts the two revolutions together. Both revolutions are generated by mathematicians for mathematics. They impose no change in the new mathematical practices. As Russell put it, "one of the chief triumphs of modern mathematics consists in having discovered what mathematics really is" (*MM*, p. 75). And, he should add, what it is not! It is *not* about abstract particulars. In Russell's view, mathematicians are, in fact, doing cp-Logic (i.e., studying relational structures grounded in the mere existence of relations, exemplified or not) when they do mathematics.

Indispensability arguments for metaphysical abstract particulars (numbers, sets/classes, triangles, propositions, and the like) and specialized kinds of non-logical necessity stem, in Russell's view, from metaphysicians working from an impoverished conception logic that blinds them to *logical form* (i.e., the relational structures involved). In so doing, they

---

5 The original title was "Recent Italian Work on the Foundations of Mathematics."

6 At the time, Russell knew only of Peano's quantification theory of propositions. Impredicative comprehension was emulated in it by uniformly substituting *wffs* for free proposition variables in the axioms of quantification theory.

introduce metaphysical "muddles" into mathematics. Amusingly, Russell wrote (HWP, p. 829):

> Then came George Cantor, who developed the theory of continuity and infinite number. "Continuity" has been, until he defined it, a vague word, convenient for philosophers like Hegel who wished to introduce metaphysical muddles into mathematics. Cantor gave a precise significance to the word, and showed that continuity, as he defined it, was the concept needed by mathematicians and physicists. By this means, a great deal of mysticism, such as that of Bergson, was rendered antiquated.

This passage is rich with important connotations against the metaphysicians of mathematics whose confusions concerning the nature of the limit, functions and continuity, introduced infinitesimals, and dynamic conceptions of change through time, all of which became, with the help of Weierstrass, antiquated. *Principia*'s task is to reveal the needed relational structures involved, and it was precisely because of its centrality that Russell proclaimed in 1914 that "logic is the essence of philosophy."

To be sure, the only known successful semantics that validates *Principia*'s formal cp-Logic of (impredicative) simple types is one that is Realist—that is, one that accepts that there are, among universals, those that are scaffolded by simple types. Russell would certainly have wished it were otherwise. And he tried hard several times without success to emulate its structure and evade a robust Realism of simple types. He never wavered from his conviction that the *formalism* of simple types was broadly correct and likely able to be emulated by some analysis yet to be found. His entire career in mathematical logic, from the time of his 1903 book *The Principles of Mathematics* (whose proposed second volume was to be in collaboration with Whitehead) to the 1910–1913 *Principia Mathematica* and into its second edition of 1925, he tried to emulate the formalism of simple type theory. He failed. But no matter. The introduction to the first edition (with its failed nominalistic semantics and ramified recursive definition of senses of "truth") and the radical Introduction to the second edition, which imagined a new grammar and Wittgensteinian extensionality, were *never* components of the formal work. Whitehead never endorsed either (especially railing against the latter) and always imagined a Realism of simple types to be the natural intended semantic interpretation. Let's embrace it. Let's not be misled by Church's "ramified type theory" (*r*-types), which imposes an untenable axiom of reducibility couched in a syntax of *r*-type regimented bindable predicate variables— some "predicative" (whose order is that of its simple type) and other non-predicative (whose order is above the order of the simple type). Such a syntax is nowhere to be found in *Principia*'s formal grammar. Ramified types is an invention of Church's genius, not Russell's. It was a theory he invented and proffered as what *Principia*'s formal theory *ought* to have been had it been in alignment with its introduction's discussion of a ramified hierarchy of senses of "truth" and "falsehood." The introduction was not part of *Principia*'s formalism, which was that of simple types and impredicative comprehension axiom schemas. *Principia* reveals that *every* kind of relational structure studied in *any* branch of mathematics can be studied within a simple type scaffolding of universals. Simple type scaffolding imposes no impediments whatsoever for the working mathematician. Church's ramification destroys mathematics altogether.

Practicing mathematicians are agreed on the revolution. Mathematics studies kinds of structure; it has no patience for the metaphysical muddles of abstract particulars. The Logicism of *Principia* and its agenda against abstract particulars in mathematics has been slow to be understood.[7] There are many reasons. But for the present chapter, I should emphasize that metaphysicians of mathematics often object that logic cannot enjoy a privileged status—its ontology and epistemology are, they imagine, compromised by controversial metaphysical commitments of its own.[8] The problem here lies in their conception of logic. For example, Williamson is concerned that various rival logics on the market today are themselves embroiled in metaphysics of abstract particulars (e.g., possible worlds). He says (Williamson, 2013, p. 147):

> readers may prefer to use the word "logic" differently. . . . but whatever advantages may accrue to their way of using "logic," they will not include isolating some claims that are in principle metaphysically uncontroversial. There are none.

Principia does not deserve inclusion in such a sweepingly negative evaluation that Williamson exemplifies. The unacceptable metaphysical claims Russell targeted are only just the claims of metaphysicians that there are special abstract particulars and kinds of necessity governing them. Moreover, *Principia* is not merely itself one among "rival" logics (such as intuitionistic, modal, relevant, paraconsistent, etc.). *Principia* is a synthetic *a priori* science of structure, and it studies every structure by studying the

---

7 Klement (2015) has noticed it. But he imagines that the techniques of *Principia*'s emulation of simple-types of class (and relation-in-extension) can serve as a general technique for eliminating abstract objects in the fields of mathematics. In my view, the technique simply offers a way to build extensional contexts out of intentional contexts. It offers no method whatsoever for undermining the many indispensability arguments from metaphysicians in favor of abstract particulars in the fields of mathematics. For example, it offers no technique for undermining the metaphysician's charge that *Principia* is guilty of If-thenism or the claim that it is arithmetically necessary that there are infinity of natural numbers as abstract particulars. See Section 5 of this chapter.

8 See Haack (1996) for an excellent introduction to "deviant" logics.

way the relations involved order the fields. There are many structures one can study, and the various structures (just as the various geometries) are "rivals" only insofar as they have rival merits and flaws in their applications. All the so-called "rival logics" are to be subsumed into *Principia*'s general study (just as were the rival geometries) as studies of different *invariant truth* in a structure. Relevant entailment systems, alethic modal systems (S4, S5, and so on), and intuitionistic systems are simply relational structures and invariances emerge over such structures. They pose no more challenge to *Principia*'s Logicism than do the non-Euclidean geometries. There are no special rival *metaphysical necessities* than there are special rival *geometric necessities* (Euclidean, etc.). There is only logical necessity (grounded in the facts composed wholly of universals). For example, the fact (which is an abstract particular) that the relation of *one-one-correspondence* has the property of being *transitive* consists only of universals. As with any metaphysics, the Platonic existence of such universals (an existence independent of their exemplification) is no doubt subject to controversy—but it is a controversy that purges abstract particulars from the subject matter of mathematic and cp-Logic. These fields study universals independently of their exemplification. In this way, mathematics and logic remain above the civil wars between metaphysicians of specialized abstract particulars and accompanying specialized kinds of necessity (metaphysical, causal, biological, psychological, etc.).

The ontological ground of *Principia*'s mathematical logic lies in facts composed *solely* of universals, not in any invariance notions of *necessary truth*. Indeed, according to the Russellian view, logical necessity stands apart even from the Tarski formal semantical notion of *logical truth* (*logical necessity*), which is a notion of an invariance on a kind of structure (a language structure). The results of *Principia*'s investigations into kinds of structures are certainly *not* always statements that are true by virtue of their structure. Instances of the axiom schema of impredicative comprehension are, for example, are not true by virtue of their structure. The results are often truths about the existence of relations that determine kinds of relational structure.

As we can see, to understand the logicism of *Principia*, one must understand its unique conception of cp-Logic and the primacy of impredicative comprehension which is the source of its being synthetic *a priori*. Impredicative comprehension is what is essential for the revolutionary view of mathematics as the study of kinds of structures. The unique non-Fregean logicism of *Principia* aims to capture this revolution *within* mathematics that remakes it as a science of relations. In this respect, it is very unfortunate that Frege's logicism has been set up as if it were the paradigm of what a logicism should be. It is not. Metaphysicians wedded to a conception of mathematics that shackles it to a favored metaphysics of abstract particulars are naturally apt to reject the quest for a Fregean logicism in favor of a theory of uniquely mathematical abstract particulars that are numbers. Frege's theory of abstract particulars (value-ranges) is, after all, inconsistent due to a version of Russell's paradox. This attitude is sometimes supported by appeal to the fact that so many working mathematicians came to use ZFC (Zermelo-Frankel set theory with the axiom of choice) as an instrument for their studies. Zermelo believed that abstract particulars are indispensable to mathematical ontology, and he sought to find a theory of sets as *mathematical* abstract particulars. In 1908, his *Aussonderung* axiom paired with other axioms seemed to capture the practical needs of the metaphysicians of mathematics wedded to abstract particulars. The intuitions were at first strange, since *Aussonderung* allows no universal set, and accepting that for any two sets there is a set which is it union, it allows no set to have an absolute complement. Later these strange intuitions about sets came to feel familiar since they stem from the notions of set building by a power-set operation and since they permit Cantor's results. Paradoxes took center stage. Russell's paradoxes of classes, Cantor's paradox of the greatest cardinal, and the Burali-Forti paradox of the greatest ordinal *had* to be solved; and *Aussonderung* seems to avoid them while minimizing technical drudgery. Alas, one must also admit that confusions about semantic paradoxes such as the Berry, Richard, and König/Dixon also played a role—even for some mathematical experts like Poincaré, who, unlike Zermelo, was clouded by his interest in marginalizing Cantor's work. But inasmuch as they are, as Russell and Peano originally pointed out in 1906, couched in confusions of equivocation on notions of "naming," "denoting," "defining" (which are unintelligible without fixing linguistic devices so that we have "naming-in-*L*," "denoting-in-*L*," and "defining-in-*L*"). They ought to have played *no* role at all (see Landini 1998).

In all this dissonance, which is still ongoing today, *Principia*'s cp-Logic and its unique logicist agenda *against* abstract particulars tend to be lost. *Principia* showed how to emulate a language of classes as a notational convenience while accepting the revolutionary view among mathematicians that abstract particulars are wholly irrelevant to any branch of mathematical study. Only kinds of relational structures matter to mathematics. Neither Zermelo-sets nor any other abstract particulars matter to the ontology of mathematics. Thus, when the classification concerns whether a figure is with, or against, the revolution in mathematics, we find that both Frege and Zermelo belong in the camp *against* the revolution. Frege's concern was to *argue* that numbers are logical objects (particulars that are value-ranges) correlated with functions and are thereby recognizable, unlike classes, sets, and the like, as logical abstract particulars. Zermelo's concern was to consistently *accept* that sets are distinctly mathematical and non-logical abstract particulars. The dispute between figures such as Zermelo and Frege is what abstract particulars one ought to identify as mathematical entities. It is, therefore, orthogonal to the interests of Whitehead and Russell, who, in stark contrast, are in the revolutionary camp of those who reject abstract particulars in mathematics altogether.

Whitehead and Russell offered *Principia* as the flagship of the revolution in mathematics. Frege and Zermelo (and their intellectual descendants, Gödel, Quine, Putnam, etc.) are all metaphysicians who thought that abstract particulars are indispensable to mathematics. Hence, they reject the revolution within mathematics that Whitehead and Russell so lauded. Philosophers of mathematics, then as today, must evaluate whether the revolution in mathematics is alive or dead.[9] I think it is very much alive.

3. How to Read *Principia Mathematica*

The formal grammar of *Principia* is just that of simple (impredicative) type theory. The primitive signs of the language of *Principia* are ∨, ~, (,), and ∃. The symbol ∀ for universal quantification is due to Gentzen's work in the 1930s and is not used in *Principia*. A type symbol of the simple type theory is recursively defined (in modern times) as follows:

i. o is a type symbol.
ii.   If $t_1, \ldots, t_n$ are type symbols, then $(t_1, \ldots, t_n)$ is a type symbol.
iii. There are no other type symbols.

The notion of order appropriate to a simple type symbol is useful in comparing rival interpretations. It can be recursively defined as follows:

i.   The type symbol o has order 0.
ii.   A type symbol $(t_1, \ldots, t_n)$ has order $m+1$ if the highest order of any of the type symbols $t_1, \ldots, t_n$ is m.[10]

The *individual* variables of *Principia*'s formal simple type theory are x_1^t, $\ldots, x_n^t$, and among these are the individual variables of the lowest simple type, namely $x_1^o, \ldots, x_n^o$ (informally x^o, y^o, z^o). Predicate variables are those individual variables whose simple type symbol is not o. The only

---

9 Those that think it is dead may have the attitude that what sealed its fate is its failure of finding a consistent system of logic that enables a theorem that there are infinity of logical abstract particulars. In section 5, however, we shall see that such an attitude is as question-begging against the revolution as were the obstinate objections to non-Euclidean geometries. The revolution in mathematics reveals that some of the metaphysician's most cherished intuitions about mathematical necessity are misguided. There is nothing in the revolutionary mathematics of Cantorian finite cardinals (i.e., natural numbers as relational properties of one-to-one correspondence) that assures that there are infinity many—nor indeed that Hume's Principle should always hold.

10 There is a parenthesis counting algorithm for the order of a simple type symbol. Count parentheses from left to right, adding one for each left and subtracting one for each right. The order of the simple type symbols is the highest reached in the counting process. (See Hatcher, 1982, p. 106.)

terms in *Principia* are the individual variables of whatever simple type. Atomic well-formed formulas are of the form:

$$x^{(t_1,\ldots,t_n)} (x_1^{t_1},\ldots,x_n^{t_n}).$$

Where $t \neq 0$, instead of the predicate variables x^t, y^t, z^t, it is convenient to use letters (informally t, t, t, f^t, g^t) as predicate variables. Accordingly, using a predicate variable $\varphi^{(t_1,\ldots,t_n)}$, the previous is this:

$$\varphi^{(t_1,\ldots,t_n)} (x_1^{t_1},\ldots,x_n^{t_n}).$$

The practice of typical ambiguity is to suppress type indices on the variables with conventions of restoration understood. With typical ambiguity in place, however, it is vitally important to distinguish bindable object-language predicate variables from schematic letters. In *Principia*, type ambiguity uses $\varphi!$, $\psi!$, $\chi!$, $f!$, $g!$ for its bindable object-language predicate variables, and they always come with the exclamation (shriek !). When letters occur *without* the shriek, as with φ, ψ, χ, f, g, they are schematic for *wffs* (well-formed formulas).[11] Thus, with typical ambiguity, an atomic *wffs* of some or other simple type is expressed as follows:

$$\varphi!(x_1,\ldots,x_n).$$

This, as we noted, must not be conflated with

$$\varphi(x_1,\ldots,x_n).$$

This is schematic for a *wff* in which that free object-language variables $x_1,\ldots,x_n$ occur. A schematic letter φ for a *wff* obviously cannot be bound in a quantification.

The formulas (*wffs*) are the smallest set K containing all atomic *wffs* and which is such that if φ and ψ are *wffs* in K and x^t is an individual variable free in formula ψ then $\sim(\varphi)$, $(\varphi \vee \psi)$, and $(x^t)(\psi x^t)$ are *wffs*. Where p, q, and r are schematic for quantifier-free formulas and where φ, ψ are schematic for all *wffs*, quantifier free or otherwise, the axiom schema are as follows:

$*1.2\ p \vee p .\supset. p$
$*1.3\ q .\supset. p \vee q$

---

11 Readers should not expect to find consensus on the grammar of *Principia* in the literature. A discussion of Church was well aware that he was offering a non-historical reconstruction of *Principia* as a ramified type theory. This will be addressed anon. I am offering my interpretation of the historical *Principia*.

*1.4 $p \vee q \mathrel{.\supset.} q \vee p$

*1.5 $p \vee (q \vee r) \mathrel{.\supset.} q \vee (p \vee r)$[12]

*1.6 $q \supset r \mathrel{.\supset.} p \vee q \mathrel{.\supset.} p \vee r$

*9.1 $\varphi x^t \supset (\exists x^t)\varphi x^t$

*9.12 $\varphi x^t \vee \varphi y^t \mathrel{.\supset.} (\exists x^t)\varphi x^t$

*12.1n $(\exists f^{(t_1,\ldots,t_n)})(x_1^{t_1}, \ldots, x_n^{t_n})(f^{(t_1,\ldots,t_n)}(x_1^{t_1}, \ldots, x_n^{t_n}) \equiv \varphi(x_1^{t_1},\ldots,x_n^{t_n})),$

where $f^{(t_1,\ldots,t_n)}$ is not free in the *wff* $\varphi(x_1^{t_1}, \ldots, x_n^{t_n})$. Written with the suppression of type-symbols under the convention of typical ambiguity, these are as follows:

*9.1 $\varphi x \supset (\exists x)\varphi x$

*9.12 $\varphi x \vee \varphi y \mathrel{.\supset.} (\exists z)\varphi z$

*12.1n $(\exists f)(x_1, \ldots, x_n)(f!(x_1, \ldots, x_n) \equiv \varphi(x_1,\ldots,x_n)),$

where $f!$ is not free in the *wff* φ. *Principia* writes the following instances of *12.1n

*12.1 $(\exists f)(x)(f!x \equiv_x \varphi x),$

where $f!$ is not free in the *wff* φx.

*12.11 $(\exists f)(f!(x, y) \equiv_{x,y} \varphi(x, y)),$

where $f!$ is not free in the *wff* $\varphi(x,y)$. There is no doubt that the comprehension axiom schemas *12.1.11 and so on are impredicative, since they impose no restrictions whatsoever on which *wffs* are instance of the schematic φx and φxy occurring in them. Note that upon restoration of type-indices under the conventions of typical ambiguity, "φ" would not get a simple type index because it is a schema for a *wff*.

   Where φ and ψ are any *wffs*, quantifier free or otherwise, the inference rules are just the following three:

*1.1 *Modus Ponens*: From φ and $\varphi \supset \psi$, infer ψ

*10.1 *Universal Generalization*: From φx^t, infer $(x^t)\varphi x^t$

Switch: From $(x^{t_1})(\exists y^{t_2})\varphi(x^{t_1}, y^{t_2})$ infer $(\exists y^{t_2})(x^{t_1})\varphi(x^{t_1}, y^{t_2}),$

where there is a logical particle in the *wff* φ on one side of which all free occurrences of x^{t_1} occur and on the other side of which all free occurrences of y^{t_2} occur.[13]

---

12 This was shown to be redundant.

13 This rule is implicit. We know of it by noting that it is explicit in the system of *8 of *Principia*'s 1925 second edition, a system which was offered as a replacement for *9.

Definitions in *Principia* are always stipulative. They introduce notional conveniences which may be everywhere replaced by appeal simply to definition. They include the following:

*1.01 $p \supset q = df \sim p \lor q$
*2.01 $p \bullet q = df \sim(\sim p \lor \sim q)$
*4.01 $p \equiv q = df (p \supset q) \bullet (q \supset p)$

where p and q are any *wffs*, quantifier free or otherwise. I use dots symmetrically for convenience, with the greater number of dots indicating the main connective.[14] In addition, the quantification theory employs the following definitions of Section *9. For the present, I will continue to write out type indices for clarity. Where p is quantifier free and not containing x^t, *Principia*'s definitions include the following:

*9.01 $\sim(x^t)\varphi x^t = df (\exists x^t) \sim \varphi x^t$
*9.02 $\sim(\exists x^t)\varphi x^t = df (x^t) \sim \varphi x^t$

Where x^t does not occur free in the quantifier-free formula p, *Principia* has:

*9.03 $(x^t)\varphi x^t \lor p = df (x^t)(\varphi x^t \lor p)$
*9.04 $p \lor (x^t) \varphi x^t = df (x^t)(p \lor \varphi x^t)$
*9.05 $(\exists x^t) \varphi x^t \lor p = df (\exists x^t)(\varphi^{x^t} \lor p)$
*9.06 $p \lor (\exists x^t) \varphi x^t = df (\exists x^t)(p \lor \varphi x^t)$

Where the *wff* φx^{t_1} does not contain the variable y^{t_2} free and the *wff* ψy^{t_2} does not contain x^{t_1} free, *Principia* has the following definitions for pulling quantifiers to an initial position in pairs:

*9.07 $(x^{t_1})\varphi x^{t_1} \lor (\exists y^{t_2})\psi y^{t_2} = df (x^{t_1}) (\exists y^{t_2})(\varphi x^{t_1} \lor \psi y^{t_2})$
*9.08 $(\exists x^{t_1}) \varphi x^{t_1} \lor (y^{t_2})\psi y^{t_2} = df (\exists x^{t_1}) (y^{t_2})(\varphi x^{t_1} \lor \psi y^{t_2})$
*9.0x $(x^{t_1})\varphi x^{t_1} \lor (y^{t_2})\psi y^{t_2} = df (x^{t_1}) (y^{t_2})(\varphi x^{t_1} \lor \psi y^{t_2})$
*9.0y $(\exists x^{t_1})\varphi x^{t_1} \lor (\exists y^{t_2})\psi y^{t_2} = df (\exists x^{t_1}) (\exists y^{t_2})(\varphi x^{t_1} \lor \psi y^{t_2})$.

The omission of *9.0x and *9.0y seems to be an oversight. This completes the system. The axiom schema of the convenient quantification theory of Section *10 are derived. They are:

*10.1 $(x^t)\varphi x^t \supset \varphi y^t$,

---

14 *Principia* used dots asymmetrically when viable, and it ordered the conjunction sign "*p.q*" as having wider force over all other of the previous logical particles. This often enabled it to avoid using dots in connection with the conjunction sign $\cdot$, which was written as small as a punctuation dot.

where y^t is free for x^t in the *wff* φ.

*10.1 $(x^t)(\varphi x^t \vee p) .\supset. (x^t)\varphi x^t \vee p,$

where x^t is not free in p.[15] As we can see, it is of utmost importance to understand that the letters φ and ψ without the exclamation are schematic for *wffs*. They are not bindable predicate variables.

Principia sets out its grammar and its axiom schemas without writing in its simple type indices. It uses the technique of typical ambiguity. This naturally leads to opportunities for misunderstanding its grammar. This must be kept in mind in reading Section *9, where we find the following:

Primitive Idea: Individual. We say x is "individual" if x is neither a proposition nor a function. The passage is concerned with grammar, but it is stated under the convention of typical ambiguity. The intent is that *individual* variables "x," "y," and so on will never appear in predicate positions as do *predicate variables* φ!, ψ!, and so on. And they are distinct from schematic letters φ, ψ, p, q, and so on that stand in for *wffs*. Of course, once one removed typical ambiguity and assigned type indices to the variables, the individual variables whose simple type is not o can indeed occur in predicate positions. But the point is being made with the technique of typical ambiguity in mind. Similarly, at Section *12, we are introduced to the grammar of bindable predicate variables. We find:

> "Matrix" or "predicative function" is a primitive idea.
> The fact that a function is predicative [i.e., the fact a sign is a bindable predicate variable] is indicated . . . by a note of exclamation after the functional letter.

This tells us that the bindable predicate variables, of whatever simple type, are recognizable by the exclamation—for example, φ!, ψ!, and so on. Whitehead and Russell write "the only functions [predicate variables] which will be so used [i.e., used as bindable variables] will be predicative functions" (p. 165). In short, the formal grammar of *Principia* is the grammar of simple (impredicative) type theory. There is no way to confuse individual variables that serve as object language bindable predicate variables with individual variables of the lowest simple type, which do not.

---

15 The formal proof that every instance of these schemas are derivable from *9, were it to have been rigorously conducted, would have required meta-linguistic induction on the length of a *wff*.

The following definitions are of interest, especially the last, which stipulates notation conveniences for the use of definite descriptions as quantifiers:

*13.01 $x = y = df (\varphi)(\varphi!x \supset \varphi!y)$.
*13.02 $x \neq y = df \sim(x = y)$
*14.01 $[\iota x \varphi x][\psi(\iota x \varphi x)] = df (\exists x)(\varphi y \equiv_y y = x. \bullet. \psi x)$.

Though *Principia* does not adopt it, the following is also quite convenient:

**14.01 $[\iota x \varphi x][\psi x] = df (\exists x)(\varphi y \equiv_y y = x. \bullet. \psi x)$.

The benefit of stipulative definition *14.01 is that it enables Russell to drop the scope marker $[\iota x \varphi x][. . .]$ in favor of $\psi(\iota x \varphi x)$ under the strict convention that the scope intended is the smallest possible in the context in question. It should be noted as well that one must use *14.01 (or **14.01) before using *13.01 and *13.02. (The mark ** indicates this is a correction or addition to *Principia*.) A definition framed with the genuine terms that are individual variables cannot apply to expressions of definite descriptions which are not genuine terms at all. Thus, one cannot apply *13.02 to a clause such as "$\iota x \varphi x \neq y$." A convention is adopted on dropping scope markers for smallest scope which assures that "$\sim(\iota x \varphi x = y)$" means "$\sim[\iota x \varphi x][\iota x \varphi x = y]$." Thus one must not think that "$\iota x \varphi x \neq y$" and "$\sim(\iota x \varphi x = y)$" are equivalent, unless of course E!$(\iota x \varphi x)$.[16]

The difference between a schema and an object language bindable predicate variable is important to the contextual definition *14.01 as well. When, in dropping a scope marker and writing $\psi(\iota x \varphi x)$, what is intended is some *wff* ψ in which the scope marker is to be restored in the smallest scope possible. For example, an instance of the schema $\psi(\iota x \varphi x)$ is

$f !(\iota x \varphi x) \supset p$.

Since $f!$ is a genuine (object language bindable) predicate variable, this *has* to be interpreted as

$[\iota x \varphi x][f!(\iota x \varphi x)] \supset p$.

That renders the smallest scope. Notice that we also have the instance

$f (\iota x \varphi x) \supset p$.

---

16 The difference plays a role at *Principia* *96.48.

The scope here cannot be determined because f is schematic for a *wff* and we don't know which.

Beware of the special temporary convention that is briefly used in Section *14 and not used in the work in general—the temporary convention to help the reader understand scope was that it meant a primary scope in writing $\psi(\iota x \varphi x)$ and a secondary scope when in writing $f\{\psi(\iota x \varphi x)\}$. Unfortunately, this temporary convention is incompatible with the formal convention on dropping scope markers. For example:

   *14.21 $\vdash \psi(\iota x \varphi x) \supset E!(\iota x \varphi x)$.

This is not a proper theorem schema because it uses the temporary convention.[17] Taken without the temporary convention, it has instances that are quite obviously false. For example, there is this:

   $\sim(\exists y)(y = \iota x \varphi x) \supset E!(\iota x \varphi x)$.

The following are proper theorems:

   **14.21 $\vdash [\iota x \varphi x][\psi(\iota x \varphi x)] \supset E!(\iota x \varphi x)$
   !*14.21 $\vdash \psi!(\iota x \varphi x) \supset E!(\iota x \varphi x)$.

The latter works because $\psi!$ is an object-language bindable predicate variable. Note that the following are also provable:

   $\iota x \varphi x = \iota x \varphi x \supset E!(\iota x \varphi x)$
   $\iota x \varphi x \neq \iota x \varphi x \supset E!(\iota x \varphi x)$.

That is because the *smallest* scopes for the antecedents (respectively) yield these:

   $[\iota x \varphi x][\iota x \varphi x = \iota x \varphi x] \supset E!(\iota x \varphi x)$.
   $[\iota x \varphi x][\iota x \varphi x \neq \iota x \varphi x] \supset E!(\iota x \varphi x)$.

When the scope marker is dropped and a definite description expression repeats itself in a *wff*, they are connected together, as you see above. To indicate a different scope, one can simply change the individual variable used, as in $\psi!(\iota x \varphi x, \iota y \varphi y)$ as opposed to $\psi!(\iota x \varphi x, \iota x \varphi x)$.

   The case of *14.21 has to be corrected. But contrast the important case of *14.18, which is quite proper as it stands; we find:

   *14.18 $\vdash E!(\iota x \varphi x) .\supset. (x)\psi x \supset \psi(\iota x \varphi x)$.

---

17 See Landini (2013).

In the case of *14.18, we find that *Principia's* formal convention makes
ψ(ιxφx) secondary scope. But all is well. Consider, for example, the following case:

$$\vdash E!(ιxφx) .\supset. (x)(\exists y)(y = x) \supset (\exists y)([ιxφx][y = ιxφx]).$$

In the proof of *14.18, however, the scope taken is primary. That proof
is of these:

!*14.18 $\vdash E!(ιxφx) .\supset. (x)ψ!x \supset ψ!(ιxφx).$
ι*14.18 $\vdash E!(ιxφx) .\supset. (x)ψ!x \supset [ιxφx][ψ!(ιxφx)].$

The proper proof of *14.18 which allows secondary scopes would require
an induction in the metalanguage on the length of a *wff*. The base case of
that induction is the proof of !*14.18. The secondary scope is needed to
justify the following comments (*PM*, vol. 1, p. 180):

> The above proposition shows that, provided ιxφx exists, it has (speak-
> ing formally) all the logical properties of symbols which directly rep-
> resent objects. Hence when ιxφx exists, the fact that it is an incomplete
> symbol becomes irrelevant to the truth-values of logical propositions
> in which it occurs.

Whitehead and Russell are justified in this comment.

When it comes to definite descriptions, Whitehead and Russell make
the further comment at the end Section *14 that is also worth quoting
indicating that a primary scope always logically entails the secondary.
The write (p. 186):

> It should be observed that the proposition in which ιxφx has the
> larger scope always implies the corresponding one in which it has the
> smaller scope.

The general statement of the equivalence of primary and secondary scope
is this:

*14.3 $\vdash (p \equiv q ._{p,q}. fp \equiv fq).\bullet. E!(ιxφx) :\supset:$
$f\{[ιxφx][χ(ιxφx)]\} \equiv [ιxφx][f\{χ(ιxφx)\}],$

But before citing this "theorem," Whitehead and Russell write:

> In this proposition, however, the use of propositions as apparent
> variable involves an apparatus not required elsewhere, and we have
> therefore not used this proposition in subsequent proofs.

Then, afterward, they continue to apologize,

> *14.3 introduces propositions (*p*, *q*, namely) as apparent variables, which we have not done elsewhere, and cannot do legitimately without the explicit introduction of the hierarchy of propositions with a reducibility-axiom such as *12.1.

This comment, is accurate, excepting the clause "without explicit introduction." That clause is likely a vestige (again) of Russell's earlier 1903–1908 conception of logic as a theory of entities he called *propositions* and that required $x \supset y$ as a *wff* and "$\supset$" as a relation sign. (We noticed similar vestiges in Section *2.) What is proper is this:

**14.3　$E!(\iota x \varphi x) . \supset . f\{[\iota x \varphi x][\chi(\iota x \varphi x)]\} \equiv [\iota x \varphi x][f\{\chi(\iota x \varphi x)\}],$

where f is a truth-functional context. The notion of a context f being *truth-functional* has to be expressed in the meta-language. It is this:

$$p \equiv q . \supset . fp \equiv fq,$$

for all *wffs* p and q. Whitehead and Russell admit this, in effect. We shall soon see that there is a rather more austere analog for class expressions.

For expressions of classes of *individuals* of some or other simple type, *Principia* has the following definitions. I have restored the scope marker $[\hat{x}(\psi x)][\ldots \hat{x}(\psi x) \ldots]$ to *20.01 as is warranted by the comments of *Principia*'s introduction (p. 80). It must be understood that occurrences of $\hat{x}(\psi x)$ without the determining scope marker are to be taken in *smallest scope*. Moreover, class expressions never have simple type indices since their expressions are *façon de parler*. There are no classes. To emphasize this, I shall restore some simple type indices. In *Principia*, we find:

*20.01　$[\hat{x}(\psi x)][f\{\hat{x}(\psi x)\}] = df (\exists \varphi!)(\varphi! x \equiv_x \psi x . \bullet . f(\varphi!))$
*20.02　$x \in \varphi! = df \varphi! x$

If we restore simple type indices to the individual variables besides $\varphi!$, these are (respectively):

$[\hat{x}^t(\psi x^t)][f\{\hat{x}^t(\psi x^t)\}] = df (\exists \varphi!)(\varphi! x^t \equiv_{x^t} \psi x^t . \bullet . f(\varphi!))$
$x^t \in \varphi! = df \varphi! x^t$

Of course, a full restoration of simple type indices would replace $\varphi!$ with $\varphi^{(t)}$ and, in turn, replace that with an *individual* variable of appropriate type, say, $y^{(t)}$. This yields:

$[\hat{x}^t(\psi x^t)][f\{\hat{x}^t(\psi x^t)\}] = df (\exists y^{(t)})(y^{(t)}(x^t) \equiv_{x^t} \psi x^t . \bullet . f(y^{(t)}))$
$x^t \in y^{(t)} = df y^{(t)}(x^t)$

The *definiendum* of the last looks a bit like a type theory of classes as genuine entities that are abstract particulars. But it is not. No *individuals* in *Principia* are classes. There are no classes. Now, when free lowercase Greek α, β, μ, and so on occur *free*, they stand in for the expressions $\hat{z}(\psi z)$ or for $\hat{a}\psi a$. Hence,

*20.06 $x \notin \alpha = df \sim (x \in \alpha)$
i.e., $x \notin \hat{z}(\psi z) = df \sim (x \in \hat{z}(\psi z))$
i.e., $x^t \notin \hat{z}^t(\psi z^t) = df \sim (x^t \in \hat{z}^t(\psi z^t))$

In the previous, we know to use $\hat{z}(\psi z)$ rather than $\hat{a}\psi a$ because of the individual variables. Bound lowercase Greek is defined in *Principia* by the following:

*20.07 $(\alpha)f\alpha = df (\varphi) f\{\hat{x}(\varphi!x)$
*20.071 $(\exists a)fa = df (\exists \varphi) f\{\hat{x}(\varphi!x)\}.$

Restoring simple type indices on some individual variables, we get the following:

$(\alpha)fa = df (\varphi) f\{\hat{x}^t(\varphi!x^t)\}$
$(\exists a)fa = df (\exists \varphi) f\{\hat{x}^t(\varphi!x^t)\}.$

But, of course, none of these definitions apply generally to classes of classes (of some or other relative type). The definitions for bound lowercase Greek need to be repeated:

c*20.07 $(\alpha)fa = df (\varphi) f\{\hat{a}(\varphi!a)$
c*20.01 $(\exists a)fa = df (\exists \varphi) f\{\hat{a}(\varphi!a)\}.$

The decision as to which of the definitions to use is contextual. *Principia*'s definitions for classes of classes of some or other relative type are the following:

*20.072 $[\iota a\varphi a][\psi a] = df (\exists a)(\varphi \beta \equiv_\beta \beta = a.\bullet. \psi a)$
*20.08 $[\hat{a}\psi a][.f (\hat{a}\psi a)] = df (\exists \varphi)(\varphi!a \equiv \hat{a}\psi a.\bullet. f(\varphi!))$
*20.081 $a \in \psi! = df \psi!a$

These definitions do not apply to classes of *individuals* of any simple type.
    It is of utmost importance to note the use of schemata in the definitions and the convention that omission of the scope marker for class expressions indicates that the smallest (most secondary) scope is intended. One cannot (as Church's influential reading does) properly read *f* in *20.07 as a predicate variable (i.e., a non-predicative bindable object language predicate variable). It would destroy the no-classes theory. Church's

reading would force a primary scope, and yet the intent is that the scope be secondary. A primary scope in *20.07 is this:

$$(\alpha)f\alpha = df (\varphi)[\hat{x}\varphi!x][f(\hat{x}\varphi!x)].$$

But the definitions won't work properly unless the scope is secondary. For example, consider

$$(\alpha)(\varphi!\alpha \equiv \psi\alpha).$$

Since the scope marker is absent, the intended scope on both sides is the most secondary possible. The only scope possible for $\varphi!(\hat{x}g!x)$ is the primary scope $[\hat{x}g!x][\varphi!(\hat{x}g!x)]$ since $\varphi!$ is a bindable predicate variable. The scope of $\psi(\hat{x}g!x)$ is unknown until after the schema ψ is assigned a *wff*. The proper application of *20.07 is therefore this:

$$(g)([\hat{x}g!x][\varphi!(\hat{x}g!x)] \equiv \psi(\hat{x}g!x)).$$

This secondary scope reading is essential to the viability of theorem c*20.3 for classes of classes. This reveals that Church's interpretation of *Principia* is mistaken. Church viewed the letters φ, ψ, χ, f, and g without the exclamation as genuine non-predicative predicate variables of an *r*-type (ramified-type) grammar. Church's account would mistakenly take *20.08 as if it intended the following scope:

$$(\alpha)(\varphi!\alpha \equiv \psi\alpha) = df (g)([\hat{x}g!x][(\varphi!(\hat{x}g!x) \equiv \psi(\hat{x}g!x)].$$

This undermines the no-classes theory.

The fundamental theorem of Section *20 is the following, which is the simple type replacement for what has come to be called "naïve abstraction." Note that the replacement does not render types of classes! Again, there are no classes. The no-classes theory, couched in the theory of simple types, is what saves it from the Russell contradiction. To illustrate the role of the comprehension axiom *12.1, it is useful to look at a proof.

*20.3 ⊢ $x \in \hat{z}\psi z \equiv \psi x$
i.e., ⊢ $(\exists\varphi!)(\varphi!z \equiv_z \psi z .\bullet. \varphi!x) \equiv \psi x$

1. ⊢ $\varphi!z \equiv_z \psi z .\bullet. \varphi!x :\supset: \psi x$ *10
2. ⊢ $(\varphi)(\varphi!z \equiv_z \psi z .\bullet. \varphi!x :\supset: \psi x)$ 1, ug
3. ⊢ $(\exists\varphi)(\varphi!z \equiv_z \psi z) .\bullet. \varphi!x) :\supset: \psi x$ 2, dR(*10.23)
4. ⊢ $x \in \hat{z}(\psi z) \supset \psi x$ 3, *df* (*20.01)
5. ⊢ $\psi x .\bullet. \varphi!z \equiv_z \psi z :\supset: \varphi!z \equiv_z \psi z .\bullet. \varphi!x$ *10
6. ⊢$(\varphi)(\psi x .\bullet. \varphi!z \equiv_z \psi z :\supset: \varphi!z \equiv_z \psi z .\bullet. \varphi!x)$ 5, ug.
7. ⊢$(\exists\varphi)(\psi x .\bullet. \varphi!z \equiv_z \psi z) :\supset: (\exists\varphi)(\varphi!z \equiv_z \psi z .\bullet. \varphi!x)$ 6, dR(*10.28)

8. ⊢ ψx • (∃φ)(φ!z ≡_z ψz) :⊃: (∃φ)(φ!z ≡_z ψz .•. φ!x) 7, dR(*10)
9. ⊢ (∃φ)(φ!z ≡_z ψz) *12.1
10. ⊢ ψx .⊃. ψx • (∃φ)(φ!z ≡_z ψz) 9, dR(*3)
11. ⊢ ψx ⊃ (∃φ)(φ!z ≡_z ψz .•. φ!x) 8, 10, dR(*syll*).
12. ⊢ ψx ⊃ x ∈ ẑψz 11, df(*20.01)
13. ⊢ x ∈ ẑφz ≡ ψx 4, 12, dR(*3.2, *5.1)

This is for classes of individuals (of any simple type). *Principia* does not
carry out the analogous proof for classes of classes. But it is worth doing
to show the role of *12.1 in that proof as well. Using the lowercase Greek
letter for classes, the proof is largely in parallel until it comes to the line
for #9. We have:

c*20.3 ⊢ β ∈ α̂ψα ≡ ψβ
i.e., ⊢ (∃φ)(φ! ≡ ψα .•. φ!β) ≡ ψβ

1. ⊢ ! ≡ ψα .•. φ!β :⊃: ψβ *10
2. ⊢ (φ)(φ! ≡ ψα .•. φ!β :⊃: ψβ) 1, ug
3. ⊢ (∃φ)(φ! ≡ ψα) .•. φ!β) :⊃: ψβ 2, dR(*10.23)
4. ⊢ β ∈ α̂(ψα) ⊃ ψβ 3, *df* (*20.01)
5. ⊢ ψβ .•. φ! ≡ ψα :⊃: φ! ≡ ψα .•. φ!β *10
6. ⊢(φ)(ψβ .•. φ! ≡ ψα :⊃: φ! ≡ ψα .•. φ!β), 5, ug.
7. ⊢(∃φ)(ψβ .•. φ! ≡ ψα) :⊃: (∃φ)(φ! ≡ ψα .•. φ!β) 6, dR(*10.28)
8. ⊢ ψβ • (∃φ)(φ! ≡ ψα) :⊃: (∃φ)(φ! ≡ ψα .•. φ!β) 7, dR(*10)
9. ⊢ (∃φ)(φ! ≡ ψα) *12.1
10. ⊢ψβ .⊃. ψβ • (∃φ)(φ! ≡ ya) 9, dR(*3)
11. ⊢ ψβ ⊃ (∃φ)(φ! ≡ ψα .•. φ!β) 8, 10, dR(*syll*).
12. ⊢ ψβ ⊃ β ∈ α̂ψα 11, *df*(*20.01)
13. ⊢ β ∈ α̂ψα ≡ ψβ 4, 12, dR(*3.2, *5.1)

As we can see, line #9 has to be proved from *12.1. The needed proof, for
classes of classes of individuals (of any simple type), is given in *Principia* at
*20.112. The proof doesn't involve itself with *20.08, but the application
of that definition to the result is clear. We find:

*20.112 ⊢ (∃g)(g!(ẑφ!z) ≡ ψ(ẑφ!z)).

What is wanted, however, is the more general case for classes of classes
of any relative type:

cls*20.112 ⊢ (∃g)(g!α ≡ ψα).

1.   ⊢ (∃g)(g!(φ!) ≡ ψ(φ!)) *12.1

    i.e., (∃g^{(t)})(g^{(t)})(φ^t) ≡ ψ(φ^t)

2.   ⊢ $g!(\varphi!) \equiv [\hat{a}\varphi!a][g!(\hat{a}\varphi!a)]$
3.   ⊢ $\psi(\varphi!) \equiv \psi(\hat{a}\varphi!a)$
4.   ⊢ $(\exists g)([\hat{a}\varphi!a][g!(\hat{a}\varphi!a)] \equiv [\hat{a}\varphi!a][\psi(\hat{a}\varphi!a)])$ 1, 2, 3, dR(*10.)

One may have an initial worry about line #3 above. The worry is that $\psi(\hat{a}\varphi!a)$ has a secondary scope, depending on what *wff* ψ is. *Principia*'s proof assumes a primary scope. The inference is sound, but the proof of this line as a theorem has to be conducted as a metalinguistic induction on the length of a *wff*.

The same point holds for many theorems in Section *20. What is found there takes only the primary scope and thus the base of the needed metalinguistic proof by strong induction on the length of a *wff*. In particular,

> ⊢ $(\alpha)f\alpha \supset f\beta$.
> *20.61 ⊢ $(\alpha)f\alpha \supset f(\hat{x}\Gamma x)$
> cls*20.61 ⊢ $(\alpha)f\alpha \supset f(\hat{a}\Gamma\sigma)$.

Consider, for example, the base case of such a strong induction:

> ⊢ $(\alpha)f!\alpha \supset f!(\hat{x}\Gamma x)$
> i.e., ⊢ $(\varphi)f!(\hat{x}\varphi!x) \supset [\hat{x}\Gamma x][f!(\hat{x}\Gamma x)]$

1.   ⊢ $\varphi!z \equiv_z \Gamma z .\supset. (\exists\psi)(\varphi!z \equiv_z \varphi!z .\bullet. f!(\psi!)) \supset (\exists\psi)(\varphi!z \equiv_z \Gamma z .\bullet. f!(\psi!))$
2.   ⊢ $\varphi!z \equiv_z \Gamma z .\supset. f!(\hat{x}\varphi!x) \supset [\hat{x}\Gamma x][f!(\hat{x}\Gamma x)]$ 1, df(*20.01)
3.   ⊢ $(\varphi)(\varphi!z \equiv_z \Gamma z .\supset. f!(\hat{x}\varphi!x) \supset [\hat{x}\Gamma x][f!(\hat{x}\Gamma x)])$ 2, ug
4.   ⊢ $(\exists\varphi)(\varphi!z \equiv_z \Gamma z) .\supset. f!(\hat{x}\varphi!x) \supset [\hat{x}\Gamma x][f!(\hat{x}\Gamma x)]$ 3, dR(*10.23)
5.   ⊢ $(\exists\varphi)(\varphi!z \equiv_z \Gamma z)$ *12.1
6.   ⊢ $f!(\hat{x}\varphi!x) \supset [\hat{x}\Gamma x][f!(\hat{x}\Gamma x)]$ 4, 5, mp
7.   ⊢ $(\varphi)f!(\hat{x}\varphi!x) \supset f!(\hat{x}\varphi!x)$ *10.1
8.   ⊢ $(\varphi)f!(\hat{x}\varphi!x) \supset [\hat{x}\Gamma x][f!(\hat{x}\Gamma x)]$ 6, 7, dR($syll$)
9.   ⊢ $(\alpha)f!\alpha \supset f!(\hat{x}\Gamma x)$ 8, *20.07

Now compare a secondary scope case such as this:

> ⊢ $(\alpha)\sim f!\alpha \supset \sim f!(\hat{x}\Gamma x)$
> i.e., ⊢ $(\varphi)\sim f!(\hat{x}\varphi!x) \supset \sim [\hat{x}\Gamma x][f!(\hat{x}\Gamma x)]$

1.   ⊢ $\varphi!z \equiv_z \Gamma z .\supset. (\exists\varphi)(\varphi!z \equiv_z \varphi!z .\bullet. f!(\psi!)) \equiv (\exists\psi)(\varphi!z \equiv_z \Gamma z .\bullet. f!(\psi!))$
2.   ⊢ $\varphi!z \equiv_z \Gamma z .\supset. f!(\hat{x}\varphi!x) \equiv [\hat{x}\Gamma x][f!(\hat{x}\Gamma x)]$ 2, df(*20.01)
3.   ⊢ $(\varphi)(\varphi!z \equiv_z \Gamma z .\supset. f!(\hat{x}\varphi!x) \equiv [\hat{x}\Gamma x][f!(\hat{x}\Gamma x)])$ 2, ug
4.   ⊢ $(\exists\varphi)(\varphi!z \equiv_z \Gamma z) .\supset. f!(\hat{x}\varphi!x) \equiv [\hat{x}\Gamma x][f!(\hat{x}\Gamma x)]$ dR(*10.23)
5.   ⊢ $(\exists\varphi)(\varphi!z \equiv_z \Gamma z)$ *12.1
6.   ⊢ $f!(\hat{x}\varphi!x) \equiv [\hat{x}\Gamma x][f!(\hat{x}\Gamma x)]$ 4, 5, mp
7.   ⊢ $\sim f!(\hat{x}\varphi!x) \supset \sim[\hat{x}\Gamma x][f!(\hat{x}\Gamma x)]$ 6, dR(*4.11)

8. ⊢ (φ)~f!(x̂φ!x) ⊃ ~f!(x̂φ!x) *10.1
9. ⊢ (φ)~f!(x̂φ!x) ⊃ ~[x̂Γx][f!(x̂Γx)] 7, 8, dR(*syll*)
10. ⊢ (α)~f!α ⊃ ~f!(x̂Γx) 9, *20.07

As we can see, a secondary scope poses no difficulty. Only base case scopes are proved in Section *20 of *Principia*, but secondary scopes are justified by metalinguistic strong induction on the length of a *wff*.

The general principle governing the logical equivalence of primary and secondary scopes of a definite description for a class is the same as it is at **14.3. Thus, we have the expected version

Clsdesc *14.3 E!(ιαφα) .⊃. f{[ιαφα][χ(ιαφα)]} ≡ [ιxφx][f{χ(ιαφα)}],

where *f* is a truth-functional context. However, it is quite important to note that when it comes to class expressions, the following is *not* a proper analog:

E!(x̂φx) .⊃. f{[x̂φx][χ(x̂φx)]} ≡ [ιxφx][f{χ(x̂φx)}],

where *f* is a truth-functional context. Of course, in *Principia*, there is no expression E!(x̂φx). But one could imagine a definition

E!(x̂φx) = *df* (∃g!)(g!x ≡ₓ φx).

In any case, *12.1 would ensure such an antecedent always holds. Nonetheless, it would be quite mistaken to imagine that the following holds:

f{[x̂φx][χ(x̂φx)]} ≡ [ιxφx][f{χ(x̂φx)}],

where *f* is a truth-functional context. The proper analog is more austere. It is this:

cls *14.3 f{[x̂φx][χ(x̂φx)]} ≡ [x̂φx][f{χ(x̂φx)}],

where *f* is a truth-functional context and χ is an extensional context. A context χ is an extensional context just in case we have:

(φ, ψ)(φ!x ≡ₓ ψ!x .⊃. χ(φ!) ≡ χ(ψ!)).

Consider the following case of identity, which is not an extensional context. We *don't* have:

(φ, ψ)(φ!x ≡ₓ ψ!x :⊃: φ! = Γ!. ≡. ψ! = Γ!).

Now by *12.1, we have $(\exists g!)(g!x \equiv_x \varphi x)$, and yet the following are not logically equivalent

$[\hat{x}\varphi x][\sim(\hat{x}\varphi x = \Gamma!)]$
i.e., $(\exists g!)(g!x \equiv_x \varphi x \ .\bullet.\ \sim(g! = \Gamma!))$

$[\hat{x}\varphi x][\sim(\hat{x}\varphi x = \Gamma!)].$
i.e., $\sim(\exists g!)(g!x \equiv_x \varphi x \ .\bullet.\ g! = \Gamma!)).$

Thus, the clause requiring extensionality is important.

Just as *Principia* is a no-classes theory, it is also a no-relation-e theory. Expressions for relations-e (relations in extension) are *façon de parler*, just as assuredly as are class expressions. Relations-e are not construed as classes of ordered *n*-tuples. Moreover, relations-e notations are defined in terms of relation (in intension) notations, which are primitive expressions such as $\varphi!xy$, $\varphi!xyz$, and so on. [Alternatively, $\varphi!(x, y)$ $\varphi!(x, y, z)$ etc.] A relation may well be non-homogeneous, holding between different simple types of *individuals*. So also a relation-e such as in $\alpha R\beta$ may be non-homogeneous, holding between a class α of one relative type and a class β of a higher or lower relative type. A relation-e such as in αTR, is between a class α and a relations-e such as R. or between individuals (of some or other simple type) and classes or relation-e. Moreover, there are classes of relations-e as well. Accordingly, many different definitions are needed. *Principia* only gives a few that are frequently used.

The (dyadic) relations-e have contextual definitions which also distinguish relation-e of individuals (of whatever simple type) from contextual definitions of (dyadic) relations-e of relation-e. The contextual definitions parallel those of classes:

*21.01 $[\hat{x}\hat{y}(\psi xy)][f\{\hat{x}\hat{y}(\psi xy)\}] = df\ (\exists\varphi)(\varphi!xy \equiv_{x,y} \psi xy \ .\bullet.\ f(\varphi!))$
*21.02 $x\{\varphi!\}y = df\ \varphi!xy$

If we restore the simple type indices on individual variables besides $\varphi!$, these are, (respectively):

$[\hat{x}^{t_1}\hat{y}^{t_2}\ (\psi(x^{t_1}, y^{t_1}))][f\{x^{t_1}y^{t_2}\ (\psi(x^{t_1}, y^{t_1})\}] = df$
$(\exists\varphi)(\varphi!(x^{t_1}, y^{t_1}) \equiv_{x^{t_1},y^{t_2}}\psi(x^{t_1}, y^{t_1}) \ .\ .\bullet.\ f(\varphi!))$
$x^{t_1}\{\varphi!\}\ y^{t_2} = df\ \varphi!(x^{t_1}, y^{t_1}).$

Had a membership sign been used, *21.02, which is the analog of *20.02, would have looked like this:

$<x^{t_1}, y^{t_2}> \in \varphi! = df\ \varphi!(x^{t_1}, y^{t_2})$

There is an important letter of 27 April 1905 from Whitehead to Russell where he praises Veblen's work and goes on to raise the concern about the limitations, as is found at *21.02 of writing the variables on either side for relations-in-extension sign R:

> He proves that Descriptive Geometry is the study of the properties of a *single three-term* relation, and the points are the field of this relation. Of course he does not quite know that this is his point of view, but it is the gist of it, and it throws a flood of light on the whole subject. Now this advance makes it urgent that we produce a notation suitable for three-term relations. In fact since four-term relations occur (harmonic relations etc.) we want a notation suitable for relations with any number of terms. . . . I should propose to keep xRy as a simplification in this instance of the general form, *but otherwise* use the new symbolism?

Of course, *Principia* has $\varphi!(x_1,...,x_n)$ for relations (in intension) of any *adicity* whatsoever. But Whitehead is explaining the limitations of writing xRy and $x\{\hat{x}\hat{y}\varphi!(x,y)\}y$. In contrast, writing $<x, y> \in R$ and $<x, y> \in \hat{x}\hat{y}\varphi!(x, y)\}$ sets the stage for relations-in-extension expressions of any number of places. It allows expression such as $<x, y, z> \in R$ and also $<x, y, z> \in \hat{x}\hat{y}\hat{z}\varphi!(x, y, z)$. *Principia* writes the typically ambiguous expressions "xRy" and also "$\alpha R\beta$," which indicates that "R" is a relation-e sign and is not a genuine dyadic predicate variable such as is found in the expressions "$\varphi!xy$" and "$\varphi!\alpha\beta$". Some of the needed definitions are these:

*21.07 $(R)fR = df (\varphi^{(t_1, t_2)}) f \{\hat{x}^{t_1}\hat{y}^{t_2}(\varphi^{(t_1, t_2)}(x^{t_1}, y^{t_1})\}$

*21.071 $(\exists R)fR = df (\exists\varphi^{(t_1, t_2)}) f \{\hat{x}^{t_1}\hat{y}^{t_2}(\varphi^{(t_1, t_2)}(x^{t_1}, y^{t_1})\}$

*2.072 $[\iota R\varphi R][\psi(\iota R\varphi R)] = df (\exists R)(\varphi S \equiv_s S = R .\bullet. \psi R)$

*21.08 $[\hat{R}\hat{S}\psi(R,S)][\hat{R},\hat{S}\psi(R,S)] = df (\exists\varphi)(\psi(R,S) \equiv_{R,S} \varphi!(R,S) .\bullet. f(\varphi!))$

*21.081 $P\{\psi!(R,S)\}Q = df \psi!(P,Q).$

As before, *21.081 might better have been written as:

$<P, Q> \in \psi! = df \psi!(P,Q).$

The primitive dyadic predicate variables are easily distinguished from incomplete relation-e signs, since (as with all predicate variables) they have the distinctive exclamation, as in $\varphi!xy$ or $\varphi!\alpha\beta$. Further relations-e contextual definitions are introduced at *21 for relations-e of classes (of any relative type), and, as before, these must be given using typical ambiguity. Moreover, still further new definitions would be needed for the treatment relation-e of relations-e. The patterns are nonetheless clear.

Definitions for union, intersection, and complement are too obvious to mention here. But it is important to take note that these definitions must be applied before applying the definitions for class and relations-e. A definition that is very useful and worth mentioning is this:

*24.03 $\exists! \alpha = df (\exists x^t)(x^t \in \alpha)$
cls*24.03 $\exists! \alpha = df (\exists \gamma)(\gamma \in \alpha)$.

Note that definition *24.01 can be applied to free lowercase Greek α and any class expressions of the form $\hat{x}^{t_1} \psi x^t$ or $\hat{a} \psi \alpha$. It cannot be applied in cases such as $\exists! \iota \alpha \varphi \alpha$. One must first apply definition *2.072, eliminating the definite description. This is because definite description expressions are not genuine terms. There is also the important notation:

*30. 01 $[R \text{ '}y][\psi(R \text{ '}y)] = df [(\iota x)(xRy)][\psi\{(\iota x)(xRy)\}]$.

This enables *Principia* to use the notion $\hat{R}\text{'}x = y$ to emulate the mathematical function notation $fx = y$.

The definition has an analog for classes as follows:

cls*30.01 $[R \text{ '}\alpha][\psi(R \text{ '}\alpha)] = df [(\iota\sigma)(\sigma R\alpha)][\psi\{(\iota\sigma)(\sigma R\alpha)\}]$.

These notions are very often used in the work.

It is very important to understand that one cannot speak in *Principia*'s object-language about simple types of individuals. *Principia*'s object language can express the notion of *relative types* of classes (and of relation-e). Of course, no class or relation-e symbols ever get type indices. The definitions are given at *63–*65. Thus, for instance, one can write:

*63.01 $t \text{ '}x = df \hat{y}(y = x \lor y \neq x)$
i.e., $t \text{ '}x^t = df \hat{y}^t(y^t = x^t \lor y^t \neq x^t)$.

Similarly, there are definitions such as these

*65.01 $_x = df \alpha \cap t\text{'}x$, i.e., $\alpha \cap \hat{y}(y = x \lor y \neq x)$.
cls*65.01 $= df \alpha \cap t \text{ '}\beta$, i.e., $\alpha \cap \hat{\sigma}(\sigma = \beta \lor \sigma \neq \beta)$

And also, there is the important notation $R(\beta)$, where R is a relation-e whose domain consists of classes that are equal to β or not equal to β. This is important for clarifying the relation-e expression "Nc" involved in the definite description "Nc 'α" (i.e., "the cardinal number of α"). We find the following:

Nc '$\alpha = df (\iota\gamma)(\gamma \text{ Nc } \alpha)$.

To fix the relative type at β, Whitehead used the notation Nc(β) 'α [i.e., it is the class of all classes σ (such that σ = β ∨ σ ≠ β) such that σ is *similar* to α].

Observe that ambiguity arises here in many ways. There are non-homogeneous (descending and ascending) *similarity* relations and thus non-homogeneous *similarity* relations-e. Thus Whitehead must distinguish the *homogeneous* similarity relation-e, N_oc, from the *ascending* similarity relation-e, that is, $N^i c$, and also from the *descending* similarity relation-e, that is, N_ic. In a descending cardinal $N_ic(β)'α$ we have a class of classes of the same relative type as b (which is lower than that of α) that are similar to α. This is a descending cardinal as opposed to an ascending or homogeneous cardinal. Thus, we get:

⊢ ∃!N^ic 'α
⊢ ∃!N_oc 'α.

Ascending and homogeneous cardinals are never empty. It is not so for descending. Hence:

⊬ ∃!N_ic 'α
⊬ ∃!Nc 'α

Very serious issues arises with "Nc 'α" because of its uniquely unruly ambiguity in relative type. Whitehead discovered that "Nc 'α" is unstable in meaning even in distinct occurrences of it in the same *wff*. (Mind you, this is not an issue of *simple type* ambiguity of *Principia*'s individual variables, which is quite uniform.) The ambiguity of "Nc 'α" may make it seem that the rule of Modus Ponens admits of exception! Whitehead offers the following outrageous remark (*PM*, vol. II, p. 34):

> It is a fallacy to infer ⊢ ∃!Nc'α from the true propositions ⊢ α ∈ Nc 'α ⊃ ∃!Nc'α and ⊢ α ∈ Nc 'α.

This is obviously just an infelicitous way to make the point, which is that

α ∈ Nc 'α ⊃ ∃!Nc'α

may at first appear as though it should be a theorem, since, as Whitehead put it, it is "true whenever significant." All the same, some of its disambiguations are not theorems. The following is a theorem:

α ∈ Nc(α) 'α ⊃ ∃!Nc(α) 'α

The following disambiguation is not a theorem:

α ∈ Nc(α) 'α ⊃ ∃!Nc(β) 'α.

Since the last is *not* a theorem, there is no valid application of Modus Ponens using it with $\alpha \in Nc(\alpha)$ 'α. The horrors of relative types of classes (not simple types of individuals) led Whitehead to delay the publication of *Principia*'s volume 2 for a year until he could sort out needed conventions.[18] Whitehead made extensive alterations to volume 2 in 1911. Russell had nothing to do with it. When in 1925 there was an opportunity to address in *Principia*'s second edition some problems of Whitehead's emendments in volume 2, the two had a discussion about it. Whitehead's comment was this (Lowe, 1990, p. 276):

> I don't think "Types" [i.e., relative types] are quite right. They [are] "tending toward the truth," as the Hindoo said of his fifth lie on the same subject. But for heaven's sake, don't alter them in the text.

Apparently, neither had the will to reexamine the matter.

4.   Unique Results of the Revolution *Within*

Principia's conception of mathematics is thoroughly a study of the kinds of relational structures—which are given by relations independently of whether they are exemplified.[19] The transformation of the field holds many important surprises, not only in its embracing the existence of non-Euclidean geometry. Metaphysicians clinging to abstract particulars are unprepared for such surprises and have often interpreted them if they were failings of *Principia*'s Logicism. This is *question begging*. It is important to put their confusions to rest.

In *Principia*, Cantor's power-class theorem is expressed in terms of homogenous cardinals. We find the following theorem:

$$h^*117.66 \ N_oc \ 'Cl \ '\alpha > N_oc \ '\alpha$$

This says that the *homogeneous* cardinal number of Cl 'α, that is, the power-class of all subclasses of a class α, is larger than the homogeneous

---

18 See Landini (2017).

19 There are, of course, existence theorems in *Principia* for individuals of all sorts of types. But none of its existence theorems ensure that such an *individual* (whose type t is not o) is exemplified. It should be noted that even individuals of lowest type o are not to be regarded as particulars. Such a semantic interpretation of the formal language is one that Russell experimented with in 1925, but it is quite foreign to his original intentions. Indeed, there is solid evidence in *Principia*'s vol. 2 that Whitehead imagined a fully realist semantics according to which all individuals are universals ontologically regimented by simple types. Whitehead's expression of the theorem relies on his illicit definitions:$^*117.02 \ \mu > Nc \ '\alpha = df \ \mu > N_oc \ '\alpha, ^*117.03 \ Nc \ '\alpha > \mu = df \ N_oc \ '\alpha > \mu$. The definitions are illicit because the *definiendum* involved in each has already been defined.

cardinal number of α. And it should be noted that, according to *Principia*, the power-class Cl 'α is of higher relative type than that of α. So the previous is a comparison of cardinals of *different* relative type. We are familiar with Cantor's diagonal theorem

$$\text{h*}116.72 \vdash N_o c \text{ 'Cl '}\alpha = 2^{N_o c'\alpha}$$

It is homogeneous cardinals that are needed for the theorem.[20] Moreover, it is only where $2^{N_o c'\alpha}$ is of higher relative type than that of $N_o c'\alpha$. that the substitution of identicals applied to *117.66 and *116.72 yield the following:

$$\vdash 2^{N_o c'\alpha} > N_o c \text{ '}\alpha.$$

Thus, there is a problematic ambiguity in using the substitution of identicals and writing:

$$\text{*117.661} \vdash \mu \in N_o C \text{ induct } 2 > \mu.$$

This is found in *Principia*, but it should be:

$$\text{**117.661} \vdash \mu_\xi \in N_o C \text{ induct } .. (2)_{(\xi)} > \mu_\xi.$$

The subscripts for relative type of classes reveal that in the case in question of $(2)_{(\xi)}$ we have a higher relative type of class involved. In fact, in *Principia*, we have only the following:

$$2^{2''} \neq \Lambda . \supset . \ \mu < 2'' < 2^{2''}.$$

Here the relative type of μ is the *same* as that of $2^{2''}$ and $2''$. Whitehead eventually sorted out some conventions and notations clarifying such issues arising from ambiguity of relative type. The issue must be watched with care.

All this concerns relative types of classes. What is rarely appreciated is that Whitehead recognized in the opening of *Principia*'s volume 2 that Cantor's diagonal method, paired with non-homogeneous relations between *individuals* of different types, yields the following

$$Cantor^{indiv} : \ \sim (\exists \beta)(\beta_{x'} \approx V_{x^{(t)}}).$$

This says that no class $\beta_{x'}$ of *individuals* (each member of which is equal to or not equal to the individual $\beta_{x'}$) is similar to the universal

---

20 It is a small point, but one must be on the lookout for such issues.

class $V_{x^{(t)}}$ of *individuals* (each member of which is equal to or not equal to the individual $x^{(t)}$). You won't find the expression of *Canto-r^{Indiv}* anywhere in *Principia*. Obviously, its proof requires that simple type indices be restored to the individual variables. All the same, we find that Whitehead also knew of the provability in *Principia* of the following theorem:

$$N_0c' V_{x^{(t)}} > N_0c' V_{x^t}.$$

Again, you won't find this as a theorem in *Principia* because Whitehead and Russell were loathe to restore simple type indices. But the derivability remains. The homogeneous cardinal number of the universal class $V_{x^{(t)}}$ of *individuals* of simple type (t) is larger than the homogeneous cardinal number of the universal class V_{x^t} of *individuals* of simple type t. This is not about classes of classes. It is about classes of individuals of different simple types. Here is the revealing quote (*PM*, vol. 2, vii):

> We often speak as though the type represented by small Latin letters were not composed of functions. It is, however, compatible with all we have to say that it should be composed of functions. It is to be observed, further that, given the number of individuals, there is nothing in our axioms to show how many predicative functions of individuals there are, i.e., their number is not a function of the number of individuals: we only know that their number is $\geq 2^{Nc'Indiv}$, where "Indiv" stands for the class of individuals.

What Whitehead is saying is that, in the case of individuals, we have:

$$\vdash N_0c\,`V_{x^{(t)}} \geq 2^{N_0c'V_{x^t}}.$$

This says that the cardinal number of the universal class $V_{x^{(t)}}$ of individuals of simple type (t) is at *least* as large as 2 to the power of the cardinal number of the universal class V_{x^t} of individuals of the next lower type t. It might well be greater. Whitehead's view is that it is epistemically possible that the universal class V_{x^t} of individuals of a lower simple type t may be finite while the universal class $V_{x^{(t)}}$ of individuals (not classes) of next higher simple type is infinite. To say this, and to prove it, we need to restore simple type indices. So you won't find it said explicitly in *Principia*, and it has gone largely unnoticed.

Now Whitehead does go on to formally prove his discovery that when *descending* non-homogeneous relations of "similarity" (sm) are involved, Cantor's power-theorem reveals that some descending cardinals are empty. He arrives at the following:

*105.26 $\vdash N_1c\,`t`\alpha = \Lambda.$

Note that t 'α is the class $\hat{\mu}(\mu = \alpha \lor \mu \neq \alpha)$, which, of course, is the universal class of classes of the same relative type as α. The theorem proves that N_1c 't'α, which is a cardinal number, equals the empty class Λ. Since N_1c 't'α is a descending cardinal, theorem *105.26 says that the empty class Λ is a descending cardinal number. This is quite important, too. It entails that Hume's Principle, according to which classes are similar *iff* they have the same cardinal number, has false instances! That is a strike to the core of any neo-Fregean "logicist" for whom Hume's Principle is supposedly the essence of the notion of *cardinal number* itself. Now where homogeneous cardinals are concerned, all is well. *Principia* has:

$$\vdash \alpha \; sm \; \beta \equiv N_o c \; '\alpha = N_o c \; '\beta.$$

But neo-Fregeans cannot be content with this. In spite of calling them-selves "Fregean," they work from Hume's Principle as essential to the notion of cardinal number and neglect Frege's higher levels of functions. In volume 2 of *Principia*, however, Whitehead explains that Hume's *Principle* is false—that is, it has exceptions once levels (or analogously simple types) are involved. He writes (*PM*, vol. 2, p. 15):

*100.321 $\vdash \alpha \; sm \; \beta \supset Nc \; '\alpha = Nc \; '\beta$

Note that Nc 'α = Nc 'β ⊃ α sm β is not always true. . . . For if the Nc concerned is descending, and α and β are sufficiently great, Nc 'α and Nc 'β may both be Λ.

That is, since N_1c 't'α is a descending cardinal, theorem *105.26 says that the empty class Λ is a descending cardinal number. (Of course, nothing here says that it is an inductive cardinal.) Applying this, where instead of t 'α we have the universal class $V_{x(t)}$ of *individuals*, we get:

**105.26 $\vdash Nc(\beta_{x^t}) \; 'V_{x(t)} = \Lambda.$

This follows from the theorem *Cantor*[Indiv].

Note that Whitehead speaks of a descending cardinal Nc 'α where α is "sufficiently great" so that it may be empty. He leaves it open when it is sufficiently great. In light of this, one can imagine adding a new axiom **105.1 that is motivated by appeal to Whitehead's *105.26, or more exactly **105.26. It applies that result to the universal class $V_{x(t)}$ of *individuals* and extends the result.

I propose the following new axiom:

**105.1 $\Lambda = Nc(\beta_{x^t}) \; '\alpha \supset \alpha \approx V_{x(t)}.$

It applies that result to the universal class $V_{x(t)}$ of *individuals* and extends the result. Axiom **105.1 maintains that the *only* case where α is sufficiently great for the descending cardinal to be empty is when α is similar to a universal class. The only case where $\Lambda = \mathrm{Nc}(\beta_{x^t})\,\text{'}\alpha$ for such a descending cardinal number (as a class of classes that are each equal to or not equal to β_{x^t}) is when $\alpha \approx V_{x(t)}$. That is, the *only* case in which the empty class Λ is equal to the descending cardinal number $\mathrm{Nc}(\beta_{x^t})\text{'}\alpha$ is when α is similar to the universal class $V_{x(t)}$. In transposition, the axiom is this:

$$\alpha \neq V_{x(t)} = df\,(\exists\beta)(\beta_{x^t} \approx \alpha).$$

This says that, if not similar to $V_{x(t)}$, then there is some class β_{x^t} of individuals of the next lower type that is similar to it.

I believe that Whitehead would be sympathetic to new axiom **105.1 because it secures the infinity of natural numbers if we go up sufficiently high in relative type. The new axiom yields:

$$\vdash_{**105.1} \mathrm{Inf}\ V_{x((t))}.^{21}$$

With axiom **105, we can see that if we go high enough in simple type, we can avoid having

Infin ax

as an hypothesis on theorems concerning inductive cardinals. We get

$$\vdash_{**105.1} \textit{Infin ax}\ (x^{((t))}).$$

The infinity of natural numbers (inductive cardinals) at this relative type (and at any higher) is provable by virtue of new axiom **105.1. That is a nice result, perhaps even too nice for Whitehead to resist.[22]

I hasten to add, however, that new axiom **105.1 does *not* yield the infinity of V_{x^o}. That is quite important. The lowest one can go is:

$$\vdash_{**105} \mathrm{Inf}\ V_{x((o))}.$$

---

21 For convenience of exposition, I omit the proof. See Landini (2022), forthcoming.
22 Indeed, it completely obviates his appeal to his notion of "NC ind," which relies on his untenable doctrine of "true-whenever significant" whereby we are instructed to find a relative type high enough to make true a *wff* he puts down as if a theorem. Even in the presence of this strange doctrine (which Russell seems never to have vetted), he cannot evade difficulties arising with unruly relative typical ambiguity.

One cannot use axiom **105.1 to arrive at the infinity of $V_{x(o)}$. From axiom **105.1, one *can* arrive at:

$$\vdash {}_{**105} (\exists x^o, y^o)(x^o \neq y^o) \supset V_{x(o)}.$$

Accordingly, we get

$$\vdash {}_{**105} (\exists x^o, y^o)(x^o \neq y^o) \supset \textit{Infin ax } (x^{(o)}).$$

If there are (contingently) at least two individuals of the lowest simple type, then we are assured contingently of the infinity of $V_{x(o)}$, that is, the universal class of individuals of simple type (o). The empirical census of individuals of the lowest type that would be needed to ensure that Inf $V_{x(o)}$ is quite a bit easier to conduct than Whitehead thought. Much to the shock and dismay of a great many unaware of the revolution within mathematics (or rejecting it), Whitehead and Russell concluded that nothing ensures that there are infinitely many natural numbers. Nothing, that is, ensures it given the revolutionary Cantorian notion of *natural number* as a finite (inductive) cardinal (understood in terms of one-to-one correspondence relations). Not all the Peano-Dedekind postulates for natural numbers are legitimately *mathematical* postulates. The question of the number of natural numbers is not, according to *Principia*, in the province of mathematics seen from within the revolution against numbers as abstract particulars. My addition of an axiom **105.1 is friendly since it does *not* yield the infinity of V_{x^o}. This is as it should be once one accepts the revolution in mathematics.

Principia's cp-Logic endeavors to remove abstract particulars from the subject matter of the branches of mathematics. It leaves it epistemically open, however, whether any abstract particulars may be subject matter of logic. None are assumed as part of *Principia*'s formal cp-Logic. This comes clear when we examine Whitehead and Russell's comments in *Principia* concerning their so-called "axiom" of infinity. There is, mind you, no "axiom" of infinity of abstract particulars adopted in *Principia*. The entire point of the work, as we have taken pains to point out, was to upturn the metaphysician's indispensability argument for abstract particulars (e.g., numbers, spatial geometrical figures, sets, etc.) in any branch of mathematics. *Principia* does not hide from the fact that *nothing* in the Cantorian revolutionary conception of *cardinal* number assures that there are infinitely many cardinals. It is a conclusion arrived at from *within* mathematics as the science of relational structure, not from outside of it as a metaphysics of special abstract particulars governed by special arithmetic or metaphysical essences. That is to say, Whitehead and Russell reached the conclusion working under what they regarded as Cantor's revolutionary conception of mathematics. The question of whether there are infinitely many natural numbers (finite cardinals) is *not* a question for mathematics to decide.

As we can see, keeping the revolution within mathematics in mind is important to combating the objection that *Principia*'s Logicism fails to capture arithmetic as a part of logic. Often the objection is that it offers an "If-thenism," that is, the thesis that from the axioms governing the entities in a given field of (abstract) *particulars* only the logic of quantification theory is needed to derive the theorems of that field. From the axioms of Euclidean geometry governing abstract particulars (lines, points, etc.), nothing more than quantification theory is needed to derive the theorems. From the Peano/Dedekind axioms for natural numbers as abstract particulars, nothing more than quantification theory is needed to derive elementary arithmetic. If-thenism applies to axiomatic astronomy as much as to mathematics. It is certainly antithetical to Fregean logicism, which endeavors to show that numbers are logical abstract particulars.

This objection, pioneered by Putnam (1967) and Boolos (1994), among others, is very misguided. It entirely misses (or dismisses) the revolution within mathematics. It ignores the important point that Whitehead and Russell accepted the revolution in mathematics which made it a science of relational structures and not a science of abstract particulars.[23] That is, If-thenism requires that one write axioms governing particulars. Quantification theory (with bindable predicate variables or without) is wholly inadequate for such an elimination of abstract particulars, and thus the transformation of any theory of abstract particulars into a theory of relational structures must be rejected by every If-thenism.[24] The import of the revolution within mathematics is that the infinity of the natural numbers (in lowest type) as Cantorian inductive cardinals (i.e., as special relational properties of one-to-one correspondence) is not a matter for mathematics to decide. Hence, it is no longer regarded as a mathematical result that there are infinitely many primes. The logical truth is that if there are infinitely many natural numbers, then there are infinitely many primes. That result is no more a form of empty If-thenism than is the discovery that it is not a geometric necessity that every equilateral triangle is equiangular and that instead the proper mathematical result is that if the metric is Euclidean, then every equilateral triangle is equiangular.

Principia's agenda is to reveal that the study of *every* kind of structure, including the study of relations that are "Progressions," can be conducted without the metaphysician's assumption of abstract particulars. The property that a relation has when it is a Progression can be fully studied independently of whether any relation that is a progression is exemplified. And *Principia* explicitly leaves open whether any relation that is a progression is exemplified. Whitehead and Russell write (*PM*, Vol. 2, Cp. 245):

> It is not convenient to *define* a progression as a series [serial relation] which is ordinally similar to that of the inductive cardinals, both because this definition only applies if we assume the axiom of infinity,

---

23 Compare Coffa (1980) and Griffin (1980), Gandon (2014), Kraal (2014).

24 Ramsification, that is, the eponym for the position named after Ramsey, according to which changing every predicate constant into a variable and universally closing the result, is inadequate to the task. It is cp-Logic that is absolutely essential.

and because we have in any case to show (assuming the axiom of infinity) the series of inductive cardinals has certain properties which can be used to afford a direct definition of progressions.

As we see from this passage, they do not follow Cantor in defining the notion of a *progression* by appeal to an assumption of infinitely many inductive cardinals (natural numbers). For a similar reason, they stray from Cantor's definition of aleph null. They write (PM, Vol. 2, p. 260):

> Cantor defines $\aleph_0$ as the cardinal number of any class which can be put in one-one relation with the inductive cardinals. This definition assumes that $\nu \neq \nu + {}_c 1$, when ν is an inductive cardinal; in other words, it assumes the axiom of infinity; for without this, the inductive cardinals would form a finite series, with a last term, namely Λ. For this reason among others, we do not make similarity with the inductive cardinals our *definition*. We define $\aleph_0$ as the class of those classes which can be arranged in progressions.

In light of such clear evidence from *Principia*, we can see the work accepted that it is simply *not* the business of mathematics to decide whether any progression relation is exemplified.[25]

It remains true, however, that Whitehead and Russell thought that it may (epistemically) be a logical truth that there are infinitely many inductive cardinals. They write (*PM*, Vol. 2, p. 183):

> It is important to observe that, although that axiom of infinity cannot (so far as appears) be proved *a priori*, we can prove that any given inductive cardinal exists in a sufficiently high type.

The comment "so far as appears" is important. Whitehead and Russell accepted that *Principia*'s formal system did not render a logical theorem of the infinity of logical particulars (of the lowest type). It certainly does not follow that they thought it is *not* a logical necessity. There is no special epistemic access to what is or is not a logical truth. We can be wrong, thinking something is logically true when it is not—which is what happened with Russell's paradoxes of attributes. And, moreover, we can be wrong, thinking something is not logically true when it is—which happened with $2 + 3 = 5$. Whitehead and Russell knew well enough that *Principia*'s formal axiomatic system was incomplete and is not the last word on the subject of logic. That is, they knew that there may well be logical truths not captured

---

25 The general theory of relation arithmetic as well as the Reals (as classes of well-ordering relations) are also developed in *Principia* independently of whether there are infinitely many natural numbers. Some important theorems in these fields, of course, require antecedent clauses expressing infinity. But, again, this is not at all an *If-thenism* because these theories are not at all theories of abstract particulars.

in it as theorems. So it may (epistemically) be that it is logically necessary that there are infinitely many logical abstract particulars of the lowest type.

Russell's Logicism leaves it open that there may be logical necessities that are not mathematical necessities. If there are infinitely many logical abstract particulars, then it is a logical necessity that there are infinitely many inductive cardinals. Even so, according to *Principia*, this is an issue in the field of logic, and it is not an issue in the field of mathematics. The question of the infinity of the natural numbers is *not* a matter of mathematics—the metaphysicians of mathematics notwithstanding. Mathematics is not a study of abstract particulars; it is the study of relations, and it is conducted quite independently of the contingencies of their exemplification. On page 498 of volume 2, we find a note concerning Part V Section C, which reads as follows:

> Section C . . . is concerned with convergence and the limits of functions and the definition of a continuous function. Its purpose is to show how these notions can be expressed, and many of their properties established, in a much more general way than is usually done, and without assuming that the argument or values of the functions concerned are either numerical or numerically measurable.

Principia evades adopting rationals as abstract particulars. Whitehead and Russell write (PM, vol. 3, p. 234 of second *300):

> it seems improper to make [the study of relations grounding] the theory of (say) $\frac{2}{3}$ depend on the assumption that the number of objects in the universe is infinite.

Though a great many theorems in the theory of the Reals cannot be similarly free of having *Infin ax* as an antecedent clause, the same stand is taken against the Reals as abstract particulars. On page vi of volume 3, we find:

> (3) We have developed a theory of ratios and real numbers which is prior to our theory of measurement, and yet is not purely arithmetical, i.e. does not treat ratios as mere couples of integers, but as relations between actual quantities such as two distances or two periods of time. (4) In our theory of "vector families," which are families of the kind to which some sort of measurement is applicable, we have been able to develop a very large part of their properties before introducing numbers.

In short, what *Principia* makes plain is that there are no numbers as abstract particulars and that mathematics gets along perfectly well. The metaphysician's indispensability arguments in favor of numbers as abstract particulars are mistaken.

According to *Principia*, nothing in logic militates against the existence of purely logical individuals. It is just that we have no epistemic access to them. This point is quite important for understanding the conception of logically necessary truth that Whitehead and Russell envisioned in the *Principia*. Whitehead clearly has a realist semantics in mind when he views universals (which are not particulars) as being among the individuals of any relative type, lowest or otherwise. He explicitly notes in volume II of *Principia* that the number of individuals of a given logical type may well outrun the number of classes that can be emulated in that type. Nothing in logic prohibits their being infinitely many logically existing individuals (universals) in every logical type. Nonetheless, in *Principia*, Whitehead held that full generality and truth *may* (epistemically) be sufficient for logically necessary truth. At the same time, it is far from epistemically certain that logic ensures an infinity of logical individuals. So far as *Principia* is concerned, the science of cp-Logic must do without.[26]

Particulars are nowhere to be found in the subject matter of mathematics or of *Principia*'s cp-Logic. Of course, the ontological ground of cp-Logic lies in facts composed solely of universals! (*Universals* are just those entities that have the capacity to unify a fact; *particulars* lack that capacity.)[27] But the ontological ground of cp-Logic is not the subject matter of cp-Logic. So all is well. Indeed, *Principia* maintains that we do know *a priori* the following existence theorem

$$\vdash (\exists x)(\varphi)(\varphi!x \supset \varphi!x).$$

It follows rather immediately by *9.1 from the theorem $\vdash (\varphi)(\varphi!x \supset \varphi!x)$. An analogous existence theorem holds for every simple type of *individuals* no matter how high in the hierarchy. Note, however, that this is not a commitment to any particulars (abstract or concrete). Universals with which we are acquainted are themselves *individuals* (of the lowest type) and are certainly not particulars. Individuals of higher simple type (in a Realist semantics for *Principia*) are universals, too. So nothing here conflicts with Russell's view that particulars (abstract or concrete) are not the subject matter of its cp-Logic. In *Principia*, there is no theorem of infinity for any type of individuals. However, if there are *n*-many individuals of a given type, it is a theorem of *Principia* (and thus we know *a priori*) that the next higher type contains *at least* 2^n many individuals. Moreover, nothing in *Principia* interferes with its being *possible* (epistemically) that it is a logical matter that there are infinitely many individuals in every type. There is, of course, a strong logical intuition that there are infinitely many individuals (universals). If universals are included in the lowest simple type, then there surely is a strong intuition that there is an

---

26 Russell changed his mind in his 1919 book *Introduction to Mathematical Philosophy* and concluded, contrary to the epistemic uncertainty expressed in *Principia*, that he does know that it is not a logical truth that there are infinitely many logical individuals.

27 If one disagrees, the objection should be put by saying that there are no *universals* and *particulars* in Russell's sense of the terms.

infinity of individuals in every simple type (including the lowest). But Whitehead and Russell held that our epistemology of logic does not rule out the epistemic possibility of *monism* in the lowest type. The epistemic possibility of *monism* is mentioned in *Principia* at least four times (vol. 1, *24, p. 216; vol. 2, p. 8, 325). It will appear shocking to those unaware or hostile to the fact that a transformation *within* mathematics has taken place.[28] But it is not a deficit—a failing. It is a substantive feature of the revolution and is not a consequence of *Principia*'s logicism itself.

One might recall, however, Russell's own later comments in the 1912 book *The Problems of Philosophy* which suggest that *all* existence is known *a posteriori*. Likely, Russell meant only to speak of the existence of (abstract or concrete) of particulars—his example being the Emperor of China, which is a bit of a joke, since his cause célèbre is obviously the status of ontological proofs of the existence of God. But in his book *Introduction to Mathematica Philosophy* (1919), Russell wrote: "There does not even seem any logical necessity why there should even be one individual—why, in fact there should be any world at all" (*IMP*, p. 203). In a footnote, he continues: "The primitive propositions in *Principia Mathematica* are such as to allow the inference that at least one individual exists. But I now view this as a defect in logical purity" (*Ibid.*). What explains away this comment is that under the influence of Wittgenstein, in 1919, Russell had ceased to hold that universals have an *individual* nature. Russell was exploring the idea that universal words must stay in predicate positions. This demanded from him a new semantic interpretation of *Principia*'s formal theory. But it is a significant departure from his acquaintance epistemology, since any universal with which one is acquainted must have an individual nature. Happily, it may be ignored.

Curiously, in his introduction to the 1925 second edition *Principia*, he didn't even mention the issue. The only amendment to the work was to replace section *9 with a new section *8, which accepted its definition of subordinate quantifiers in terms of quantifiers initially placed but offered a technical system of deduction without free variables.[29] As applied to *9, the method of deduction without free variables would take the following universal closure of every instance of *9 as an axiom schema:

$$\text{Closure}*9.1 \ (\varphi_1, \ldots, \varphi_n) \ (y_1, \ldots, y_m)(x)(\varphi x \supset (\exists z)\varphi z),$$

where $\varphi!_1, \ldots, \varphi!_n$ and $y_1, \ldots, y_m$ are all the variables occurring free in the *wff* φ. The new system of *8 was perhaps the first quantification theory without

---

28 See Landini (2007, 2011) for details about Whitehead's heavily emended volume 2 of *Principia*.
29 See Landini (2005). Russell wrote the new Introduction and appendices to the second edition without the collaboration of Whitehead. In these experiments, Russell explored some ideas he attributed to Wittgenstein, and found them only of limited success. There is good evidence that Whitehead did not want these experiments included in the second edition. See Lowe (1985, 1990).

free variables, but it does not avoid existential theorems—not even those framed with individual variables of the lowest type. The technical problem of conducting deduction to avoid existential theorems involving individual variables of the lowest type is made complicated by the fact that Russell does not accept vacuous quantification and that in both of his systems of *9 and *8, subordinate occurrences of quantifiers are defined in terms of quantifiers initially placed. Even when p is quantifier free, the system cannot accept, $p \mathbin{.} \supset. \ (x)\varphi x \supset p$, since it is defined as the existential:

$$(\exists x)(p \mathbin{.} \supset. \ \varphi x \supset p)$$

(Compare Quine 1951). In any case, a concern against free variables is one thing, a blanket rejection of existence theorems known *a priori* is quite another.

Principia's philosophy of mathematics—the revolution within mathematics against the metaphysicians's abstract particulars—is worthy of revival. We need a Realist semantic interpretation of *Principia*'s bindable variables in order to validate its simple type-regimented impredicative comprehension. Under such an interpretation, type-free universals with which we may be acquainted are values of its individual variables of the lowest type. So there is no need to offer a repair of its quantification theory to avoid free variables even of the lowest type. Since we have *a priori* knowledge of every instance of comprehension, we are forced to accept that there is *a priori* knowledge of the existence of universals (some are type free and values of the individual variables of the lowest simple type and some regimented by simple types).[30] In this matter, it was the Rationalists who were correct. *Principia* cannot be successfully criticized by those outside the revolution within mathematics—and those outside it include Frege, Zermelo, Quine, Gödel, and every metaphysician embracing abstract particulars, empiricist or rationalist alike. Indeed, we do well to take stock. Does Gödel's famous first incompleteness theorem apply to the historical *Principia*? I wonder! There being no abstract particulars, does Gödel's diagonal function exist in the ontology accepted by the revolution? We cannot let the metaphysicians imposing abstract particulars on the branches of mathematics determine the agenda for what would have to be emulated were the revolutionaries to be right. It is high time to see *Principia Mathematica* and its unique non-Fregean Logicism as it was intended—as the flagship of revolution *within* mathematics.

References

Boolos, G. (1994). "The Advantages of Theft over Honest Toil," in Alexander George, ed., *Mathematics and Mind* (Oxford: Oxford University Press), pp. 27–44.

---

30 Recall that Russell hoped to emulate comprehension by offering a nominalistic semantic interpretation of *Principia*'s bindable predicate variables. Their intended nominalistic interpretation failed to validate comprehension, and thus we have accepted an objectual (Realist) semantics in its place.

Carnap, R. (1931). "The Logicist Foundations of Mathematics," *Erkenntnis*, pp. 91–121.

Church, A. (1976). "Comparison of Russell's Resolution of the Semantical Antinomies with That of Tarski," *Journal of Symbolic Logic* 41, pp. 747–760.

Coffa, A. (1980). "Russell and Kant," *Synthese* 45, pp. 43–70.

Dewey, J. (1920). *Reconstruction in Philosophy* (New York: Henry Holt & Co.). Pagination is to the edition: New York: Mentor, 1950.

Gandon, S. (2014). *Russell's Unknown Logicism: A Study in the History and Philosophy of Mathematics* (London: Palgrave MacMillan, 2014).

Griffin, N. (1980). "Russell on the Nature of Logic (1903–1913)," *Synthese* 45, pp. 117–188.

Heck, R. (2012). *Reading Frege's Grundgesetze* (Oxford: Oxford University Press).

Klement, K. (2015). "A Generic Russellian Elimination of Abstract Objects," *Philosophia Mathematica* 25, pp. 91–115.

Kraal, A. (2014). "The Aim of Russell's Early Logicism," *Synthese* 7, pp. 1493–1510.

Landini, G. (1998). *Russell's Hidden Substitutional Theory* (Oxford: Oxford University Press).

———. (2005). "Quantification Theory in *8 of *Principia Mathematica* and the Empty Domain," *History and Philosophy of Logic* 25, pp. 47–59.

———. (2007). *Wittgenstein's Apprenticeship with Russell* (Cambridge: Cambridge University Press).

———. (2011). "Logicism and the Problem of Infinity: The Number of Numbers," *Philosophia Mathematica* 19, pp. 167–212.

———. (2013). "Typos of *Principia Mathematica*," *History and Philosophy of Logic* 34, pp. 306–334.

———. (2017). "Whitehead's (*Badly*) Emended *Principia*," *History and Philosophy of Logic* 37, pp. 114–169.

———. (2022) "Infinity $V_{x(t)}$ in *Principia Mathematica*," forthcoming.

Lowe, V. (1985). *Alfred North Whitehead: The Man and His Work, Vol I, 1861–1910* (Baltimore: Johns Hopkins University Press).

———. (1990). *Alfred North Whitehead: The Man and His Work, Vol II 1910–1947*, edited by J. B. Schneewind (Baltimore: Johns Hopkins University Press).

Putnam, H. (1967). "The Thesis That Mathematics Is Logic," in R. Schoenman, ed., *Bertrand Russell, Philosopher of the Century* (London: Allen & Unwin), pp. 273–303.

Quine, W. V. O. (1951). "On Carnap's Views on Ontology," *Philosophical Studies* 5, pp. 65–72.

———. (1963). "Russell's Theory of Types," from *Set Theory and Its Logic* (Cambridge: The Belnap Press), pp. 241–265.

Russell, B. (1901). "Mathematics and the Metaphysicians," in *Mysticism and Logic and Other Essays* (New York: Longmans, Green and Co.), pp. 74–96.

———. (1912). *The Problems of Philosophy* (London: Home University Library).

———. (1919). *Introduction to Mathematical Philosophy* (London: Allen and Unwin).

Whitehead, A. N. and Russell, B. (1910). *Principia Mathematica* (Cambridge: Cambridge University Press). Vol. 1, 1910; vol. 2, 1912; vol. 3, 1913.

Williamson, T. (2013). *Modal Logic as Metaphysics* (Oxford: Oxford University Press).

14 The Ontology of Abstraction, From Neo-Fregean to Neo-Dedekindian Logicism[*]

Fiona Doherty[†]

1. Scottish Neo-Fregean Logicism

Hale and Wright's neo-Fregean logicist programme inherits Frege's logicist ambition to establish the conjecture that the truths of arithmetic are truths of logic, such that the subject matter of these truths are *logical objects* (see Hale & Wright 2001). Inspired by the result that the Dedekind-Peano axioms for second-order arithmetic can be derived from Hume's Principle (HP) in full second-order logic, they preach that the stone that Frege rejected can become the cornerstone: HP can provide those most-sought-after of Fregean objects (the cardinals) in such a way that establishes their logicality, thus providing an answer to Frege's question in §62 of *Grundlagen*: "How are numbers to be given to us (if we cannot have any ideas or intuitions of them)?"[1]

In short, their answer is that HP, that is,

Hume's Principle $\forall F \forall G (Nx : Fx = Nx : Gx \leftrightarrow F \approx G)$[2]

* This chapter is dedicated to Bob Hale, whose keen mind and wit, demanding curiosity and boundless generosity will always be sorely missed. In writing this chapter, I am indebted to him and also to Felipe Contatto, Joseph Dewhurst, Giovanni Merlo, Bethany Nelson, Tamara Sanchez-Kapostasy, Peter Sullivan, Lukas Skiba, Shyane Siriwardena, Robert Wilson, Cripsin Wright and everyone at the Origins and Varieties of Logicism conference in Milan, especially Francesca Boccuni, Andrea Sereni, Michael Potter, Michael Frank Hallett, Salvatore Florio and an anonymous referee for their careful attention, which led to many improvements in the chapter.

† Orcid.org/0000-0002-8493-5668. E-mail: fiona.t.doherty@gmail.com

1 The neo-Fregean logicists earn their prefix by the fact that Frege himself foregoes Hume's Principle, opting instead to define numbers in terms of extensions (Frege 1884, §63–64); although Geach (1955) and Parsons (1965) both noticed that the Dedekind-Peano axioms could be derived from Hume's Principle, the conjecture did not gain much attention until Wright (1983) carried out most of the derivation in his *Frege's Conception of Numbers as Objects*. Later, Heck showed that the only "essential" use Frege makes of his ill-fated Basic Law V in *Grundgesetze* is to derive Hume's Principle (for more details, see Heck [1993, 581–584] and Heck [2011]).

2 The second-order quantifiers here are restricted to predicates expressing non-indefinitely extensible sortal concepts (see Dummett 1993, 441). Hale and Wright conjecture such

DOI: 10.4324/9780429277894-17

(where "≈" denotes an equivalence relation on sortal concepts and "$Nx : \_$" a term-forming operator, such that "$Nx : Fx$" reads "the number of Fs") is a member of a special class of sentences called *abstraction principles*. These are principles of a common syntactic structure which use the special mechanism (detailed in the following) of implicit definition to introduce abstracta. Not only that, but they do so in such a way that *a prioricity* is preserved well enough to support a species of logicism (Hale & Wright 2001, 117).

For those unfamiliar with this special mechanism, here is a condensed explanation: HP is laid down as a definition, the *definiendum* of which is "$Nx : \_$". The surface syntax of the *definiens* and the fact that it is laid down as true work together to define "$Nx : \_$" in a complex process that goes something like this: start with a recognition of the sentence as a true biconditional; help yourself to the truth of the RHS from your acquaintance with equivalence relations; infer the truth of the LHS on truth-functional grounds; recognise some familiar faces on the LHS: the very same F & G from the RHS, and the identity relation "taken as given".[3] Finally, take the *definiendum* to mean *what it has to mean in order to contribute with the determined parts of the sentence to make the LHS—and thus the biconditional—true*. If all has gone well, the meaning of the *definiendum* "$Nx : \_$" has now been fixed as an operator which maps predicate/function-expressions ("F, G") to singular terms ("$Nx : Fx$"—i.e., "the number of Fs"). Vitally (here I skip over the detail of Hale and Wright's theory of reference), the *reference* of the *definiendum* "$Nx : \_$" has thereby been fixed as a function which maps concepts/functions (F, G) to objects ($Nx : Fx$, i.e., the number of Fs).[4]

Have the numbers been given to us? The short answer is: not directly but indirectly via the number-function which *has* been given to us, that is, $Nx : \_$ which takes F & G to the *same* object when and only when F

---

a restriction can be made good in order to "exorcise" Boolos's anti-zero; see Hale and Wright (2001, 314); Boolos (1997, 260). For a contemporary discussion of these issues, see Rumfitt (2001).

3 In Frege's own discussion of Hume's Principle, he permits beginning with the general notion of identity and from it forming the notion of *numerical identity*, so that a grasp of the identity relation is given prior to a grasp of Hume's Principle (Frege 1884, §64).

4 Abstraction principles are typically read as impredicative definitions which define abstracts by quantifying over a domain to which those abstracts belong. This permits a straightforward reading of the quantifiers, since the objects defined by abstraction fall under the quantifiers on both sides of the abstraction principle. By contrast, in a predicative reading of abstraction principles, the abstracts defined are not included in the range of the quantifiers used to introduce them, and a many-sorted logic is needed. This latter kind of definition has the advantage of avoiding Russell's paradox, but the disadvantage of being unable to deliver classical arithmetic. Thus, Potter and Sullivan conclude that if we "view Hume's principle (or Basic Law V, for that matter) predicatively, it is harmless and plausible, but quite useless as a route to substantial mathematical theories" (2005, 192). For details and further discussion, see: Potter (2004, Appendix C.); Potter and Sullivan (2005, 187–192).

& G stand in an equivalence relation. The conjecture is that such objects are the cardinals, so that the singular terms on the LHS of HP must refer to the very objects Frege asked for, bestowing them with a (curious) species of *a prioricity*, a kind of meaning-in-virtue-of-truth. Assuming such a property is preserved across logical consequence, then what is won for HP is also won for arithmetic, thanks to Frege's Theorem.[5]

Importantly, Hale and Wright reinflate Frege's logicist hope by developing this process of definition by abstraction in which a mere definition can establish reference to infinitely many objects without being ontologically extravagant—what they call their "central ontological idea" (2009, 181).

To put my cards on the table, I believe that Hale and Wright have long forgone their aim of delivering the natural numbers as Fregean objects and instead defend an account of abstraction with an implicit structuralist ontology. But here I can only hope to convince the reader of the former negative claim and leave the structuralist accusation as merely suggestive.

I believe the most straightforward way to see the implicit abandonment of Frege's ontology is to examine a seemingly innocent ontological precondition that Hale and Wright put on the objects of abstraction—namely that they may admit of no independent means of specification (§2–4). Their move away from Frege's ontology can also be seen from their preferred conception of reference and their solution to the Caesar problem (§5). This gives the result that Scottish neo-Fregean logicism's best hope of delivering a kind of logicism recognisable to Frege is to do so by establishing the logicality of objects *un*recognisable to Frege (§6).

2. The Specification Requirement

I have found that the simplest and most straightforward application of my objection is Hale and Wright's repeated appeal to be granted the assumption that the objects recovered by abstraction should not be understood as having "independent means of specification".[6] However, in order to understand what this appeal amounts to, we will have to do a fair bit of unpacking first. What counts as a means of specification independent of abstraction? For that matter, what constitutes a means of specification?

To build up an answer to these questions, I will elaborate on the general story of abstraction set out in §1. Recall that—if all goes well—HP can be used to establish reference to the natural numbers but that it does so in an indirect way: HP is used to define "$Nx : \_$", which then refers to the

---

5 For more details, see Hale and Wright (2001), especially "Implicit Definition and the A Priori" (Hale & Wright 2001, 117–150).
6 For example: Hale and Wright (2008, 204); Hale and Wright (2009, 92; 210).

number *function*; that is, the function that takes one-one concepts to the same object.[7] We infer that such objects must be the numbers and that the singular terms "$Nx : Fx$; $Nx : Gx$" on the LHS of HP must refer to them. The singular terms thus afford us a means of specification of the numbers.

It is the final step we need to put under the microscope. How do we go from defining a function expression to referring to the right objects? In short, abstraction principles (e.g., HP) establish reference to objects by securing a function (e.g., the number function) that has those objects (e.g., the natural numbers) in its value-range. This is a curious, circular, yet seemingly genuine mechanism for establishing reference. For example, (only) a philosopher might tell someone they were hired by announcing: "The new chair of logic and rhetoric is the unique object in the value-range of the one-place property *person standing in front of me right now*". Or I might tell you to "Meet me at the restaurant where we first met".

What makes abstractionism more complex than these examples is that the relevant function expression is *defined* as well as *employed* to refer to a function. The process of definition by abstraction must begin with a meaningless function expression (on the LHS of the biconditional) and imbue it with the right sense for it to refer to the relevant function from which we hope to extract the numbers in its value-range. Hale and Wright have their own analogy for this process:

> To invent a meaning, so conceived, is to fashion a concept: it is to be compared to making a mould and then fixing a certain shape-concept by stipulating that its instances comprise just those objects which fit the mould.
>
> (2001, 131)

To make this analogy explicit: an abstraction principle acts as an implicit definition of the function term on its LHS, which then expresses a mode of presentation of a determinate function with a determinate value-range. The natural numbers are stipulated to be just those objects which fit this function, that is, which fall in its value-range. This is what it means to say that abstraction principles establish reference to objects by presenting them as *those objects falling in the value-range of the defined function*. Abstraction secures a value-range as part and parcel of defining a function. We can then peer down into the function's value-range in search of the very objects that the object-demanding singular terms on the LHS must refer to. To add my own analogy, the process is akin to searching for a proximal object denoted by a suitable word (and prompts) in the game "I spy with my little eye"—granted that the process of abstraction engages something more akin to the "mind's-eye". The selected object is

---

7 For all to go well, the relevant terms must, among other things, feature in a range of true atomic sentences, as I will discuss in §5.

not named directly, but the game provides us with alternative means of specification of that object, although in the form of clues rather than as the value-range of a function. Importantly, for game to work, the object must have been there all along in our "domain" of vision.[8]

The crucial idea here of presenting objects as values of concepts can be seen as Fregean terminology for something more straightforward: the fact that we can pick out things by their properties as well as by calling them by name. In this case, natural numbers are picked out as those objects with the property of *being mapped to the same concepts when and only when those concepts are one-one.*

What this more detailed picture of abstraction shows is that defining the number-function via HP is at the same time establishing the sense of a function expression, reference to a function and reference to the objects in the function's value-range. This finally puts us in a position to make sense of what Hale and Wright mean by their repeated censure of an "independent means of specification". By this requirement, they mean to prevent any way of referring to the relevant functions or objects distinct from the process of establishing a "referential route" to the natural numbers by defining a function which they form the values of. This includes referring to the natural numbers by singular terms or presenting them as values of functions *other* than the function defined by the abstraction principle.

Having spelled out the specification requirement, I want to make one final observation. That is that the abstractionist mechanism we have examined qualifies as a Fregean mode of presentation. That is to say, defining singular terms via the value-range of a function gives a particular way of presenting them, distinct from presenting them via very different functions such as *the coveted objects needed to provide a logicist recovery of arithmetic* or *those objects which Count von Count from Sesame Street familiarised pre-school children with.* This is no controversial claim; it merely amounts to the claim that, in the best case, the numerical singular terms are given their *sense* by the implicitly defined number function (cf. Hale & Wright 2001, 131). However, it will later prove useful to have this observation on the table.

Altogether, Hale and Wright's position is that it is simply a misunderstanding of their view to think that there can be any means of specifying either the sense or referent of the function expression on the LHS or of the singular terms on the LHS independent of the process of abstraction.

3. First Way Into the Objection

Now that we have the relevant background in place, I can explain why the requirement to admit of no independent means of specification carries with it an inbuilt rejection of a Fregean ontology. I'll do so in two steps. First I'll

---

8 At least this is this case in the standard impredicative versions of "I spy" which I used to play.

argue that the requirement carries an implicit rejection of thick objects, and then I'll argue that Frege's conception of objects is thick in the relevant sense.

3.1.   *Through Thick and Thin*

The distinction between thick and thin objects is as suggestive as it is obscure.[9] Before we proceed, therefore, I want to clarify what I mean by these philosophical terms of art.

In what follows, I will employ the relevant distinction in line with the following definitions:

> **Thick objects:** *Objects which admit of modes of presentation further than those by which they have been introduced.*
> **Thin objects:** *Objects which do not admit of modes of presentation further than those by which they have been introduced.*

Intuitively, calling objects thick/thin is meant to capture a distinction between two *realist* conceptions of the individuals in the domain of first-order quantifiers. One can think of these denizens of reality as capable of surprising us such that they can turn out to have features which we didn't use to refer to them or define them. Such would be the case if it turned out that someone we knew by their given name was Banksy. People in general—and concrete objects even more generally—are good examples of objects for which most realists would adopt a thick conception.[10]

What is the alternative to this conception of objects? It is to think of the denizens of reality as being so essentially tied to the means by which they are defined and referred to that it would be a conceptual mistake to suppose their nature capable of outstripping our means of pointing them out. The cases for which it is most intuitive to adopt this conception of objects occur with respect to theoretical objects defined for a specific purpose. One would diagnose confusion if a student claimed to have met the person involved in your imaginary counterexample or claimed that you had incorrectly switched the names of two arbitrarily labelled vertices of a triangle you had just drawn.

Philosophical accounts of thin objects are understandably more obscure among realists. They have received most attention and development from structuralist accounts which want to retain a sense in which the objects of mathematics—although secondary to the structures of mathematics—can be said to exist and are straightforwardly referred to by mathematical singular terms, that is, *non-eliminative* mathematical structuralist accounts,

---

9 See Dummett's talk of full-blooded objects (Dummett 1981, 471).
10 This is the insight exploited by the MP objection; see Potter and Sullivan (1997), Potter and Sullivan (2005).

like Shapiro's *ante rem* structuralism.[11] Such accounts hold that mathematics is "the science of structure", by which they mean that the structures referred to by mathematical theories take ontological priority over the mathematical objects which are nothing more than reified positions in the theory's structure. Non-eliminative structuralism provides a textbook case of a thin conception of objects, since, in their view, the nature of the mathematical objects is both dependent on and exhausted by the structure of which they form a part and by virtue of which they can be referred to.

3.2. Thick Objects and the Specification Requirement

It should be obvious that one of the previous definitions is in tension with the specification requirement, but let us set out the details, due to the importance of this point.

Hale and Wright demand that the objects of abstraction may admit of no means of specification independent of abstraction. On the surface, this demand simply sounds like the demand for fair play. After all, it mitigates the risk of definitional circularity whereby an alternative means of reference to the numbers is used in order to establish the supposedly only means of reference to the numbers afforded by abstraction. This undermines the supposed advantage of abstraction principles over axiomatic stipulation and makes the impredicativity of the domain look harmful.[12]

On closer inspection, however, the requirement is not so innocent. We saw in §2 that the specification requirement amounts to the requirement that abstraction principles establish reference to their abstracts by providing a sense via the process of implicit definition. More specifically, by providing a mode of presentation of the relevant objects as falling in the value range of a function implicitly defined by abstraction. As such, the requirement amounts to the imposition that the abstracts admit of no other modes of presentation further to that which is established by the process of definition by abstraction.

This is straightforwardly in tension with the thick conception of objects defined previously, since in this conception, objects can admit of different modes of presentation beyond the means by which they have been introduced—such objects can always surprise us. Indeed, one way to understand the requirement enforced by Hale and Wright is that it serves as an arbitrator of whether a realist ontology adopts a thick or a thin

---

11 Shapiro (1997). For other recent non-eliminative structuralist accounts, see Parsons (1965), Linnebo and Pettigrew (2014) and Schiemer and Wigglesworth (2018).

12 Hale and Wright argue that the impredicativity of the domain is harmless if the domain can be specified in such a way that doesn't employ any particular reference to its members, that is, a range-unspecific specification. The use of an antecedent means of reference to the numbers, however, renders the possibility of a range-unspecific specification of the domain a moot point (Hale & Wright 2001, 242).

conception of the objects in its domain. To put it starkly: alignment with the specification requirement entails a thin conception of objects, and alignment with its negation entails a thick conception of objects.

3.3. Frege's Realism

The final piece of the puzzle, then, is the observation that the thick conception of objects is Frege's own. Like Dummett, I take it to be a characteristic of Frege's ontology that objects can always admit of different modes of presentation beyond (i.e. independent of) those by which it has been introduced, or in Dummett's terms, that objects are full blooded.[13] In consequence, Frege's conception of objects qualifies as a thick conception of objects in line with our definition. More to the point, this means Frege's conception is incompatible with the specification requirement.

Frege's commitment to a thick ontology can be seen in several central aspects of his work. To begin with, there are two related defining features of Frege's conception of objects: one is that they are the referents of singular terms, something that Hale and Wright heavily exploit in their account. The other is that such singular terms "are to be understood as standing for *self-subsistent* objects" (Frege 1884, §62, emphasis mine). I read Frege's latter claim to be in direct tension with Hale and Wright's specification requirement. The view that numbers are self-subsistent seems, on the surface, at least, to imbue numbers with a life of their own outwith their abstractionist characterisation.

Against this, however, Hale and Wright could object that Frege's requirement that numbers be self subsistent is the assertion of a straightforward kind of realism and nothing more. Furthermore, they could rightly point out that in the preceding sections of *Grundlagen*, Frege argues that number statements inherently contain "an assertion about a concept" (Frege 1884, §55). They might infer from this that Frege's realism about numbers is entirely compatible with holding that numbers can only be introduced by means of their essential characterisation, since any grasp of them must presuppose a grasp of "Nx : _" in HP. This would seem to exonerate the specification requirement.[14]

In response to this line of objection, I first note my wholehearted agreement that the specification requirement is in no tension with Frege's realism. I am happy to grant Hale and Wright that, even in the worst case, their programme can safeguard some kind of realism. My contention is rather with the kind of realism they are able to safeguard and, in particular, the fact that they cannot deliver a non-thin ontology of objects characteristic of Frege's realism.

---

13 See Dummett (1981, 471–511); Dummett (1998, 63; 153).
14 Thanks to an anonymous referee for this point.

Likewise, I agree that, for Frege, it is essential to numbers that they be numbers *of* concepts and that this provide a sense in which numbers are dependent on being introduced via the numerical operator. However, this dependency is in no tension with the claim that they are independent in another sense; that even if it is epistemically necessary that any numerical definition be formulated so as to capture this essential characteristic of number, metaphysically speaking, neither their existence nor their nature depends on their being introduced so. It is this latter metaphysical claim, concurring with my working definition of thick objects, which I take Frege to have in mind when he argues for the independence of numbers and objects more generally.

The true point of contention, then, is whether Frege's conception of numbers as self subsistent, and his foundational work in general, indicate a thick conception of objects which aligns with the definition provided previously. For, if it does, then we can draw the conclusion that the specification requirement—which is in tension with *any* thick ontology—is incompatible with Frege's realism. For this reason, I devote the next section to the examination of textual evidence establishing the implausibility of denying that Frege's conception of objects is thick, rather than thin.

3.4. Frege's Conception of Objects

Let us begin by examining the context of Frege's requirement that numbers are self subsistent in order to faithfully reconstruct what Frege means by this.

The relevant claim forms the subject matter of *Grundlagen* IV §§55–69, entitled with the lemma "Every individual number is an self-subsistent object". In that discussion, Frege claims that the *reason* he identifies numbers as self-subsistent is, "to preclude the use of such [numerical] words as predicates or attributes, which alters their meaning" (1884, §60). In other words, Frege means to establish numbers not as concepts but as first-order objects in the Fregean hierarchy. To do this, we "use the definite article to register 1 as an object", referring to numbers with singular terms like "the number 1" (Frege 1884, §57).

Were this all Frege had to say on the matter of the self-subsistence of number, then, contra to my interpretation, Hale and Wright could rightly insist that by it, Frege only meant a rather neutral ontological categorisation belying his realism about first-order entities. However, Frege goes on to draw out a revealing conclusion from the fact that numbers must be conceived of as objects: that this implies a certain class of sentences must have a sense, namely those which "express our recognition [of a number as the same again]" (Frege 1884, §62). This is because:

> If we are to use a symbol *a* to signify an object, we must have a criterion for deciding in all cases whether *b* is the same as *a*, even if it is not in our power to apply this criterion.
>
> (ibid)

The fact that Frege concludes that we must define an identity criterion for numbers *purely* from the basis that numbers are objects is revelatory of his inherent conception of objects more generally. In particular, it shows that Frege takes the objects of the domain to be such that they can be given to us in different ways. Or, in other words, that they admit of distinct modes of presentation which necessitate the requirement of a means of determining, in principle, when singular terms expressing distinct modes of presentation determinate the same referent.[15] Therefore, Frege's ontology is incompatible with the specification requirement, since the specification requirement is incompatible with a thick ontology and Frege demonstrates a thick conception of objects in the sense we have defined—that is, objects can be determined by senses further to those senses by which they have been introduced.

To see this crucial point in a little more detail, let us first remind ourselves what the specification requirement amounts to: namely, the restriction that the numbers admit of no mode of presentation independent of that which is given in abstraction. In other words, for abstraction to work, the only sense permitted to determine the numbers is the sense given by the number function implicitly defined in abstraction. If Frege's realism were compatible with the specification restriction, there would be no substance to his claim that numbers can be "recognised as the same again", since the requirement entails that objects can only be given in one way (Frege 1884, §56).

It must be admitted, however, that although there would be no *substance* to this claim, there would still be a *sense*, albeit a trivial one. For Hale and Wright could maintain that the very means of recognising an object as the same again could issue from a recognition of whether those objects were introduced by the same abstractionist mechanism.

Against this, Frege anticipates and explicitly rejects any such strategy where, later in the discussion, he considers it for the analogous case of directions. He protests that if objects are identified on the basis of their means of introduction:

> then the way in which the object . . . is introduced would be treated as a property of it, which it is not.
>
> (Frege 1884, §67)

Frege's objection here seems to be that, as we already inferred, if objects were to be distinguished by their definitional introduction, this would

---

15 I take it to be an uncontroversial reading that where Frege says that an object can be given in a different way, he does not *merely* mean that it can be referred to by two distinct singular terms expressing the same sense. Especially given that where Frege explicitly distinguishes sense from reference, he does so by a demonstration that distinct *senses* can determine the same reference, per his solution to Frege's Puzzle (Frege 1892, 210).

"presuppose that an object can only be given in one single way", so that all identities between numbers would amount to the principle that: "whatever is given to us in the same way is to be recognised as the same" (1884, §67). Frege's noted criticism regarding such a principle is that it is "so unfruitful that it is not worth stating". This is because "no conclusion could ever be drawn here that was different from any of the premises" (ibid). Frege claims this goes against every purposeful use of "meaningful" equations involving numbers, which he says depend "on the fact that something can be re-identified even though it is given in a different way" (ibid).

Thus, in his argument for rejecting the strategy which must appear tempting to Hale and Wright, Frege ultimately relies on his prior claim that numbers can be given in different ways. To digress for a moment, I take it to be entirely consistent with a *neo*-logicist programme that neo-logicists do not wholesale adopt Frege's framework, in particular Frege's rejection of Hume's Principle, of which the previous objection forms a part. After all, it is the conjecture of their programme that any scruples that Frege harboured in that regard can be overcome. So I want to make clear before I proceed that by pressing this objection of Frege's, I do not mean to suggest otherwise but only to use it as part of a excavation of the realism which underlies Frege's foundational programme and which Hale and Wright claim to safeguard by the resuscitation of Hume's Principle towards this aim.

To return to the findings of said excavation: Frege's reliance on the claim that numbers can be given in different ways is significant for various reasons. The first is that it demonstrates that Frege does not consider discriminating between objects by means of their definitional origins to honour his claim that different objects can be recognised as the same again.

Second, it is revelatory of the importance of this claim to Frege. It clarifies beyond doubt that the conception of numbers as inherently capable of admitting distinct modes of presentation is not an incidental or optional thesis. Nor is it a throwaway means of indicating a neutral kind of realism. Rather, the way in which Frege employs this claim solidifies his adoption of a thick ontology of objects.

Third, it gives us insight as to *why* Frege took this claim to be beyond dispute. Namely, because it facilitates the very "fruitfulness" of mathematics—something as close to Frege's heart as the logicality of mathematics. Indeed, that Frege's *Begriffsschrift* is designed to preserve the generative nature of mathematics is what Frege points to in distinguishing the aim and power of his conceptual notation from the closed "sterile" systems that have gone before—mainly Aristotelian logic and Boolean logic (Frege 1881, 15; 46).[16]

16 Here Frege claims that his concept-script is fruitful because his function-argument conception enables thoughts to admit of novel decompositions distinct from their constituent senses. However, the key idea here is the same, that is, that his logic is a truly generative Leibnizian *lingua characteristica* in a way that the Boolean *calculus* is not (Frege 1881).

3.5. Summary

We can conclude from these considerations that it is central to Frege's conception of an object that it can be presented under different modes, that is, referred to by means of distinct senses. This also secures the result that Frege's conception of objects is thick in the sense that his conception of the independent individuals falling in the domain are such that they must be able to be presented by different modes of presentation, further to the means by which they have been introduced. This feature of objects is essential to Frege, since it underlies the inherent fruitfulness of mathematics.

This is significant because we already established that an ontology of thick objects is incompatible with the specification requirement. This is because, if the objects that Hale and Wright recover by abstraction are required not to admit of any mode of presentation other than the one created by the process of definition by abstraction, then the only kind of numbers this process could hope to recover would not be recognised by Frege as objects. Thus, the mere adoption of the specification requirement amounts to an implicit rejection of a Fregean ontology.[17]

4. The Requirement Underpins the Central Ontological Idea

The question that remains to be addressed is why Hale and Wright would insist on this problematic requirement. Unfortunately, this is no oversight—rather—it is required if they are to defend the ability of abstraction principles to deliver their central ontological idea.

The basic point here is that if the objects of abstraction are independent of the means by which they are introduced, then they could go rogue in various ways: the objects of abstraction could fail to exist, the function defined by the abstraction principle could fail to exist, the objects of abstraction could have other aspects to them such that the abstraction principle would fail as a definition. Granting the object's independence makes all of these things a possibility and thus forcibly requires any successful introduction of them to provide a means of ruling out these possibilities. In other words, to define independent objects, abstraction principles would have to be supplemented with a guarantee of characterising the essential properties of an object and with an existential proviso. But such supplementation directly undermines the central ontological

---

17 This is not to deny that there are some distinct modes of presentation of numbers secured by Hume's Principle such as '1+1' and 'the successor of 1'. Nor is it to deny that Hume's Principle can underpin new arithmetic theorems. The relevant point is that such limited modes of presentation are entirely derivative from the successor function and the resources of Hume's Principle cannot provide all the features of numbers Frege recognised, in particular, their sheer potential to have further modes of presentation which go beyond their means of introduction by Hume's Principle. Many thanks to Eric Reck for this point.

idea: the idea that an implicit definition can establish a route to infinitely many objects while avoiding metaphysical and epistemic debt.

Thus, the requirement of a single mode of presentation of an object— that is, that we may have—indeed there may be possible—no prior independent way of conceiving of the objects—is needed to shore up the central ontological idea that abstraction principles are able to deliver the numbers. Otherwise, additional assumptions would be required beyond the abstraction principle to the effect that the objects/functions are actually *there* and that they are how the definition says they are.

It should be stressed that Hale and Wright are not unaware of the important link between the central ontological idea and their requirement that objects recovered by abstraction admit only that mode of presentation. Take, for example, where they say:

> So when Boolos asks, what reason do we have to think that there is any function of the kind an abstraction principle calls for, it is to skew the issues to think of the question as requiring to be addressed by the adduction of some kind of evidence for the existence of a function with the right properties that takes elements from the field of the abstractive relation as arguments *and objects of some independently available and conceptualisable kind as values.*
>   (Hale & Wright 2009, 92, emphasis mine)

Here they diagnose Boolos as having failed to adopt a conception which upholds both the idea that no independent ontological collaboration of the abstraction principle is needed (the central ontological idea) and their requirement that the object-values admit no independent means of specification (the specification requirement). The point is that *both* of these elements are required for a conception that can hope to deliver logicism via abstraction.

Or, take where they summarise their position against Potter and Sullivan by concluding that:

> Our position remains, unrepentantly, that even in the impredicative case, *the newly introduced terms may provide our sole underived means of reference to the objects in question,* and that warrant to regard those terms as referring need be sought no further afield than in the abstraction principle itself, along with our ordinary and antecedently available means of appraising the truth of instances of its right-hand side.
>   (Hale & Wright 2008, 206, emphasis mine)

The placement of this (emphasised) appeal again shows that Hale and Wright are quite aware of the important role this requirement plays in their defensive strategy. They are quite right to insist on the importance of being granted the requirement that the objects of abstraction admit of a single mode of presentation. The issue is that this important requirement is, as we have seen, inherently anti-Fregean.

Seen in this light, it becomes clear that Hale and Wright's specification requirement essentially builds an ontological presupposition into the process of implicit definition by abstraction. Abstraction is intended to recover objects which admit of a single mode of presentation and may have possible no independent means of specification external to the abstraction principle. This renders abstraction principles theoretical tools which are only suitable for establishing reference to objects of a certain kind, namely thin objects.

5.   What Kind of Numbers Should Be Given to Us?

I want to take a step back in this final section to observe where these considerations leave the project of Scottish neo-Fregean logicism.

Taking Stock

What we have seen is that the very same thing that enables Hale and Wright to protect the central ontological idea that they have forgone a Fregean ontology in which the natural numbers feature as thick objects.

The first thing I want to strongly emphasise is that I take the abandonment of this central feature of Frege's conception of objects to beset many canonical elements of Hale and Wright's project, not merely their specification requirement. I have used their claim that there is no means of specification of the relevant objects independent of abstraction because I think this is the easiest and shortest way into my objection, but it is by no means the only one. To show this, I want to briefly illustrate two further routes into this same ontological sleight of hand: Hale and Wright's stance on a theory of reference and their solution to the Julius Caesar problem.

5.1.   *The Preferred Conception of Abstraction*

Hale and Wright accuse Potter and Sullivan of saddling them with a conception of reference which they want to reject at all costs—the "Lockean" theory of reference—upon which there is an element of blind pointing involved in our attempts to refer to things in the world (Hale & Wright 2009, 180).[18] This being so, the purported referents of our expressions can always let us down by being different, by not being unique or by not being there at all. This doesn't bode well for an account of definition by abstraction which is bent on doing without any ontological supplementation. Instead Hale and Wright present a theory of reference which they call "minimalism". They develop their theory partly by analogy to

---

18 This occurs in the context of their dispute spanning Potter and Sullivan (1997), Potter and Sullivan (2005), Hale and Wright (2008) and Hale and Wright (2009).

Aristotle and partly by analogy to the abundance theory.[19] However, Hale and Wright make it clear that their preferred minimalism is distinct from both views, for, like the abundance theory, minimalism holds that establishing the sense of an expression is enough to guarantee reference on account of a plentiful ontology—but unlike the abundance theory, minimalism restricts this ontological bounty to a metaphysically conservative headcount of the functions and objects in the domain of second-order logic. Like Aristotle, minimalism requires that expressions be true of something—but unlike Aristotle, minimalism requires singular terms, not predicates, to be true of something (in a range of atomic statements) (2009, 208).

Putting all this together, we get minimalism's distinctive thesis: all that is needed for a watertight assurance of successful reference is that the term in question is "reference demanding", that is, is a referring expression, and that it features in a range of true atomic sentences.[20] Unpacking this campaign slogan a little: Hale and Wright take such an atomic context to be given by the identity statements on the LHS of abstraction principles, so that, in at least some cases, the task of establishing the truth of an expression in an atomic sentence is realised by the process of abstraction, that is, by the success of the bi-conditional as an implicit definition which bestows sense on an expression on its LHS as well as establishing it as a syntactically reference demanding expression and—via the uncontroversial worldly assist of the truth of the RHS—the truth of the atomic context on the LHS. All together, this establishes the reference of the senseful reference demanding expression on the LHS (e.g., "$Nx : \_$").

The details of this slogan are important to spell out, because it means that the bottom line of Hale and Wright's endorsement of the minimalist theory is that implicitly defining an expression via an abstraction principle is enough to establish the existence and nature of the referent of that expression. This must be appreciated in order to understand that it is Hale and Wright's minimalism that enables them to preserve their central ontological idea. For, under minimalism, the process of definition by abstraction *internally* provides the guarantee that the objects exist and are exactly as the definition stipulates them to be (Hale & Wright 2009, 207). As such, abstraction can no longer be thought to require any independent metaphysical assist or appeal.

However, the thesis that abstraction can provide such an internal guarantee also marks the point of departure between minimalism and

---

19 According to the abundance theory, whatever property can exist does exist (Lewis 1983, 346). As such, every well-formed predicate will refer to a property. This contrasts with a sparse theory according to which a well-formed predicate might fail to refer to a property simply because the property doesn't exist; Hale and Wright's appeal to Aristotle concerns his claim that predicates be "true of something" (Hale & Wright 2009, 208).

20 See Hale and Wright (2009, 208; 209) and Hale and Wright (2001, 22).

Lockeanism. For Lockeanism, the truth of an atomic context, or any context, is not sufficiently established by a successful implicit definition by abstraction. Why not? In short, because minimalism assumes an ontology of thin objects, exhausted by their means of introduction, such that the process of abstraction provides a sufficient ontological guarantee, while Lockeanism assumes an ontology of thick objects which of necessity outstrip their means of introduction. As such, no species of implicit definition could provide the necessary assurance of their existence or nature.

So Hale and Wright are right to insist that minimalism is incompatible with Lockeanism and convincingly demonstrate that abstraction principles can fulfil minimalism's demands (2009, 205). However, this unexpectedly implies that minimalism is also incompatible with *Frege's* conception once it is appreciated that thin objects have no place in Frege's ontology.[21]

5.2.   *The Partial Solution to Julius Caesar*

Finally, let us look at an entirely different angle from which the ontological switch-out can be witnessed: Hale and Wright's solution to the Julius Caesar objection. As is familiar, they appeal to the sortal inclusion principle:

> **Sortal Inclusion Principle** (SI#): Some Fs are Gs only if F and G are each sub-sortals of one and the same category.[22]

By this means, Hale and Wright partition the domain into categories: a special kind of sortal concept (2001, 389). Hale and Wright then claim that the sortals *number* and *people* must be associated with different identity conditions. This is because Hume's Principle derives the identity criterion for the natural numbers by way of an appeal to equinumerosity relations between concepts and a criterion that personal identity can never be determined by appeal to equinumerosity relation between concepts. This is made even more intuitive by their claim that we are dealing with pure sortals—that is, the most general sortals which characterise the "essence" of an object. The thought is that the intrinsic properties of personhood can never be established by an appeal to concepts and relations between them. Thus, the number 2 and Julius Caesar must belong to distinct categories, and cross-categorical identity statements involving them must be false or as problematically indeterminate as cross-categorical statements in general (Hale & Wright 2001, 394–396).

---

21 I take the ontological alignment of Lockeanism and Frege's conception to be no coincidence, since I read Potter and Sullivan to understandably assume that Hale and Wright mean to recover Frege's ontology proper.
22 Hale and Wright (2001, 396).

The issue here is that Frege, at a stretch, would have admitted only two categories: objects and functions. More precisely, this is to say that in his ontology, he takes there to be fundamentally only two *kinds of things*. Further, Frege would have admitted only *one* category of objects, since, for him, all objects are of level 0.[23] Thus, where Hale and Wright classify "abstract object" and "concrete object" as two distinct categories, Frege would subsume them as sortals under the single category "object". The most substantial argument against subsuming these sortals into this more general sortal in Hale and Wright's discussion is the fact that the abstract/concrete divide promises to provide a temporary solution to the Julius Caesar problem.[24] However, I have always taken it to be the case that the Julius Caesar problem occurs *precisely because*, for Frege, Julius Caesar and the number two *are both objects which fall under sub-sortals of the same category*, that is, the category *object*.

The point of importance here is that the difference in categorical categorisation constitutes a substantive ontological difference in bifurcating Frege's unified domain of objects. Let me be clear, I am not claiming that Hale and Wright's categories overturn Frege's ontological division between objects and concepts; this is not the case. What they do overturn is the *lack* of division between objects so as to signal a departure from Frege, not by tampering with the hierarchy but rather by establishing a parallel ontological classification system which does not remain faithful to all of the ontological principles underlying and informing the hierarchy.

In this way, Hale and Wright have moved away from a universal domain and a Fregean conception of an object as a member of that domain to a collection of equivalence classes (or categories) into which fall objects whose nature is exhausted by their falling into an equivalence class. This produces an ontology more akin to non-eliminative structuralism than to Fregean logicism. Again, this move is not easily revocable by Hale and Wright because upon it rests the solution to the very problem by which Frege himself points to the untenability of treating Hume's Principle as a foundational basis for the natural numbers, that is, the Caesar Problem.

6. Interesting Future Directions

I have focused on an argument which amounts to the charge that Hale and Wright's account is not a neo-Fregean species of logicism and furthermore that Hale and Wright cannot revise their unFregean thesis that objects

---

23 Note that if we understand categories as dividing up logically, then actually Frege would have admitted infinitely many kinds of things and their associated categories, since there are infinitely many logical functions typed by their arguments and values. However, this would still give only a single category for level 0.

24 I say temporary because it doesn't yet stop Caesar coming back as an *abstractum*, that is, as any non-numerical object falling in the same category as the natural numbers.

admit one mode of presentation on account of the fact that it would invalidate their central ontological idea. Although this might initially be seen as an unwelcome result for abstractionism, I think it can also be seen in a positive light: the ontology is not revisable precisely because the unFregean ontology *does* support the central ontological idea. What we are left with is a better understanding of the requisite metaphysics of the best-developed account of abstraction available. The logicist hope is not dead yet.

For reasons that I have not detailed, I would sooner call this position structuralist logicism. After all, abstraction principles may not be fit for the delivery of thick objects, but in light of what we have discussed, they appear very well suited for the recovery of thin objects such as objects which are anyway merely reifications of positions in a structure—that is, non-eliminative structuralist objects.[25] A logicism founded on the recovery of thin objects is reminiscent of elements in Zalta's proposal of providing a modal route to the natural numbers,[26] in particular Zalta's encoding relation which holds between an abstract object and its properties is such that "[t]he properties that an abstract object encodes are constitutive of its nature, and as such, are essential to its identity as an object" (1993, 396). This comes very close to the definition of thin objects we have been using. So we might say that abstraction principles, when successful, establish an encoding relation between an *abstractum* referred to on its LHS and its properties, by virtue of its RHS. Such an account is thus consistent with a broader structuralist project and perhaps even a (non-eliminative) structuralist logicism.

More generally, the attempt to provide a distinctively structuralist logicism goes back to Dedekind. Dedekind is often pointed to as an early proponent of mathematical structuralism.[27] But there is a growing recognition that Dedekind was also a logicist, as argued by Ferreirós (forthcoming), Klev (2017), Reck (2013) and Shapiro and Hellman (2019) in the second chapter of their recent book. Indeed, Frege himself identifies Dedekind as a logicist in *Grundgesetze*: "Dedekind too is of the opinion that the theory of numbers is a part of logic" (1893, 196).

I believe there is great promise in a view we might call *neo-Dedekindian logicism*—a marriage of elements of Dedekind's approach and Hale and

---

25 Landry has suggested that we should replace logicism with structuralism so we can best account for "the objectivity required for the justification of the truth of mathematical propositions" (2001, 79). This offers a contrast to the current proposal of bringing them together via the infrastructure of Hale and Wright's account of abstraction.

26 See Zalta (1993, 1999). Note that Zalta doesn't classify his approach as a logicist one—perhaps on account of the controversial question of the logical status of modal logic.

27 For further discussion of Dedekind's structuralism, see Shapiro (2005), chapter 5. Hellman (1989, vii) and Shapiro (1997, 14) both identify him as probably the first mathematical structuralist. See also Reck (2003), Sieg and Morris (2018).

Wright's account of abstraction. The latter bolsters their central ontological idea by means of Dedekind's structuralist ontology. The former finds a means of being absolved of his most enduring criticism: that the "free creation" of his infamous Theorem 66 in *Was sind und was sollen die Zahlen?* is fundamentally a mental process which thus infects the real numbers and natural numbers defined—at best establishing something like "psychologism" for arithmetic (Dedekind 1888, 357).

Of course, Dedekind's notion of abstraction is quite different from Hale and Wright's. It begins with a system that exemplifies the target structure and "abstracts" the structure as a so-called free creation—a kind of abstraction that Frege criticises at length.[28] My suggestion is that we could fruitfully refine this suggestive process of free creation with the neologicist process of definition by abstraction. After all, the central ontological idea is precisely to defend the fact that there is no *illegitimate* free creation involved in the process of abstraction. Furthermore, defenders of Dedekind like Ferreirós argue that the process of free creation should be understood logically, bringing it closer to Hale and Wright's view:

> it is not even a matter of creating new individuals, but merely of regarding given individuals merely from the point of view of an axiomatically characterised structure to which they belong, abstracting from any other of their properties. Taking into account the purely logical character ascribed by Dedekind to the "general laws" reigning over sets and mappings, and the fact that the "creation" of numbers is strictly regulated by such laws, it seems to me that his talk of numbers as "free creations of the mind" does not cause any tension with a crisp form of logicism.[29]
>
> (Ferreirós forthcoming, 10–11)

This perspective brings out how the conceptions of abstraction at play might be different, but the ontological result is the same: the objects of each form of abstraction are guaranteed to be objects which are exhausted by their characterisation of—and inherit their logicality from—the axioms/abstraction principles which define them.

I hope for this to demonstrate that I take my objection to Hale and Wright's account to serve as a diagnostic refinement. Of course, far more needs to be said to properly develop a position like neo-Dedekindian logicism. However, I will finish on the promising note that recently the potential of a structuralist logicism has begun to gain serious attention in the literature. Foremost is Linnebo's (2018) admirable book on thin objects, which devotes most of its chapters to setting out and developing what he

---

28 See Frege (1884, §13: 34) and Frege (1906/1971, 125).
29 See also Tait (1986), Yap (2009), Shapiro and Hellman (2019), Reck (2003, 2017).

calls *dynamic abstraction*. This could be easily utilised to develop a position like neo-Dedekindian logicism, especially because Linnebo's work also draws from the wealth of Hale and Wright's project to provide an account of abstraction, not to mention the fact that his account of abstraction is specifically designed to recover thin objects.[30] There is also some useful work by Boccuni and Woods, who provide a development of versions of structuralist logicism in which mathematical singular terms refer arbitrarily. Their idea is to use abstraction principles to establish the sense and (arbitrary) reference of such singular terms together with structuralism in the sense that "we treat only the arithmetical properties—those following from facts about the equivalence relation of equinumerosity—as properties of the numbers" (Boccuni & Woods 2018, 3).[31]

7. Concluding Remarks

The real heart of my diagnostic objection comes not from the fact that there exists a range of objects which abstraction principles are impotent to introduce; rather, it comes from exposing the fact that the best cases of abstraction principles which fully characterise (thick) abstracts amount to sheer coincidences: coincidences in the sense that we require additional assurance that the characterisation of the objects given by abstraction is indeed complete. This assurance amounts to a demonstration that the abstracts are not lacking in essential properties. To carry this out, however, we require independent access to the objects characterised by abstraction in order to assure ourselves that the abstraction has delivered the desired objects in their full glory, that is, with all the properties which we antecedently know those objects to have. Any such a metaphysical supplementation flies in the face of the neo-logicist's central ontological idea.

Therefore, the only way to preserve the central ontological idea is to insist that the kind of objects that abstraction principles can establish reference to must admit of no specification independent of the abstraction (which would invalidate the central ontological idea), because they necessarily have all and only those properties provided by the abstraction

---

30 Interestingly, Linnebo argues he can avoid Russell's paradox by drawing inspiration from the iterative conception of sets, using iterated domain extensions such that problematic abstracta fall outside the domain (Linnebo 2018, 55–63).

31 By arbitrary reference, they mean without unique individuation either epistemically or, as far as I understand them, in the sense of proofs which introduce an assumption *without loss of generality*; Boccuni and Wood's account is meant to be classified as *structuralist* neo-logicism because it treats abstraction principles as logical while adopting a structuralist ontology—so that abstraction principles are implicit definitions which establish (genuine though arbitrary) reference to "objects . . . which could serve to play the role of numbers" (2018, 19).

process. In short, abstraction principles are only apt to recover thin objects.

The problematic aspect of this subtle metaphysical shift in the background ontology of abstraction is that it is inherently anti-Fregean. As we have seen, for Frege, numbers are objects and objects are thick. Therefore, the very same ontological manoeuvre that is required to shore up Hale and Wright's delivery of the central ontological idea that definition by abstraction can be made to work also entails a stark departure from Frege's metaphysics which strips the project of its claim to providing even a *neo*-Fregean result.

However, this need not be the end of the line for their logicist program if Hale and Wright are willing to embrace the ontology that they have implicitly shown to facilitate the process of definition by abstraction. For their logicist project—whether Fregean or not—constitutes the most refined and intricate alternative to brute axiomatic stipulation available and remains an essential resource in the quest to account for exactly how theoretical objects are given to us such that we can have knowledge of them and determine their nature.

References

Boccuni, F. & Woods, J. (2018). "Structuralist Neologicism", *Philosophia Mathematica* **1093**(10), 1–21. https://doi.org/10.1093/philmat/nky017.

Boolos, G. (1997). "Is Hume's Principle Analytic?", in Richard G. Heck Jnr, ed., *Language Thought and Logic: Essays in Honour of Michael Dummett*, Oxford: Clarendon Press, pp. 245–261.

Dedekind, R. (1888). *Was sind und was sollen die Zahlen?*, Braunschweig: Vieweg. English translation "The Nature and Meaning of Numbers" by W.W. Beman, pp. 29–115 in W.W. Beman, ed. (1901) *Essays on the Theory of Numbers*, Chicago: Open Court Publishing Company.

Dummett, M. (1981). *Frege: Philosophy of Language*, 2nd ed., London: Duckworth.

Dummett, M. (1993). *The Seas of Language*, Oxford and New York: Oxford University Press.

Dummett, M. (1998). *Frege: Philosophy of Mathematics*, London: Duckworth.

Ferreirós, J. (forthcoming), 'On Dedekind's Logicism', in A. Arana & C. Alvarez, eds., *Analytic Philosophy and the Foundations of Mathematics*, Palgrave Macmillan.

Frege, G. (1881). "Boole's Logical Calculus and the Concept–Script", in F. Meiner, H. Hermes, F. Kambartel & F. Kaulbach, eds., *Posthumous Writings: Gottlob Frege*, Chicago: University of Chicago Press, pp. 9–52. English translation by F. Meiner, H. Hermes, F. Kambartel & F. Kaulbach, 1979.

Frege, G. (1884). *Grundlagen der Arithmetik*, Breslau: W. Koebner. Translation by J. L. Austin pp. v–viii in Austin (1974) *The Foundations of Arithmetic. A Logic–mathematical Enquiry into the Concept of Number*, Oxford: Blackwell.

Frege, G. (1892). "Über Sinn und Bedeutung", *Zeitschrift für Philosophie und philosophische Kritik* **100**, 25–50. Translated as 'On Sense and Meaning' in M.

Black and P. Geach, eds., *Translations from the Philosophical Writings of Gottlob Frege*, 3rd ed., pp. 56–78. Oxford: Blackwell.

Frege, G. (1893). *Grundgesetze der Arithmetik: Begriffsschriftlich abgeleitet*, Jena: Verlag Hermann Pohle. Translation by P. Ebert, M. Rossberg & C. Wright, *Basic Laws of Arithmetic*, Volume I (2013), Oxford: Oxford University Press.

Frege, G. (1906/1971). "On the Foundations of Geometry: Second Series", in E. H. W. Kluge, ed., *On the Foundations of Geometry and Formal Theories of Arithmetic*, New Haven & London: Yale University Press, pp. 49–112.

Geach, P. (1955). "Class and Concept", *Philosophical Review* **64**(4), 561–570.

Hale, B. & Wright, C. (2001). *Reason's Proper Study: Essays Towards a Neo-Fregean Philosophy of Mathematics*, Oxford: Oxford University Press.

Hale, B. & Wright, C. (2008). "Abstraction and Additional Nature", *Philosophia Mathematica* **16**(2), 182–208.

Hale, B. & Wright, C. (2009). "The Metaontology of Abstraction", in *Metametaphysics: New Essays on the Foundations of Ontology*, Oxford: Oxford University Press, pp. 178–212.

Heck, R. (1993). "The Development of Arithmetic in Frege's" *Grundgesetze der Arithmetik*', *Journal of Symbolic Logic* **58**(2), 579–601.

Heck, R. (2011). *Frege's Theorem*, Oxford: Clarendon Press.

Hellman, G. (1989). *Mathematics Without Numbers: Towards a Modal-Structural Interpretation*, Oxford: Clarendon Press.

Klev, A. M. (2017). "Dedekind's Logicism", *Philosophia Mathematica* **25**, 341–368.

Landry, E. (2001). "Logicism, Structuralism and Objectivity", *Topoi* **79**(20), 79–95.

Lewis, D. (1983). "New Work for a Theory of Universals", *Australasian Journal of Philosophy* **61**(4), 343–377.

Linnebo, Ø. (2018). *Thin Objects: An Abstractionist Account*, Oxford: Oxford University Press.

Linnebo, Ø. & Pettigrew, R. (2014). "Two Types of Abstraction for Structuralism", *Philosophical Quarterly* **64**, 267–283.

Parsons, C. (1965). "Frege's Theory of Number", in M. Black, ed., *Philosophy in America*, Ithaca, NY: Cornell University Press, pp. 180–203.

Potter, M. D. (2004). *Set Theory and Its Philosophy: A Critical Introduction*, Oxford: Oxford University Press.

Potter, M. D. & Sullivan, P. (1997). "Hale on Caesar", *Philosophia Mathematica* **5**(2), 135–152.

Potter, M. D. & Sullivan, P. (2005). "What Is Wrong with Abstraction?", *Philosophia Mathematica* **13**(2), 187–193.

Reck, E. H. (2003). "Dedekind's Structuralism: An Interpretation and Partial Defense", *Synthese* **137**(3), 369–419.

Reck, E. H. (2013). "Frege, Dedekind, and the Origins of Logicism", *History and Philosophy of Logic* **34**(3), 242–265.

Reck, E. H. (2017). "Dedekind's Contributions to the Foundations of Mathematics", in Edward N. Zalta, ed., *The Stanford Encyclopedia of Philosophy* (Winter 2020 Edition), URL = <https://plato.stanford.edu/archives/win2020/entries/dedekind-foundations/>.

Rumfitt, I. (2001). "Hume's Principle and the Number of All Objects", *Noûs* 35(4), 515–541.

Schiemer, G. & Wigglesworth, J. (2018). "The Structuralist Thesis Reconsidered", *British Journal for the Philosophy of Science*, 1–26.

Shapiro, S. (1997). *Philosophy of Mathematics: Structure and Ontology*, Oxford: Oxford University Press.

Shapiro, S. (2005). *The Oxford Handbook of Philosophy of Mathematics and Logic*, Oxford: Oxford University Press.

Shapiro, S. & Hellman, G. (2019). *Mathematical Structuralism*, Cambridge: Cambridge University Press.

Sieg, W. & Morris, R. (2018). "Dedekind's Structuralism: Creating Concepts and Deriving Theorems", in E. H. Reck, ed., *Logic, Philosophy of Mathematics, and their History: Essays in Honor of W. W. Tait*, London: College Publications, pp. 251–301.

Tait, W. (1986). "Truth and Proof: The Platonism of Mathematics", *Synthese* 69(3), 341–370.

Wright, C. (1983). *Frege's Conception of Numbers as Objects*, Aberdeen: Aberdeen University Press.

Yap, A. (2009). "Logical Structuralism and Benacerraf's Problem", *Synthese* **171**, 157–173.

Zalta, E. N. (1993). "Twenty-Five Basic Theorems in Situation and World Theory", *Journal of Philosophical Logic* 22(4), 385–428.

Zalta, E. N. (1999). "Natural Numbers and Natural Cardinals as Abstract Objects: A Partial Reconstruction of Frege's *Grundgesetze* in Object Theory", *Journal of Philosophical Logic* 28(6), 619–660.

15 Frege's Theorem and Mathematical Cognition

Lieven Decock

1. Counting and Cardinality

At the end of the 19th century, our understanding of the natural numbers was greatly enhanced by the development of formal systems characterising them. The most common axiomatisation was first presented by Peano in 1889 in his *Arithmetices principia, novo methodo exposita* (1989). A contemporary form[1] of the axioms is:

1. $N0$
2. $Nx \land Pxy \rightarrow Ny$
3. $Nx \land Pxy \land Pxz \rightarrow y = z$
4. $Nx \land Ny \land Pxz \land Pyz \rightarrow x = y$
5. $Nx \rightarrow \sim Px0$
6. $Nx \rightarrow \exists y \, (Pxy)$
7. $F0 \land \forall x \forall y \, (Nx \land Fx \land Pxy \rightarrow Fy) \rightarrow \forall x \, (Nx \rightarrow Fx)$.

Central in Peano's conception of the natural numbers N is the relation between successive numbers, the predecessor relation P in the previous axioms. Often the inverse relation S, the successor relation, is used instead. In Peano's original formulation of the axioms, the successor of a natural number n was denoted by $n + 1$, but this notation has the disadvantage of suggesting that addition is more fundamental than the predecessor or successor relation.

An alternative formalisation was developed by Frege, informally in the *Grundlagen der Arithmetik* (1980), and formally in the *Grundgesetze der Arithmetik* (2013). The central idea in Frege's conception of number is Hume's Principle. It was given its name by Boolos (1987, 6) after the following passage in Hume's *Treatise of Human Nature*

---

1 I heavily draw on Heck (2011, 156–179) for formalism and notation, as Heck's article will be the starting point for my philosophical discussion on the relevance of Frege's Theorem for mathematical cognition.

DOI: 10.4324/9780429277894-18

(1978, bk. 1, pt. 3, ch. 1 §5), which is mentioned in Frege's *Grundlagen* (1980, §63):

> We are possest of a precise standard, by which we can judge of the equality and proportion of numbers; and according as they correspond or not to that standard, we determine their relations, without any possibility of error. When two numbers are so combin'd, as that the one has always an unite answering to every unite of the other, we pronounce them equal.

The central concept in Hume's Principle is the relation 'equinumerosity': two collections are equinumerous in case the objects of the two collections can be matched pairwise. The set of forks is equinumerous with the set of knives in the case where each fork can be paired with one and only one knife, and vice versa. Equinumerosity is a primitive relation and does not presuppose the concept of number (despite its name). Another name for the relation is one-to-one correspondence (also called bijection); two collections are equinumerous if and only if there is a one-to-one correspondence between the collections.

In Frege's *Grundlagen* (§62–63), it is invoked as a 'general criterion for the identity of numbers,' and in the *Grundgesetze*, it follows from the notorious Basic Law V. This law states that concepts F and G have the same extension if and only if $\forall x\, Fx \leftrightarrow Gx$. Shortly before the publication of the *Grundgesetze*, Russell pointed out that a contradiction, which became known as Russell's paradox, can be derived in Frege's system. Due to Russell's discovery of an inconsistency, Frege's conception of number remained for a long time unattractive.

The fate of Hume's Principle changed radically with Crispin Wright's (1983) discovery that Basic Law V can be sidestepped. Wright argued that Frege's *Grundgesetze* contains the essential ingredients[2] for the derivation of the laws of arithmetic. The second-order Peano postulates (PA) can be derived in a consistent second-order system with only one non-logical axiom HP, introducing the extra-logical predicate N (is the number of):

$$N x : Fx = N x : Gx \leftrightarrow \mathrm{Eq}_x\,(Fx, Gx).$$

This system has been called Frege Arithmetic (FA).

In this definition '$Nx : Fx$' is to be read as 'the number of Fs,' and Eq_x (Fx, Gx) expresses that the concepts F and G are in a one-to-one correspondence. This result became known as Frege's Theorem.

---

2 For a careful analysis of the extent to which Frege's Theorem can be reconstructed from parts of the *Grundgesetze*, see Heck (2011, 9–11; 40–68).

The existence of two[3] alternative conceptions of number leads to questions about their relation and priority. In recent decades, a detailed logical analysis has been made of the relative 'strength' of PA and FA. In order to make the comparison, some extra definitions are required (Heck 2011, 161):

DZ $0 = Nx : x \neq x$
DP $Pmn \equiv \exists F \, \exists y \, [n = Nx : Fx \wedge Fy \wedge m = Nx : (Fx \wedge x \neq y)]$
DN $Nx = P^{*=}0x,$

whereby $R^{*=}$ is the weak ancestral of R, defined as

$$R^{*=}ab \equiv \forall F \, [Fa \wedge \forall x \, \forall y \, (Fx \wedge Rxy \rightarrow Fy) \rightarrow Fb].$$

DZ states that 0 is the number of the concept of non-self-identity; obviously no objects fall under this concept. DP states that m precedes n if by eliminating one object from a concept F with number n, we obtain a concept with number m. DN states that by iteratively taking the predecessor of . . . the predecessor of a natural number n, at some point, the number 0 will be obtained.

Frege's Theorem shows that second-order Peano arithmetic (PA) is interpretable in Frege Arithmetic (FA), given the appropriate translations DZ, DP, and DN. Conversely, FA is not interpretable in PA, and hence, HP is 'stronger' than strictly necessary for the derivation of PA from a logical point of view. At first glance, FA looks more basic from a logical point of view. A more fine-grained analysis is possible, though. For example, if one considers a finite version of Hume's Principle (HPF) that states that Hume's Principle holds for finite concepts F and G, the axioms of PA are derivable from HPF and DZ, DP, and DN (Heck 2011, 162; 242). In this case however, HPF can be derived from PA in combination with Frege's definitions on the (rather uncontentious) condition that we add an axiom stating that every predecessor of a natural number is a natural number. Instead of full second-order logic, more restrictive logics can be used in the study of Frege's Theorem (see Heck 2011, 267–296). Moreover, an independent analysis of the derivation of each of the seven axioms of Peano can be given. Axioms (1), (2), and (7) are independent of Hume's Principle and follow from the definitions DZ, DP, and DN only. Axiom (6) is redundant, as it can be derived from axiom (3). The axioms essentially relying on Hume's Principle are axioms (3)–(5).

---

3 Other concepts of number are possible. In Decock (2008), I briefly indicate that category theory may yield a different concept of number; for a discussion of Hume's Principle in category theory, see Logan (2017).

From a logical point of view, the comparison of FA and PA is straightforward. The philosophical ramifications are less clear. In 'Cardinality, Counting, and Equinumerosity,' Heck (2000, reprinted in Heck 2011) addresses the question of whether HP is in any sense more fundamental than Peano's axioms (3)–(5) from an epistemological perspective.[4] Heck's analysis[5] differs from the neo-logicist account of the epistemological role of Hume's Principle. Wright[6] famously argued that Hume's Principle should be regarded as a conceptual truth and that the logical consequences of this kind of conceptual truths, arithmetical truths, should be regarded as conceptual or analytical truths. If correct, Frege's theory of the natural numbers would not only be vindicated from a logical point of view, but also the epistemological tenets of logicism would be saved. Heck (2000) argues that matters are more complicated and raises several issues. It is not totally obvious that Hume's Principle is a conceptual truth and that the class of conceptual truths is closed under deduction in second-order logic. Even if Hume's Principle were a conceptual truth, it is not evident that Frege's Theorem implies that mathematical truths are analytic. What the theorem proves is that PA can be interpreted in FA by means of the definitions DZ, DP, and DN. This does not guarantee that propositions in FA are *about* the natural numbers, only that the statements of FA can be systematically interpreted as statements about the natural numbers. Most importantly, it is not at all clear that Hume's Principle is actually involved in our ordinary understanding of numbers.

Heck's central issue concerns our cardinal understanding of number and its relation to counting. The issue goes back to a controversy between Frege and Husserl on the role of one-to-one correspondence. In *Philosophie der Arithmetik*, Husserl objected that equinumerosity presupposes the concept of number, and in his review of the book, Frege answers:

> It is in a similar way that the author [Husserl] judges the definition of numerical equality by means of the concept of a one-to-one correlation. 'The simplest criterion of equality of number is just that the *same* number results in counting the sets to be compared' (p. 115)

---

4 One could argue that HP is more fundamental from a metaphysical point of view, and theories of fundamentality, dependence, and grounding have been popular in metaphysics in the last two decades; for a recent discussion of the metaphysical interdependence of ordinal and cardinal numbers, see Assadian and Buijsman (2018). Heck (2011, 159) is still sceptical about metaphysical fundamentality and dependence, while arguably Frege would consider 'the dependence of truths one upon another' (1980, §2). Nevertheless, even if Hume's Principle were more fundamental metaphysically speaking, still the epistemological question how we can have access to this 'mathematical truth' would already justify the epistemological analysis.

5 See also Boolos (1998).

6 See Hale and Wright (2001) for elaborate expositions of these claims.

Naturally; the simplest way of testing rectangularity is by applying a protractor! The author forgets that this counting rests itself on a one-to-one correlation, namely of the numeral 1 to *n* and the objects of the set. Each of the two sets needs to be counted. This makes the matter less simple than it is if we consider a relation that correlates the objects of the two sets without numerals as intermediaries.

(Beaney 1997, 225)

The passage illustrates that both counting, the use of a fixed series based on a successor relation, and one-to-one correspondence seem to be implicated in the understanding of numbers. The passage highlights the issue of the relative priority of the two concepts. Heck, following Parsons (1994), argues that Frege's objection is less compelling. One may argue that in counting, only an operational procedure has been mastered, but no real concept is involved. Moreover, in view of empirical work on mathematical cognition, he claims that it is implausible that counting immediately leads to an understanding of cardinality.

2. Mathematical Cognition

2.1. *Gelman and Gallistel's Cardinality Principle*

Heck's discussion (2000) is informed by results in developmental psychology, in particular the work of Gelman and Gallistel. In their seminal work on numerical cognition in developmental psychology, Gelman and Gallistel (1978) argued that mastery of counting does not immediately lead to understanding of cardinality. Young children do not immediately attribute the same number to a collection that is in an obvious one-to-one correspondence to a collection they have already counted. They claimed that five separate principles underlie children's understanding of number: the one-to-one correspondence principle, the stable-order principle, the abstraction principle, the order irrelevance principle, and the cardinality principle.[7] The central principle in counting is the stable-order principle. It states that number concepts or words come in a stable order; there is a standard series of number words that are used in counting, and this list starts with a particular number word ('one'[8] in English). The abstraction

---

7 Many authors omit the abstraction principle and the order irrelevance principle in discussions of the Gelman and Gallistel model, as they are deemed less important from a cognitive point of view. From a philosophical point of view, they are quite important.

8 In studies of numerical cognition, the starting point is never 0 as defined in DZ previously. For an epistemological account of the understanding of zero, based on Carey's bootstrapping model and absence perception, see Barton (2019). This observation conflicts with Heck's (2011, 166) claim that the ZCE-principle, stating that Zero-Concepts are Empty, is essential in our understanding of cardinality.

principle states that every collection or set of items can be counted: trees, dogs, whistles, handshakes, and so on. There are no ontological restrictions on what can be counted, nor is counting restricted to a particular sensory modality. One-to-one correspondence is used in a restricted way; in an enumeration procedure, objects are placed in a one-to-one correspondence with an initial segment of the series of number words. The more general use of the one-to-one correspondence principle in HP to compare the numerosity of two arbitrary collections directly was seldom mentioned in the psychological literature. The order irrelevance principle articulates that the order of counting does not affect the cardinality, or, in other words, every one-to-one correspondence between a collection and an initial segment of number words will determine the same cardinality. The cardinality principle bases the determination of the cardinality of a collection on an enumeration procedure; the cardinality of a collection is determined as the last member of the enumerated series of number words. In the development of a child's numerical abilities, the five principles are not mastered at the same time. Children may be able to recite the number series, thus having mastered the stable order principle, without being able to enumerate a collection. Or if a collection can be enumerated, this does not necessarily lead to a grasp of the cardinality.[9]

In this view, the comparison of the cardinality of two collections is only possible through counting out the two collections. Moreover, the evidence available in the 1990s suggests that this conception of cardinality is not easily combined with Hume's Principle. Young children, having counted a collection of objects and having paired the objects to the objects of another collection, are not able to determine the cardinality of the second collection on the basis of this information. In Heck's (2011, 165) example, infants being shown and having counted a set of dolls and having paired them with toy hats will not be able to state the number of hats, which should be quite obvious if understanding of cardinality is based on Hume's Principle; the number of hats can only be obtained by counting them again. These observations strongly support a Husserlian conception of cardinality based on counting procedures.

However, in the recent two decades, the view has been modified. New results in psychology, neurology, and the study of animal cognition have transformed the view of the first stages of the development of numerical cognition. Moreover, anthropological results have to some extent corrected some implicit cultural biases resulting from Western mathematical

---

9 It seems possible to combine the stable-order principle and the cardinality principle with an inadequate grasp of one-to-one correspondence. A case in point is the difficulty Mundurucu had with learning to count. After being taught a counting procedure, they could not match the number words on a one-to-one basis to the objects in a collection very well; see www.youtube.com/watch?v=9iXh8wte3gM.

education. In typical experiments in developmental psychology, test subjects (infants) are selected from environments in which the number sequence is inculcated at a very early age. In view of this cultural background, one would expect that the centrality of the stable-order principle in human mathematical cognition would be overrated and partly a social construct. I will analyse how some of these findings may be relevant in the assessment of the relative importance of counting and equinumerosity in numerical cognition.

2.2.   Core Systems of Numerical Cognition

It is now generally accepted among researchers on numerical cognition that humans, together with several species of animals, share two systems[10] of prelinguistic numerical abilities, which together form the 'number sense' (Dehaene 2011). These core systems enable the employment of non-symbolic numerical concepts; it is believed that numerical concepts can be discerned in humans and animals for which no symbols or words (need to) exist in the language. A first 'innate' numerical ability is subitising, derived from the Latin 'subitus,' which means 'sudden.' Humans are able to grasp the numerosity of small collections of up to three or four objects quickly and very reliably, but performance deteriorates rapidly for larger numerosities. Larger numerosities always require more processing time, but there is a discontinuity in the slope of the function that depicts response time versus numerosity, jumping from 40 to 100 ms per additional item between collections containing four and five items (Trick and Pylyshyn 1994). The error rate is minimal for numerosities up to four but goes to 20% for collections with five items and to 40% for collections with six items (Mandler and Shebo 1982). Response times are believed to be too fast for linear processing. The most plausible explanations of subitising are that it depends on brain circuits dedicated to localising and tracking objects in space or that it is related to limitations in working memory. Neither Husserl's nor Frege's view on numerical cognition sits well with these empirical findings. In subitising, the numerosity of a collection is not obtained via an enumeration process, but it is grasped at once; both phenomenal experience and neuroimaging studies (Demeyere et al. 2012; Cutini et al. 2014) indicate that subitising differs radically from counting.[11] Moreover, one-to-one correspondence is not implicated in subitising in any transparent way.[12] Admittedly, the psychological and

---

10 For a longer introduction and additional references, see Gilmore et al. (2018).
11 For a dissenting voice, see Gallistel and Gelman (1991, 1992).
12 It remains possible that the neuropsychological mechanism of subitising does invoke a one-to-one correspondence between the objects and neuropsychological entities; at present, this mechanism remains opaque.

neural processes underlying subitising are not fully understood, but processes such as (multiple-)object tracking do not in any straightforward way involve object pairing.

The second core system, the approximate number system (ANS), is used for estimations of large numerosities. The first experiments on estimations of large numbers of items were carried out in the 1950s (Minturn and Reese 1951), but the existence of the ANS became widely accepted[13] in the 1990s. Neurophysiological evidence that certain neurons respond specifically to numerosity provided strong support for the ANS, and Dehaene's (2011) description of this 'number sense' made it well known. The ANS is an inaccurate system; it does not yield the precise number of objects but can reliably approximate large numbers such as the number of people in a square or the number of birds in a flock. It has been observed in adults, infants, and a broad variety of animals, including primates and birds. In humans, there are important differences in accuracy related to age. Newborns can perform matching tasks, but only if the numerosities differ by a 1:3 ratio, while adults can typically discriminate with a 9:10 ratio (e.g., 27 from 30), but there is significant variation between subjects, in part dependent on education.

The ANS is believed to be based on the conversion of discrete quantities into an analogue magnitude on an internal continuum, the mental number line (MNL). As the estimates are not precise, in conversion models also the deviation is modelled by means of statistical curves. Contemporary research in neurophysiology aims at discovering neural correlates of these magnitudes, in particular in the parietal cortex. There is debate whether the MNL is linear (Brannon et al. 2001; Gallistel and Gelman 2000) or logarithmically scaled (Dehaene and Changeux 1993; Dehaene et al. 1993). In the linear model, the distance on the MNL of the converted magnitudes of consecutive natural numbers remains equal for all the natural numbers, while the mean standard deviation increases for larger natural numbers. In the logarithmic model, since Fechner often used in psychology for continuous magnitudes, the (larger) natural numbers are compressed so that distances on the number line represent ratios between natural numbers and the mean standard deviation remains equal.

The conversion of discrete numerosities into continuous magnitudes looks hardly compatible with either Husserl or Frege's[14] conception of the natural numbers. It has even been argued (Gallistel and Gelman 2000;

---

13 Some authors have expressed reservations, see Gilmore et al. (2018, 23–26) for an overview of the critical literature.
14 Note, though, that in one of his latest (posthumously published) writings, Frege seems to abandon his earlier views and formulates an intriguing proposal: 'The more I have thought the matter over, the more convinced have I become that arithmetic and geometry have developed on the same basis—a geometrical one in fact—so that mathematics in its entirety is really geometry' (Beaney 1997, 373).

Gallistel et al. 2005) that from a cognitive perspective, the structure of the real numbers or the continuum is more basic than the structure of the natural numbers. Humans and other animals have an innate imprecise MNL isomorphic to the continuum, onto which discrete quantities can be mapped. An obvious rejoinder is that these findings are less relevant in relation to foundational work in mathematics, since the numerical abilities with exact numbers might well be independent of the cognitive mechanisms underlying numerosity estimations. Nevertheless, there is reason for worry, as several experimental results on cognitive abilities with exact discrete quantities indicate genuine correlations with the ANS and the ANS is standardly invoked in theoretical accounts in developmental psychology of the acquisition of the understanding of cardinality. A well-known finding that illustrates the influence of the ANS on numerical judgments is the distance effect. The speed of discrimination between numerosities depends on the distance on the MNL. Discrimination between 7 and 10 dots is typically faster than between 8 and 10 dots, as they are at a larger distance on the MNL. Remarkably, this effect can also be observed in judgments based on claims involving symbolic representations of numbers only; subjects more rapidly judge '7 is smaller than 10' than '8 is smaller than 10.' Moreover, there are noticeable though small correlations between arithmetical performance and ANS acuity, though it is difficult to explain in detail how ANS acuity would enhance arithmetical performance. Alternatively, experience with numerical symbols and mathematics will sharpen ANS precision, as can be observed in studies comparing ANS precision of adults with and without mathematical education and from studies of non-Western societies with a restricted number lexicon (Piazza et al. 2013).

2.3.   Psychological Models of the Acquisition of Exact Numbers

The innate numerical abilities are commonly believed to form the basis of more advanced numerical skills: the use of symbolic numbers, exact numbers, and larger cardinalities. Subitising explains the more or less reliable use of exact numbers up to four only,[15] and the ANS can only explain the understanding of imprecise larger cardinalities but not of exact larger cardinalities. The understanding of larger exact cardinalities is strongly correlated with the use of symbolic numbers, in particular the arabic numerals. The understanding of larger symbolic exact numerosities is not easy; it takes children years to fully master numerical (and arithmetical)

---

15 For a dissenting voice, see Buijsman (2019a). Buijsman argues that mastery of the cardinal concept 'one,' together with an understanding how the cardinality changes by adding one item to a collection, lies at the basis of children's acquisition of number concepts.

abilities. Unfortunately, there is no scientific consensus on the mechanisms that link advanced numerical abilities to the two core numerical abilities; various rivalling proposals have been put forward, and opinion remains divided.

A first proposal (Gallistel and Gelman 2000; Leslie et al. 2008; Wynn 2018) is the accumulator model. An accumulator is a receptacle (e.g., a water container or a battery) that can be filled (with water or electricity) gradually. The 'content' of the accumulator can be described by means of a continuous magnitude. It is believed that some type of accumulator is built into our cognitive system and underlies our understanding of cardinality. In preverbal counting, the accumulator is filled with a fixed amount of the continuous magnitude for each item that is counted. At the end of the counting process, the amount to which the accumulator is filled reliably represents the number of items counted, and this amount can be transferred to memory. It is readily seen that Gallistel and Gelman's principles can be read into the model. One-to-one correspondence is observed, since every item increases the accumulator with a single unit quantity. The stable order principle is observed, since the filling of the accumulator mirrors the increase in numerosity. The cardinality principle is observed, as the full content of the accumulator stands for the last object that is counted and thus represents the number of the set. The accumulator model is arguably quite reliable for small numerosities and thus can explain subitising as well as preverbal arithmetical skills in infants (Wynn 1992). It provides a clear basis for learning symbolic numbers; it suffices to understand the correspondence between the verbal counting systems and nonverbal accumulator systems. However, some facts are less easily explained. If a clear and precise cognitive system is in place, it is difficult to understand why it takes children years to link the accumulator system to symbolic numbers. Moreover, the discontinuity in subitising is left unexplained, and no dissociation between subitising and the ANS is accounted for, nor is the imprecise nature of the ANS well explained.

A second set of models are the bootstrapping models. These models claim that the initial understanding of cardinality is based on innate principles and that by means of a bootstrapping procedure, understanding of larger cardinalities and the general cardinality principle is acquired. A first form was proposed by Carey and Le Corre (Carey 2004, 2009; Le Corre and Carey 2007). It is based directly on subitising and more in particular the alleged mechanism underlying it, the object tracking system (OTS). The OTS is a perceptual system that is used in multi-object tracking. Items in the visual scene are indexed, and the OTS keeps track of the individual trajectories through the scene. Multi-object tracking is reliable for a small number of objects only and hence is readily related to subitising. The number of indices is measured and can be stored in memory. Rather than the ANS,[16] these small numerosities and their mapping onto the first number words are believed to form the basis of our understanding of cardinality. Children become one-knowers, two-knowers, three-knowers, and with these numbers learn to understand the successor principle, the idea that by adding an item, the numerosity moves to the next number word. This initial understanding of number words is subsequently used in a bootstrapping procedure to achieve an understanding of the larger numerals and cardinality principle in general. At the end of the bootstrapping process, children have become cardinality principle (CP)-knowers; they understand the cardinality principle and are able to determine the exact cardinality of an arbitrary collection through counting. Spelke (2011) has argued for a similar model in which both subitising and the ANS play a role. The understanding of cardinality is based on the ANS, whereas the notion of exact number and the successor relation are based on subitising. The verbal counting list is the medium that connects both and allows bootstrapping to larger number concepts.[17] A further modification is the Merge-model proposed by vanMarle et al. (2018). On the basis of a large-scale longitudinal study, they concluded that, even though both the OTS and the ANS underlie the initial stages of the development of numerical abilities, over time, the influence of the OTS wanes in favour of the ANS so that a full understanding of the successor relation and the cardinality principle is based on the ANS only.

The role of equinumerosity in the development of numerical abilities has been neglected for a long time. In one of the first studies of numerical cognition, Piaget (1965) explicitly studied one-to-one correspondence but concluded that children under the age of five do not have an adequate understanding of equinumerosity. Only recently, new experiments have been set up, yielding more nuanced results, mildly going against Heck's claim that young children lack the concept of equinumerosity. Sarnecka and Wright (2013) tested whether 2½–4 years old can compare sets of five or six items on two cards. They found that performance is above chance for non-CP-knowers but that CP-knowers can judge equinumerosity more reliably (but not near upper threshold). Izard et al. (2014) tested the understanding of exact equality in children between 32 and 36 months. They presented the subject with 'rubber puppets' that could

---

16 Carey et al. (2017) argue explicitly against the role of the ANS in learning the meaning of number words early in development.
17 Marshall (2018), critical of the explanatory power of neuropsychological and cognitive accounts concerning the acquisition of numerical concepts employed in mathematical practice, highlights the epistemic immediacy of multi-digit decimal numerals and regards this representation as more important than the verbal counting system and the ANS. Also, in Buijsman (2019b), the syntactic structure of the numeral system is essential for grasping the successor axiom. Note that Heck (2011, 173f) also mentions the crucial role of the multi-digit decimal numerals in our understanding of cardinality.

be put on 'trees,' allowing for a visible pairing operation. Children are capable, without taking recourse to counting, of using one-to-one correspondence clues to determine the exact equality of numerosities. However, these abilities are restricted; the children are not able to process the impact of adding or removing rubber puppets from the box or branches of the tree on numerosities.[18] Results on substitution, that is, the replacement of rubber puppets in the box, are intriguing; removing and adding identical puppets does not affect the numerical judgments, while the replacement of a puppet by a nonidentical puppet results in guesses not beyond chance level. Muldoon et al. (2009) further note that there may be a difference between item-by-item matching between collections and word-item matching in counting. Children are able to use one-to-one correspondence in sharing tasks but are not immediately able to relate this ability to counting. Muldoon et al. point at the importance of error detection in sharing tasks as a catalyst in the linking word-item and item-item correspondence. In conclusion, one may conclude that mastery of the cardinality principle does not coincide with mastery of equinumerosity and that a limited understanding of equinumerosity can be observed in young children. These results have only recently been established, and theoretical underpinnings of the origins of the mastery of equinumerosity judgments have not yet been proposed in the literature. Sarnecka and Wright (2013, 1504) argue that this ability cannot be grounded in the ANS only but offer no further explanation.

2.4. Numerical Abilities of Anumerate Societies

Many of the models on the acquisition of advanced numerical abilities critically hinge on the linking process of the innate numerical abilities to a verbal counting list. It may be objected that the existence[19] of a verbal counting list, which enables an enumeration procedure to arbitrarily large numerosities, introduces a Western bias,[20] giving too much weight to counting in the description of the development of numerical abilities. The analysis of numerical abilities of people using languages with

---

18 According to Heck (2011, 166), the addition preserves cardinality (APC) and removal preserves cardinality (RPC) principles, which follow directly from HP, are essential in our understanding of the notion of cardinal number.

19 Symbolic or verbal counting lists have been constructed, and the archaeological and historical evidence may also cast light on the development of numerical abilities. The discovery of tally systems indicates the early use of enumerations systems; see, for example, Everett (2017, ch. 2) for an overview. The study of the development of symbolic notation systems for larger numerosities may yield additional clues on the development of numerical abilities with larger cardinals; see, for example, Cajori (1993) and Ifrah (1998).

20 Recently several authors have highlighted the importance of social and cultural factors in the development of numerical abilities; see De Cruz (2018), Fabry (2019), and Pantsar (2019).

a limited numerical system yields further evidence on the role of counting and one-to-one correspondence in numerical cognition. In recent years, studies of the Pirahã and the Mundurucu, two Amazonian communities, have become well known. The Pirahã use an extremely limited amount of number words, with only number words for one and two. Gordon (2004, 497) designed various nonverbal numerical tasks to test their numerical abilities and found that the numerical abilities rapidly decrease for larger numerosities. A natural interpretation of the performance of the Pirahã, confirmed by renewed experiments (Frank et al. 2008), is that tasks typically relying on enumeration procedures (fig. 1: C, E, G) are executed poorly, while one-to-one pairing tasks (fig. 1: A, D) lead to almost correct results. The Mundurucu (Pica et al. 2004) use a more evolved number vocabulary: the words most regularly used are 'pũg' for one, 'xep xep' for two, 'ebapũg' for three, 'ebadipdip' for four, 'pũg pogbi' ('one hand') for five, and 'xep xep pogbi' ('two hands') for ten. The compositionality principles behind these words are not completely transparent. Pica et al. (2004, 500) suggest that enumeration plays a role:

> The words for 3 and 4 are polymorphemic: ebapũg = 2+1, ebadipdip = 2 + 1 +1, where 'eba' means 'your (two) arms.' This possibly reflects an earlier base-2 system common in Tupi languages, but the system is not productive in Mundurukú.

Alternatively, Pica and Lecomte (2008) have argued that symmetry principles (involving one-to-one correspondences between the symmetrical parts) are central in the composition of Mundurucu number words. For example, they submit that the reduplicated 'dip' in 'ebadipdip' reflects a symmetry principle and not a double tally added to 'eba.' The words for five and ten are clearly related to the number of fingers on a hand, which is common in many languages, including Proto-Indo-European, in which '*penkwe' is etymologically related to 'finger' and 'fist.' The fingers of the hand could be regarded as what I have called a 'canonical collection' (2008), a standard reference collection that can be paired one-to-one with the objects of collections containing five objects, so that cardinal assessment can be made on the basis of one-to-one correspondence only. The alternative explanation that the fingers are used in a stable order in an enumeration procedure is unlikely, in view of the fact that the Mundurucu lack verbal and gestural counting procedures.

Additional clues can be found in communities with a slightly more evolved vocabulary, but again, there is evidence that both one-to-one correspondence and enumeration play a role in the development of numerical abilities. Epps' (2006) study of the development of numerals in the languages of the Nadahup family in the Amazon offers interesting evidence of the role of the one-to-one correspondence principle, in particular because the etymological origins of various numbers words are traceable. Several languages (Dâw, Hup, Yuhup) can be seen to use canonical collections for small numerals: the word for two is in some languages related to the word for eye; the word for the three is a cognate of the word for the rubber tree seed, a distinctive three-lobed seed or nut, which is culturally highly salient; and the word for five is again related to the word for hand. Interestingly, a pairing operation is important in some of the languages: the expression 'has a brother' functions as the word 'even,' and 'has no brother' as 'uneven,' and these expressions play a role in the numerical expressions for larger numbers and in enumeration procedures. Other anthropological evidence highlights the role of enumeration in the development of verbal counting systems. Tally systems based on body parts are common, mostly decimal (or vigesimal) systems based on the fingers (and toes). More peculiar are tally systems based on other parts of the human body, such as in the Wambon language in Irian Jaya (de Vries 1994, 543):[21]

Wambon numeral	Corresponding body part
1 sanop	little finger
2 sanopkunip	ring finger
3 takhem	middle finger
4 hitulop	index finger
5 ambalop	thumb
6 kumuk	wrist
7 mben	lower arm
8 muyop	elbow
9 javet	upper arm
10 malin	shoulder
11 nggokmit	neck
12 silutop	ear
13 kelop	eye
14 kalit	nose

Speakers of Wambon point consecutively to their body part while counting. For larger numbers, up to 27, they continue on the other part of the body, and prefix 'em-' to the previous number words, as in 'emkelop' for 15. According to one informant, it is possible to continue further by returning again and prefixing the word 'nggisikhivo' as in 'nggisikhivo-sanop' (return little finger = 28). Many languages in the region have these body-part counting systems, all with minor variations on the body parts involved (and some with sex-based variations within a language).

---

21 I thank Lourens de Vries for clarification on the use of the body parts–based number systems. Other languages in the region with similar body-based enumerative counting practices are Korowai (van Enk and de Vries 1997, 73) and Kombai (de Vries 1993, 39–40).

Cardinality judgments are manifestly made on the basis of a regular enumeration procedure in which objects are placed into a one-to-one correspondence with a fixed series of body parts in a stable order, thus supporting the Gelman and Gallistel view on the development of numerical cognition.

In conclusion, Gelman and Gallistel's empirical results mentioned by Heck (2011) are less convincing and more controversial at present. In contemporary developmental psychology, the Gallistel and Gelman model has been supplemented by findings concerning innate preverbal numerical abilities. Gallistel and Gelman have modified their earlier view, incorporating the innate mechanisms in the more recent models. However, a plethora of rivalling accounts on the acquisition of the cardinality principle has been proposed, and no consensus is emerging. Importantly, the innate core numerical mechanisms are not readily related either to enumeration or one-to-one correspondence. Moreover, anthropological evidence on anumerate societies does not cast additional light on the role of both principles; both counting and pairing play a role in the development of verbal number systems, though not uniformly.

3. Psychologism and Its Critics

The lack of scientific consensus over the precise mechanisms for the acquisition of advanced numerical abilities is not helpful for the determination of the relative contributions of one-to-one correspondence and counting in the development of numerical abilities. But the situation is actually worse. Several cognitive scientists have argued that it remains dubious whether the proposed mechanisms actually lead to a genuine understanding of arithmetic and its basic axioms, whether FA or PA. Rips et al. (2008) wrote an influential critique of current accounts of the acquisition of numerical abilities. They argued that children arrive at concepts of numbers and arithmetic not from the bottom up, from innate core principles via bootstrapping to larger natural numbers, but rather in a top-down way, by constructing mathematical schemas. An example is the acquisition of the principle of additive commutativity, based on a generalisation of the schema $a + b = b + a$ over all the natural numbers. They argue that 'children form a schema for the numbers that specifies their structure as a countably infinite sequence' (Rips et al. 2008, 637). Central in their critique is the argument that bottom-up bootstrapping doesn't provide a plausible account of the induction principle: 'For any count word n,' the next count word '$s(n)$' in the count sequence refers to the cardinality $(n + 1)$ obtained by adding one element to collections whose cardinality is denoted by 'n.' Typically children have only a finite number vocabulary; they are able to count to 10 or 100 but not beyond that. In order to acquire the adult concept of natural numbers, the compositional structure of the counting vocabulary must also be mastered, and the generalisation is not unique, or so they argue. While some details of their critique might be contentious,[22] Rips et al. do point out a genuine gap between children's acquisition of concepts of small cardinalities and the adult understanding of number concepts. A related critique has been made by Cheung et al. (2017),[23] who demonstrated that children acquire an adequate understanding of the successor relation only at the age of 5½, which is 1½–2 years after mastering the cardinality principle. They argue that this experiment jeopardises most bootstrapping accounts, as these often critically hinge on an earlier understanding of the successor relation.[24] Relaford-Doyle and Núñez (2018) argue for the stronger claim that even adults' number concepts are not compatible with Peano's axioms. They present an experiment in which university students manifestly abandon the principle of mathematical induction in a fairly straightforward visual mathematical proof and conclude that the students must have an inconsistent concept of natural numbers. In sum, there is genuine reason to doubt that at present cognitive science offers a convincing account of how humans come to understand the full[25] structure of the natural numbers.

An even more fundamental worry concerns the methodology of studying number concepts in the cognitive sciences.[26] Typically, in experiments that test whether a particular numerical concept should be

---

22 Rips et al. (2008, 632) argue, in a Kripkenstein-style argument (Kripke 1982), that different inductive generalisations are possible; children might equally well generalise in a cyclical pattern, for example, the natural numbers modulo 10 or 38 or 983. However, one might object that this would violate Peano's axiom 5 and thus would not lead to concepts of the natural numbers. Based on particular (nontrivial) assumptions concerning the structure of the mind, Piantadosi et al. (2012) present a precise formal bootstrapping model that can explain the inductive steps.

23 Cheung et al. (2017) also tested whether children having mastered counting have an adequate understanding of infinity, by testing whether they understand that there cannot be a highest natural number. Following Heck (2011), I want to sidestep the discussion of the infinity of the natural numbers. It is quite obvious that Cheung et al.'s results would be equally valid for the general understanding of the predecessor function.

24 Beck (2017) has given a more detailed analysis of bootstrapping and argued that the 'circularity challenge' can be answered.

25 The delimitation of what constitutes a full understanding of the natural numbers is not universally accepted. Some authors argue that an understanding of the concepts underlying the axiomatic systems FA or PA is not necessary for numerical or arithmetical knowledge; for example, Pantsar (2019, 2) defines arithmetic as 'a sufficiently rich discrete linear system of explicit number words or symbols with explicit rules of operations.' For present purposes, that is, a comparison of PA and FA from a cognitive perspective, this characterisation will be too weak.

26 It goes beyond the scope of this chapter to give a precise ontological account of concepts and an account of concept possession. For present purposes, it suffices to point out that most accounts of concepts would reject the claim that concepts are fully determined by

attributed to test persons, the concept must be operationalised, which can lead to a de facto modification of the concept. Often the limitations of the operationalisations are implicitly hinted at in the psychological literature. Some examples will suffice to illustrate the impact of these methodological decisions. One of the most central numerical concepts is the cardinality principle. Notwithstanding the clarity of the concepts involved, in experiments testing whether subjects are CP-knowers, specific extra procedural refinements are required. A standard test is the Give-A-Number task (Wynn 1992); subjects are asked to hand over a certain number of items from a large set. A second test is the How Many?-test; test persons are asked how many items are in a collection. While both tests clearly rely on an enumeration procedure, experimental outcomes are not exactly the same. Another example can be found in Gordon's (2004) ascription of number abilities to the Pirahã; as pointed out in Decock (2008), some tests (Gordon 200, fig. 1 A and D) are pairing tests that are more naturally interpreted as tests of mastery of the one-to-one correspondence principle than mastery of cardinality. In Cheung et al. (2017), a particular successor task is devised in order to distinguish mastery of the successor function from mastery of the cardinality principle; test subjects, that is, children between 4 and 7 years old, are presented with a box in which N objects are placed, and subsequently the box is closed so that the child can no longer count the objects. Subsequently, an object is added to the box, and the child is asked whether the box contains $N+1$ or $N+2$ objects. The authors also operationalise an infinity task to test whether children have an adequate understanding of the successor relation, that is, to test whether children think there is a largest natural number, by means of a particular set of questions. In one of the clearest contemporary studies of the role of one-to-one correspondence in the development of numerical abilities, Izard et al. (2014) present an operationalisation of exact equality by means of three principles: the identity principle, the addition/subtraction principle, and the substitution principle. Tests on exact equality comprise physical operations on collections without changing the object (e.g., shaking a box), adding or removing objects, and replacing objects with other objects in a collection. One notices that exact equality is not a perfect operationalisation of Frege's equinumerosity; physical one-to-one pairing operations are not included. In some cases, authors explicitly state that some operationalisations are dubious or wrong. Izard et al. (2014, 29) indicate that Piaget's negative results on exact equality, in particular the identity principle, in experiments with children under five years old have been contested, with

operational or behavioural criteria; Bridgman's operationalism or Carnap's verificationism have only a few followers left.

critics mainly appealing to the pragmatics of the tests. Sarnecka and Wright (2013, 1495) point out that the Compare-Sets test, employed in Sarnecka and Gelman (2004) to test children's knowledge that number concepts are specific, does test equinumerosity rather than specificity. Furthermore, extra methodological issues arise in the absence of a shared language between experimenter and subjects, complicating the separation of numerical skills from other cognitive abilities (e.g., perceptual routines), in particular in animal cognition or the interpretation of violation of expectation tests in developmental psychology, or arise from ambiguities in translation, complicating the identification of genuine number words in anumerate societies. Admittedly, in most cases, the operational procedures do reflect central characteristics of the concepts they are designed to reveal and yield important insights in the development of numerical skills, yet often the procedures have an ad hoc character and seldom lead to unambiguous results. In view of the intricacies related to these inevitable methodological choices, drawing inferences concerning the possession of number concepts or other central numerical concepts such as one-to-one correspondence or a successor or predecessor relation from numerical behaviour remains precarious.

The overview of recent studies in the cognitive sciences on the empirical basis of the foundational concepts in arithmetic leads to unsatisfactory conclusions. There is no general consensus on the acquisition mechanisms of advanced numerical abilities; it has been argued that the extant explanations do not offer an adequate account of the genuine mastery of the central concepts used in axiomatic systems, and there are serious methodological difficulties related to these studies. From a Fregean and Husserlian perspective, this would hardly be problematic, in view of Frege and Husserl's harsh critique on psychologism in the study of foundations of mathematics. As Frege wrote in the last year of his life:

> I myself at one time held it to be possible to conquer the entire number-domain, continuing along a purely logical path from the kindergarten-numbers; I have seen the mistake in this. I was right that you cannot do this if you take an empirical route. . . . So an *a priori* mode of cognition must be involved here.
>
> (Beaney 1997, 373)

Neologicists (see Hale and Wright 2001) or contemporary phenomenologists (see, e.g., da Silva 2017) are equally critical of the use of psychological studies on foundational mathematical concepts.

However, in view of developments in the cognitive sciences and epistemology in the last century, antipsychologism is deeply unattractive. It is hard to deny that human mathematical reasoning is possible only by means of the reasoning skills implemented in the human brain.[27] In view of these advances, it would hardly be possible to avoid the incredulous stare of cognitive scientists if one wants to take recourse to unexplained nonnaturalistic faculties such as Gödelian intuition or reason or if one wants to invoke a transcendental level in order to explain mathematical reasoning. As Heck (2011, 160) indicates, explaining mathematical concept formation is ultimately an empirical question. Nevertheless, while present-day empirical studies do shed light on the role of counting and Hume's Principle in the acquisition of numerical abilities, they cannot plausibly decide questions concerning the cognitive or epistemic primacy of the two concepts. We ought not exclude the possibility that future research in the cognitive sciences will provide us with a better or complete grasp of the conceptual structures underlying mathematical cognition and its acquisition, but in deliberations over the relative logical, conceptual, and philosophical merits of foundational systems in arithmetic, in particular FA versus PA, at present, we are better off relying on expert logicians and mathematicians. For this reason, a provisional modest apsychologism[28] seems advisable concerning logical and philosophical work on the foundations of arithmetic.

Acknowledgements

Earlier versions of this chapter have been presented at the Università Vita-Salute San Raffaele in Milan, at the Vrije Universiteit in Amsterdam, and at the Ruhr-Universität in Bochum, and I thank the participants for valuable comments. I thank César dos Santos, Catarina Dutilh Novaes, Regina Fabry, and an anonymous referee for detailed comments on the manuscript.

References

Assadian, B., and Buijsman, S. (2018). "Are the natural numbers fundamentally ordinals?" *Philosophy and Phenomenological Research* 99(3): 564–580. https://doi.org/10.1111/phpr.12499.

Barton, N. (2019). "Absence perception and the philosophy of zero". *Synthese* 197: 3823–3850. https://doi.org/10.007/s11229-019-02220-x.

Beaney, M. (1997). *The Frege Reader*. Oxford: Blackwell.

---

27 There is room for debate over how much mathematical cognition is located within the brain. 4E-philosophers of cognition would claim that mathematical cognition can exist only as a brain-environment interaction, whereby the mathematical inscriptions serve as cognitive tools; see, for example, Menary (2015) and Fabry (2019). I have tried to remain neutral in this debate throughout the text.

28 I thank Catarina Dutilh Novaes for suggesting this term.

Beck, J. (2017). "Can bootstrapping explain concept learning?" *Cognition* 158: 110–121.

Boolos, G. (1987). "The consistency of Frege. In J.J. Thomson (Ed.)," *On Being and Saying* (pp. 3–20). Cambridge, MA: MIT Press.

Boolos, G. (1998). "Is Hume's Principle analytic?" In R. Jeffrey (Ed.), *Logic, Logic, and Logic.* Cambridge, MA: Harvard University Press.

Brannon, E., Wusthoff, C., Gallistel, C., and Gibbon, J. (2001). "Numerical subtraction in the pigeon: Evidence for a linear subjective number scale." *Psychological Science* 12: 238–243.

Buijsman, S. (2019a). "Learning the natural numbers as a child." *Noûs* 53: 3–22.

Buijsman, S. (2019b). "Two roads to the successor axiom." *Synthese* 197(3): 1241–1261. https://doi.org/10.1007//s11229-018-1752-5.

Cajori, F. (1993). *A History of Mathematical Notations.* New York: Dover.

Carey, S. (2004). "Bootstrapping and the origin of concepts." *Daedalus* 133: 59–68.

Carey, S. (2009). *The Origin of Concepts.* Oxford: Oxford University Press.

Carey, S., Shusterman, A., Haward, P., and Distefano, R. (2017). "Do analog number representations underlie the meanings of yound children's verbal numerals?" *Cognition* 168: 243–255.

Cheung, P., Rubenson, M., and Barner, D. (2017). "To infinity and beyond. Children generalize the successor function to all possible numbers years after learning to count." *Cognitive Psychology* 92: 22–36.

Cutini, S., Scatturin, P., Basso Moro, S., and Zorzi, M. (2014). "Are the neural correlates of subitizing and estimation dissociable?" An fNRIS investigation. *Neuroimage* 85: 391–399.

Da Silva, J. (2017). *Mathematics and Its Applications.* Cham: Springer.

Decock, L. (2008). "The conceptual basis of numerical abilities: One-to-one correspondence versus the successor relation." *Philosophical Psychology* 21: 459–473.

De Cruz, H. (2018). "Testimony and children's acquisition of number concepts. In S. Bangu (Ed.)," *Naturalizing Logico-Mathematical Knowledge* (pp. 164–178). New York: Routledge.

Dehaene, S. (2011). *The Number Sense.* Oxford: Oxford University Press.

Dehaene, S., Bossini, S., and Giraux, P. (1993). "The mental representation of parity and number magnitude." *Journal of Experimental Psychology: General* 122: 371–396.

Dehaene, S., and Changeux, J.-P. (1993). "Development of elementary numerical abilities: A neuronal model." *Journal of Cognitive Neuroscience* 5: 390–407.

Demeyere, N., Rotshtein, P., and Humphreys, G. (2012). "The neuroanatomy of visual enumeration: Differentiating necessary neural correlates for subitizing versus counting in a neuropsychological voxel-based morphometry study." *Journal of Cognitive Neuroscience* 24: 948–964.

De Vries, L. (1993). *Forms and Functions in Kombai, an Awyu Language of Irian Jaya.* Canberra: Australian National University Press.

De Vries, L. (1994). "Numerals systems of the Awyu language family of Irian Jaya." *Bijdragen tot de Taal-, Land-, en Volkenkunde* 150: 539–567.

Epps, P. (2006). "Growing a numeral system. The historical development of numerals in an Amazonian language family." *Diachronica* 23: 259–288.

Everett, C. (2017). *Numbers and the Making of Us.* Cambridge, MA: Harvard University Press.

Fabry, R. (2019). "The cerebral, extra-cerebral bodily, and socio-cultural dimensions of enculturated arithmetical cognition." *Synthese* 197: 3685–3720. https://doi.org/10.1007/s11229-019-02238-1.

Frank, M., Everett, D., Fedorenko, E., and Gibson, E. (2008). "Number as cognitive technology: Evidence from Pirahã language and cognition." *Cognition* 108: 819–824.

Frege, G. (1980). *The Foundations of Arithmetic*. Transl. J. Austin. Evanston: Northwestern University Press.

Frege, G. (2013). *Basic Laws of Arithmetic*. Transl. Ph. Ebert and M. Rossberg. Oxford: Oxford University Press.

Gallistel, C., and Gelman, R. (1992). "Preverbal and verbal counting and computation." *Cognition* 44: 43–74.

Gallistel, C., and Gelman, R. (2000). "Non-verbal numerical cognition: From reals to integers." *Trends in Cognitive Sciences* 4: 59–65.

Gallistel, C., Gelman, R., and Cordes, S. (2005). "The cultural and evolutionary history of the real numbers. In S. Levinson and P. Jaisson (Eds.)," *Culture and Evolution* (pp. 247–274), Cambridge, MA: MIT Press.

Gallistel, R., and Gelman, R. (1991). "Subitizing: The preverbal counting process. In W. Kessen, A. Ortony, and F. Craik (Eds.)," *Memories, Thoughts, and Emotions: Essays in Honor of George Mandler* (pp. 65–81) Hillsdale, NJ: Lawrence Erlbaum Associates.

Gelman, R., and Gallistel, R. (1978). *The Child's Understanding of Number*. Cambridge, MA: Harvard University Press.

Gilmore, C., Göbel, S., and Inglis, M. (2018). *Introduction to Mathematical Cognition*. London: Routledge.

Gordon, P. (2004). "Numerical cognition without words: Evidence from Amazonia." *Science* 306: 496–499.

Hale, B., and Wright, C. (2001). *The Reason's Proper Study*. Oxford: Oxford University Press.

Heck, R. (2000). "Cardinality, counting, and equinumerosity." *Notre Dame Journal of Formal Logic* 41: 187–208.

Heck, R. (2011). *Frege's Theorem*. Oxford: Oxford University Press.

Ifrah, G. (1998). *The Universal History of Numbers*. London: Harvill Press.

Izard, V., Streri, A., and Spelke, E. (2014). "Toward exact number: Young children use one-to-one correspondence to measure set identity but not numerical equality." *Cognitive Psychology* 72: 27–52.

Le Corre, M., and Carey, S. (2007). "One, two, three, four, nothing more: An investigation of the conceptual sources of verbal counting principles." *Cognition* 105: 395–438.

Leslie, A., Gelman, R., and Gallistel, C. R. (2008). "The generative bias of natural number concepts." *Trends in Cognitive Science* 12: 213–218.

Logan, S. (2017). "Categories for the neologicist." *Philosophia Mathematica* 25: 26–44.

Mandler, G., and Shebo, B. (1982). "Subitizing: An analysis of its component processes." *Journal of Experimental Psychology: General* 111: 1–22.

Marshall, O. (2018). "The psychology and philosophy of natural numbers." *Philosophia Mathematica* 26: 40–58.

Menary, R. (2015). Mathematical cognition: "A case of enculturation. In T. Metzinger and J. M. Windt (Eds.)," *Open MIND* (pp. 1–20). Frankfurt am Main: MIND Group. https://doi.org/10.15502/9783958570818.

Minturn, A., and Reese, T. (1951). "The effect of differential reinforcement on the discrimination of visual number." *The Journal of Psychology* 31: 201–231.

Muldoon, K., Lewis, C., and Freeman, N. (2009). "Why set-comparison is vital in early number learning." *Trends in Cognitive Sciences* 13: 203–208.

Pantsar, M. (2019). "The enculturated move from proto-arithmetic to arithmetic." *Frontiers in Psychologoy. Theoretical and Philosophical Psychology* 10: 1454.

Parsons, C. (1994). "Intuition and number. In A. George (Ed.)," *Mathematics and Mind* (pp. 141–147). New York: Oxford University Press.

Peano, G. (1989). *Arithmetices Principia, Novo Methodo Exposita*. Reprinted in G. Peano, *Opere Scelte* (Vol. 2, pp. 20–55). Rome: Editizioni Unione Mathematica Italiana.

Piaget, J. (1965). *The Child's Conception of Number*. New York: Norton.

Piantadosi, S., Tenenbaum, J., and Goodman, N. (2012). "Bootstrapping in a language of thought. A formal model of numerical concept learning." *Cognition* 123: 199–217.

Piazza, M., Pica, P., Izard, V., Spelke, E., and Dehaene, S. (2013). "Education enhances the acuity of the nonverbal approximate number system." *Psychological Science* 24: 1037–1043.

Pica, P., and Lecomte, A. (2008). "Theoretical implications of the study of numbers and numerals in Mundurucu." *Philosophical Psychology* 21: 507–522.

Pica, P., Lemer, C., Izard, V., and Dehaene, S. (2004). "Exact and approximate arithmetic in an Amazonian indigene group." *Science* 306: 499–503.

Relaford-Doyle, J., and Núñez, R. (2018). "Beyond Peano. Looking into the unnaturalness of natural numbers." In S. Bangu (Ed.), *Naturalizing Logico-Mathematical Knowledge* (pp. 234–251). New York: Routledge.

Rips, L., Bloomfield, A., and Asmuth, J. (2008). "From numerical concepts to concepts of number." *Behavioral and Brain Sciences* 31: 623–687.

Sarnecka, B., and Gelman, S. (2004). "Six does not just mean a lot: Preschoolers see numbers as specific." *Cognition* 108: 662–674.

Sarnecka, B., and Wright, C. (2013). "The idea of exact number: Children's understanding of cardinality and equinumerosity." *Cognitive Science* 37: 1493–1506.

Spelke, E. (2011). "Natural number and natural geometry." In S. Dehaene and E. Brannon (Eds.), *Space, Time and Number in the Brain* (pp. 287–317). London: Academic Press.

Trick, L., and Pylyshyn, Z. (1994). "Why are small and large numbers enumerated differently?" A limited-capacity preattentive stage in vision. *Psychological Review* 101: 80–102.

Van Enk, G.J., and de Vries, L. (1997). *The Korowai of Irian Jaya. Their Language in Its Cultural Context*. Oxford: Oxford University Press.

vanMarle, K., Chu, F., Mou, Y., Seok, J., Roulder, J., and Geary, D. (2018). "Attaching meaning to number words. Contributions of object tracking and approximate number systeems." *Developmental Science* 21: e12495.

Wright, C. (1983). *Frege's Conception of Numbers as Objects*. Aberdeen: University of Aberdeen Press.

Wynn, K. (1992). "Addition and substraction by human infants." *Nature* 358: 749–750.

Wynn, K. (2018). "Origins of numerical knowledge." In S. Bangu (Ed.), *Naturalizing Logico-Mathematical Knowledge* (pp. 106–130). New York: Routledge.

Contributors

Patricia Blanchette: Department of Philosophy, University of Notre Dame

Paola Cantù: Université Aix-Marseille et Centre National de Recherche Scientifique, Centre Gilles Gaston Granger, Aix-en-Provence, France

Roy T. Cook: Department of Philosophy, University of Minnesota—Twin Cities

Lieven Decock: Department of Philosophy, Vrije Universiteit Amsterdam

Fiona Doherty: Faculty of Law and Philosophy, University of Stirling

Bob Hale: (formerly) University of Sheffield

Michael Hallett: Department of Philosophy, McGill University, Montreal

Mirja Hartimo: Department of Social Sciences and Philosophy, University of Jyväskylä, Finland

Kevin C. Klement: Department of Philosophy, University of Massachusetts Amherst

Gregory Landini: Department of Philosophy, The University of Iowa

Robert May: Department of Philosophy, University of California, Davis

Marco Panza: CNRS, IHPST (CNRS—Paris 1 University) and Chapman University

Erich Reck: Department of Philosophy, University of California at Riverside, USA

Georg Schiemer: Department of Philosophy, University of Vienna

Index

Taylor & Francis Group
an **informa** business

Taylor & Francis eBooks

www.taylorfrancis.com

A single destination for eBooks from Taylor & Francis
with increased functionality and an improved user
experience to meet the needs of our customers.

90,000+ eBooks of award-winning academic content in
Humanities, Social Science, Science, Technology, Engineering,
and Medical written by a global network of editors and authors.

TAYLOR & FRANCIS EBOOKS OFFERS:

A streamlined
experience for
our library
customers

A single point
of discovery
for all of our
eBook content

Improved
search and
discovery of
content at both
book and
chapter level

REQUEST A FREE TRIAL
support@taylorfrancis.com

Routledge
Taylor & Francis Group

CRC Press
Taylor & Francis Group

For Product Safety Concerns and Information please contact our EU
representative GPSR@taylorandfrancis.com
Taylor & Francis Verlag GmbH, Kaufingerstraße 24, 80331 München, Germany

www.ingramcontent.com/pod-product-compliance
Lightning Source LLC
Chambersburg PA
CBHW061615220326
41598CB00026BA/3767